Principles of Plasma Physics for Engineers and Scientists

This unified introduction provides the tools and techniques needed to analyze plasmas, and connects plasma phenomena to other fields of study. Combining mathematical rigor with qualitative explanations, and linking theory to practice with example problems, this is a perfect textbook for senior undergraduate and graduate students taking a one-semester introductory course in plasma physics.

For the first time, material is presented in the context of unifying principles, illustrated using organizational charts, and structured in a successive progression from single-particle motion to kinetic theory and average values, through to the collective phenomena of waves in plasma. This provides students with a stronger understanding of the topics covered, their interconnections, and when different types of plasma models are applicable. Furthermore, mathematical derivations are rigorous yet concise, so physical understanding is not lost in lengthy mathematical treatments. Worked examples illustrate practical applications of theory, and students can test their new knowledge with 90 end-of-chapter problems.

Umran Inan is a Professor of Electrical Engineering at Stanford University, where he has led pioneering research on very low frequency studies of the ionosphere and radiation belts, space plasma physics, and electromagnetics for over 30 years. He also currently serves as President of Koç University in Istanbul, Turkey. As a committed teacher, he has supervised the Ph.D. dissertations of 42 students and has authored two previous books that have become standard textbooks for electromagnetics courses, as well as receiving numerous awards including the Tau Beta Pi Excellence in Undergraduate Teaching Award and the Outstanding Service Award from the Electrical Engineering Department for excellence in teaching. He is a Fellow of the Institute for Electrical and Electronics Engineers (IEEE), the American Geophysical Union (AGU), and the American Physical Society (APS), and is the recipient of the 2008 Appleton Prize from the International Union of Radio Science and the Royal Society, the 2007 Allan Cox Medal of Stanford for Faculty Excellence in fostering undergraduate research, and the 2010 Special Science Award given by the Scientific and Technological Research Council of Turkey.

Marek Gołkowski is an Assistant Professor in the Department of Electrical Engineering at the University of Colorado Denver, which he joined after completing his Ph.D. at Stanford University. He has won several awards including the Young Scientists Award from the International Association of Geomagnetism and Aeronomy and the Outstanding Student Paper Award from the American Geophysical Union. His current research focuses on electromagnetics and biological applications of plasmas.

Principles of Plasma Physics for Engineers and Scientists

Umran Inan
Stanford University†

and

Marek Gołkowski
University of Colorado Denver

†Now serving as President of Koç University, Istanbul

CAMBRIDGE
UNIVERSITY PRESS

University Printing House, Cambridge CB2 8BS, United Kingdom

Cambridge University Press is part of the University of Cambridge.

It furthers the University's mission by disseminating knowledge in the pursuit of education, learning and research at the highest international levels of excellence.

www.cambridge.org
Information on this title: www.cambridge.org/9780521193726

First published 2011

A catalogue record for this publication is available from the British Library

Library of Congress Cataloguing in Publication data

Inan, Umran S.
Principles of plasma physics for engineers and scientists / Umran Inan and Marek Gołkowski.
p. cm.
Includes bibliographical references and index.
ISBN 978-0-521-19372-6 (Hardback)
1. Plasma (Ionized gases)–Textbooks. I. Gołkowski, Marek. II. Title.
QC718.I435 2010
530.4′4–dc22 2010038466

ISBN 978-0-521-19372-6 Hardback

Additional resources for this publication at www.cambridge.org/9780521193726

To my parents, my beautiful wife Elif, my dear children Ayşe and Ali, and my very special granddaughter Ayla.

USI

To my father, who taught me to appreciate physics, my mother, who taught me to appreciate writing, and my wife, who gave me the support and motivation to finish this project.

MG

I have been teaching Introductory Plasma Physics to senior undergraduates and beginning graduate students for many years, and I find the level of the presentation of material, the order that the topics are presented, and the overall length of the book to be an excellent match for my needs in a textbook.
David Hammer, Cornell University

The authors have done an excellent job in introducing the vast scope of plasma physics for basic plasma physics courses. The schematic illustrations and flow charts used are especially helpful in understanding the complexities involved in the hierarchal nature of plasmas. Mathematics is kept at just the right level for the intended readers and the descriptions of the physical processes are clear. Although this book is targeted to advanced undergraduate or beginning graduate students, it will be a good addition to the personal library of every plasma physicist.
Gurudas Ganguli, Naval Research Laboratory

This new book provides an excellent summary of the basic processes occurring in plasmas together with a comprehensive introduction to the mathematical formulation of fluid (MHD) and kinetic theory. It provides an excellent introduction to the subject suitable for senior undergraduate students or entry-level graduate students.
Richard M. Thorne, University of California at Los Angeles

Contents

Preface

This book is intended to provide a general introduction to plasma phenomena at a level appropriate for advanced undergraduate students or beginning graduate students. The reader is expected to have had exposure to basic electromagnetic principles including Maxwell's equations and the propagation of plane waves in free space. Despite its importance in both science and engineering the body of literature on plasma physics is often not easily accessible to the non-specialist, let alone the beginner. The diversity of topics and applications in plasma physics has created a field that is fragmented by topic-specific assumptions and rarely presented in a unified manner with clarity. In this book we strive to provide a foundation for understanding a wide range of plasma phenomena and applications. The text organization is a successive progression through interconnected physical models, allowing diverse topics to be presented in the context of unifying principles. The presentation of material is intended to be compact yet thorough, giving the reader the necessary tools for further specialized study. We have sought a balance between mathematical rigor championed by theorists and practical considerations important to experimenters and engineers. Considerable effort has been made to provide explanations that yield physical insight and illustrations of concepts through relevent examples from science and technology.

The material presented in this book was initially put together as class notes for the EE356 Elementary Plasma Physics course, newly introduced and taught by one of us (USI) at Stanford University in the spring quarter of 1998. The course was then taught regularly every other year, for graduate students from the departments of Electrical Engineering, Materials Science, Mechanical Engineering, Applied Physics, and Physics. Over the years, several

PhD students, including Nikolai Lehtinen, Georgios Veronis, Jacob Bortnik, Michael Chevalier, Timothy Chevalier, and Prajwal Kulkarni, contributed to the course in their work as teaching assistants. The course was co-taught by Prajwal Kulkarni and one of us (MG) in the Spring of 2008, and by Brant Carlson in the Spring of 2010. We offer our thanks to each of these colleagues for their enthusiastic help and contributions, as well as to the many students enrolled in the course who helped improve its content with their contributions.

More generally, we owe considerable gratitude to all the other researchers and students of the Very Low Frequency Group at Stanford University who have been a source of valuable feedback and expertise, and to our administrative assistants, Shaolan Min and Helen Wentong Niu, for their contributions. We would like specifically to acknowledge Dr. Prajwal Kulkarni, for his pedagogical insights that have helped shape this text, and Dr. Brant Carlson, for valuable help in editing the manuscript.

CHAPTER

1 Introduction

This text concerns the basic elementary physics of *plasmas*, which are a special class of gases made up of a large number of electrons and ionized atoms and molecules, in addition to neutral atoms and molecules as are present in a normal (non-ionized) gas. The most important distinction between a plasma and a normal gas is the fact that mutual Coulomb interactions between charged particles are important in the dynamics of a plasma and cannot be disregarded. When a neutral gas is raised to a sufficiently high temperature, or when it is subjected to electric fields of sufficient intensity, the atoms and molecules of the gas may become ionized, electrons being stripped off by collisions as a result of the heightened thermal agitation of the particles. Ionization in gases can also be produced as a result of illumination with ultraviolet light or X-rays, by bombarding the substance with energetic electrons and ions, or in other ways. When a gas is ionized, even to a rather small degree, its dynamical behavior is typically dominated by the electromagnetic forces acting on the free ions and electrons, and it begins to conduct electricity. The charged particles in such an ionized gas interact with electromagnetic fields, and the organized motions of these charge carriers (e.g., electric currents, fluctuations in charge density) can in turn produce electromagnetic fields. The ability of an ionized gas to sustain electric current is particularly important in the presence of a magnetic field. The presence of mobile charged particles in a magnetic field yields a Lorentz force $q\mathbf{v} \times \mathbf{B}$. When applied to a collection of particles this force leads to an electromagnetic body force $\mathbf{J} \times \mathbf{B}$ which can dominate the gas dynamics. As a result, the most novel and spectacular behavior of plasmas is exhibited in the context of their interaction with a magnetic field.

During the 1920s, I. Langmuir and colleagues first showed that characteristic electrical oscillations of very high frequency can exist in an ionized gas that is neutral or quasi-neutral, and introduced the terms *plasma* and *plasma oscillations*,[1] in recognition of the fact that these oscillations resembled those of jelly-like substances [1, 2]. When subjected to a static electric field, the charge carriers in an ionized gas rapidly redistribute themselves in such a way that most of the gas is shielded from the field, in a manner quite similar to the redistribution of charge which occurs within a metallic conductor placed in an electric field, resulting in zero electric field everywhere inside. Langmuir gave the name "plasma" specifically to the relatively field-free regions of the ionized gas, which are not influenced by the boundaries. Near the boundaries, typically metallic surfaces held at prescribed potentials, strong space-charge fields exist in a transition region Langmuir termed the *plasma sheath*. The sheath region has properties that differ from the plasma, since the motions of charged particles within the sheath are predominantly influenced by the potential of the boundary. The particles in the sheath form an electrical screen between the plasma and the boundary. We will find later that the screening distance is a function of the density of charged particles and of their temperature.

The plasma medium is often referred to as the fourth state of matter, since it has properties profoundly different from those of the gaseous, liquid, and solid states. All states of matter represent different degrees of organization, corresponding to certain values of binding energy. In the solid state, the important quantity is the binding energy of molecules in a crystal. If the average kinetic energy of a molecule exceeds the binding energy (typically a fraction of an electron volt), the crystal structure breaks up, either into a liquid or directly into a gas (e.g., iodine). Similarly, a certain minimum kinetic energy is required in order to break the bonds of the van der Waals forces in order for a liquid to change into a gas. In order for matter to make the transition to its fourth state and exist as a plasma, the kinetic energy per plasma particle must exceed the ionizing potential of atoms (typically a few electron volts). Thus, the state of matter is basically determined by the average kinetic energy per particle. Using water as a convenient example, we note that at low temperatures the bond between the H_2O molecules holds them tightly together against the low energy of molecular motion, so that the matter is in the solid state (ice). At room temperature, the

[1] The word "plasma" first appeared as a scientific term in 1839 when the Czech biologist J. Purkynie coined the term "protoplasma" to describe the jelly-like medium containing a large number of floating particles which make up the nuclei of the cells. The word "plasma" thus means a mold or form, and is also used for the liquid part of blood in which corpuscules are suspended.

increased molecular energy permits more widespread movements and currents of molecular motion, and we have the liquid state (water). Since the particle motions are random, not all particles have the same energy, with the more energetic ones escaping from the liquid surface to form a vapor above it. As the temperature of the water is further increased, a larger fraction of molecules escapes, until the whole substance is in the gaseous phase (steam). If steam is subjected to further thermal heating, illumination by UV or X-rays, or bombardment by energetic particles, it becomes ionized (plasma).

Although by far most of the Universe is ionized and is therefore in a plasma state, on our planet plasmas have to be generated by special processes and under special conditions. While we live in a bubble of essentially non-ionized gas in the midst of an otherwise ionized environment, examples of partially ionized gases or plasmas, including fire, lightning, and the aurora borealis have long been part of our natural environment. It is in this connection that early natural philosophers held that the material Universe is built of four "roots," earth, water, air, and fire, curiously resembling our modern terminology of solid, liquid, gas, and plasma states of matter. A transient plasma exists in the Earth's atmosphere every time a lightning stroke occurs, but is clearly not very much at home and is short-lived. Early work on electrical discharges included generation of electric sparks by rubbing a large rotating sphere of sulphur against a cloth [3], production of sparks by harnessing atmospheric electricity in rather hazardous experiments [4], and studies of dust patterns left by a spark discharge passing through the surface of an insulator [5]. However, it was only when electrical and vacuum techniques were developed to the point where long-lived and relatively stable electrical discharges were available that the physics of ionized gases emerged as a field of study. In 1879, W. Crookes published the results of his investigations of discharges at low pressure and remarked: "The phenomena in these exhausted tubes reveal to physical science a new world, a world where matter may exist in a fourth state ..." [6]. A rich period of discoveries followed, leading to Langmuir's coining of the word "plasma" in 1929, and continuing into the present as a most fascinating branch of physics.

Although a plasma is often considered to be the fourth state of matter, it has many properties in common with the gaseous state. At the same time, the plasma is an ionized gas in which the long range of Coulomb forces gives rise to collective interaction effects, resembling a fluid with a density higher than that of a gas. In its most general sense, a plasma is any state of matter which contains enough free, charged particles for its dynamical behavior to be dominated by electromagnetic forces. Plasma physics therefore encompasses the solid state, since electrons in metals and semiconductors fall into this category [7]. However, the redistribution of charge and the screening of the inner regions of a metal occur extremely quickly (typically $\sim 10^{-19}$ s) as a result of the very high density of free charges. Most applications of plasma physics are concerned with ionized gases. It turns out that a very low degree of ionization is sufficient for a gas to exhibit electromagnetic properties and behave as a plasma: a gas achieves an electrical conductivity of about half its possible maximum at about 0.1% ionization and has a conductivity nearly equal to that of a fully ionized gas at 1% ionization. The degree of ionization can be defined as the ratio $N_e/(N_e + N_n)$, where N_e is the electron density and N_n is the density of neutral molecules. (Since most plasmas are macroscopically neutral, as we will see later, the density of positive ions is equal to the density of electrons, i.e., $N_i = N_e$.) As an example, the degree of ionization in a fluorescent tube is $\sim 10^{-5}$, with $N_n \simeq 10^{22}$ m^{-3} and $N_e \simeq 10^{17}$ m^{-3}. Typically, a gas is considered to be a weakly (strongly) ionized gas if the degree of ionization is less than (greater than) 10^{-4}.

The behavior of weakly ionized plasmas differs significantly from that of strongly ionized plasmas. In a plasma with a low density of charged particles (i.e., low value of N_e), the effect of the presence of neutral particles overshadows the Coulomb interactions between charged particles. The charged particles collide more often with neutrals than they interact (via the Coulomb repulsion force) with other charged particles, inhibiting collective plasma effects. As the degree of ionization increases, collisions with neutrals become less and less important and Coulomb interactions become increasingly important. In a fully ionized plasma, all particles are subject to Coulomb collisions.

The Sun and the stars are hot enough to be almost completely ionized, with enormous densities ($N_e \simeq 10^{33}$ m^{-3}), and the interstellar gas is sparse enough to be almost completely ionized as a result of stellar radiation. Starting at about 60 km altitude the Sun bathes our atmosphere in a variety of radiations and the energy in the ultraviolet part of the spectrum is absorbed by atmospheric gas. In the process, significant numbers of air molecules and atoms receive enough energy to become ionized. The resulting free electrons and positive ions constitute the *ionosphere*. Maximum ionization density occurs in the F-region of the ionosphere at about 350 km altitude, where $N_e \simeq 10^{12}$ m^{-3}. With atmospheric density at 350 km altitude being $N_n \simeq 3.3 \times 10^{14}$ m^{-3}, the degree of ionization is $\sim 10^{-2}$. At even higher altitudes, the air is thin enough so that it is almost completely ionized, and the motion of charged particles is dominated by the Earth's magnetic field, in a region known as the *magnetosphere*.

Plasmas have various uses in technology because of their unique electrical properties and ability to influence chemical processes. Artificial plasmas are generated by application of heat or strong electric fields. Ultraviolet radiation from plasmas is used in lighting and plasma display panels. The plasma state opens a whole new regime of chemistry not typically accessible to normal gases. Plasmas play a key role in processing materials, including those related to the production of integrated circuits. Plasmas can also be used to process waste, selectively kill bacteria and viruses, and weld materials. Achieving controlled thermonuclear fusion, which holds promise as an abundant and clean energy source, is essentially a plasma physics problem. Thus plasma physics is essential both to understanding the basic processes of our planet and to advancing important technological applications.

One of the most important properties of a plasma is its tendency to remain electrically neutral, i.e., to balance positive and negative free charge ($N_e \simeq N_i$) in any given macroscopic volume element. A slight imbalance in local charge densities gives rise to strong electrostatic forces that act in the direction of restoring neutrality. This property arises from the large charge-to-mass ratio (q_e/m_e) of electrons, so that any significant imbalance of charge gives rise to an electric field of sufficient magnitude to drag a neutralizing cloud of electrons into the positively charged region. If a plasma is subjected to an applied electric field, the free charges adjust so that the major part of the plasma is shielded from the applied field. In order to be considered a plasma, an ionized gas must have a sufficiently large number of charged particles to shield itself electrostatically within a distance smaller than other lengths of physical interest. The quantitative measure of this screening distance is the so-called *Debye length*, discussed below. We will see that the screening distance is proportional to $N_e^{-1/2}$. A simple analogy can be made with a person entering a forest. Beyond a certain distance within the forest there are enough trees to screen the edge of the forest from view. However, if the trees are too far apart and the forest is too small, the person may never lose sight of the edge, in which case such a group of trees would not be called a forest.

At first thought the fourth state of matter may appear to be the simplest to study since the elementary fundamental laws of charged-particle motion are perfectly known, i.e., classical electromagnetic theory (Maxwell's equations) and the Lorentz force equation.[2] However, analyses of plasma effects are much more

[2] Gravitational forces are much smaller than electromagnetic forces on earthly scales. The momenta ($p = mv$) of free electrons and ions in typical plasmas are usually high and

complicated, for a number of reasons: (i) Although the various movements of individual particles are all governed by the electromagnetic fields in which they move, these fields are themselves often greatly modified by the presence and motion of the particles. (ii) Atomic processes such as ionization, excitation, recombination, and charge exchange come into play and compete with one another in a complicated manner, with complicated dependencies on particle energies and densities. (iii) The fact that charged particles move results in a variety of transport phenomena arising from both short- and long-range Coulomb interactions between various particles. (iv) The long-range Coulomb forces give rise to a number of collective phenomena, including electrostatic oscillations and instabilities. (v) Most plasmas, and in particular hot plasmas, are typically confined in a magnetic field, with which they are strongly coupled. We may think that in spite of these difficulties it should be possible to solve the equation of motion for each and every particle. We could then find the electric and magnetic fields as functions of space and time by solving Maxwell's equations with the source terms (charged density ρ and current density $\mathbf{J}$) specified using the position and velocity vectors of *all* particles in the system. This type of approach is the domain of the discipline known as computer simulation of plasmas. However, noting that any natural plasma environment (such as the Earth's ionosphere) may contain $>10^{25}$ particles, one has a tremendous accounting problem, with too much information to keep track of.

The fundamental equations governing the behavior of a plasma with freely mobile, non-relativistic particles can be summarized as follows:

$$
\left[\begin{array}{l}\text{Initial}\\ \text{and}\\ \text{boundary}\\ \text{conditions}\end{array}\right] \rightarrow
\left[\begin{array}{r}\nabla\times\mathbf{H}=\mathbf{J}+\frac{\partial\mathbf{D}}{\partial t}\\ \nabla\times\mathbf{E}=-\frac{\partial\mathbf{B}}{\partial t}\\ \nabla\cdot\mathbf{D}=\rho\\ \nabla\cdot\mathbf{B}=0\\ \mathbf{B}=\mu\mathbf{H}\\ \mathbf{D}=\epsilon\mathbf{E}\end{array}\right]
\begin{array}{c}\xrightarrow{\mathbf{E},\,\mathbf{B}}\\ \xleftarrow{\mathbf{J},\,\rho}\end{array}
\left[\begin{array}{l}\frac{d\mathbf{v}_i}{dt}=\frac{q_i}{m_i}(\mathbf{E}+\mathbf{v}_i\times\mathbf{B})\\ \rho=\frac{1}{\Delta V}\sum_{\Delta V} q_i\\ \mathbf{J}=\frac{1}{\Delta V}\sum_{\Delta V} q_i\mathbf{v}_i\end{array}\right]
\leftarrow \left[\begin{array}{l}\text{Initial}\\ \text{conditions}\end{array}\right]
$$

For gaseous plasmas, the medium is essentially free space (aside from freely mobile charged particles, which are separately

their densities relatively low, so that the de Broglie wavelengths ($\lambda_e = h/p$, where $h \simeq 6.6 \times 10^{-34}$ J s is Planck's constant) are much smaller than the mean interparticle distance, so that quantum effects are negligible, except for some types of collisions between particles. As an example, electrons with 1 eV energy have $\lambda_e \simeq 1.2$ nm, while the interparticle distance even for an extremely high electron density of $N_e \simeq 10^{12}$ cm^{-3} is $\sim 10^{-4}$ cm $= 10^5$ nm $\gg 1.2$ nm.

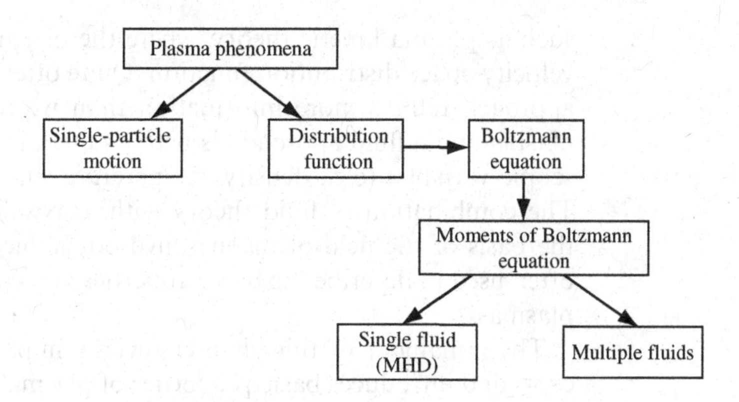

Figure 1.1 Hierarchy of approaches to plasma phenomena.

accounted for), so that we have $\mu = \mu_0$ and $\epsilon = \epsilon_0$. An equation of motion such as the one above specifying the particle acceleration $d\mathbf{v}/dt$ in terms of the fields **E** and **B**, exists for each and every positively or negatively charged particle in the plasma. For known fields $\mathbf{E}(\mathbf{r}, t)$ and $\mathbf{B}(\mathbf{r}, t)$ and specified initial conditions, these motion equations can be uniquely solved to determine the positions and velocities of every particle. However, the particle motions and locations lead to charge accumulations and current densities, i.e., ρ and **J**, which in turn modify the electric and magnetic fields. The charge and current densities are obtained from discrete charges by averaging over a macroscopically small volume ΔV, which nevertheless contains many individual particles, so that it makes sense to talk about continuous distribution of "density" of current and charge. Such averaging is appropriate, since the electric and magnetic fields in Maxwell's equations are also macroscopic fields, suitably averaged over both space and time.[3]

Given the complexity of plasma behavior, the field of plasma physics is best described as a web of overlapping models, each based on a set of assumptions and approximations that make a limited range of behavior analytically and computationally tractable. A conceptual view of the hierarchy of plasma models/approaches to plasma behavior that will be covered in this text is shown in Figure 1.1. We will begin with the determination of individual particle trajectories in the presence of electric and magnetic fields. Subsequently, it will be shown that the large number of charged particles in a plasma facilitates the use of statistical techniques

3 Such averaging is assumed to be done over spatial scales that are microscopically large (i.e., contain many individual particles) but which are nevertheless much smaller than any other relevant dimension in the context of a given application, or over time periods much shorter than the resolution of any measuring instrument.

such as plasma kinetic theory, where the plasma is described by a velocity-space distribution function. Quite often, the kinetic-theory approach retains more information than we really want about a plasma and a fluid approach is better suited, in which only macroscopic variables (e.g., density, temperature, and pressure) are kept. The combination of fluid theory with Maxwell's equations forms the basis of the field of magnetohydrodynamics (MHD), which is often used to describe the bulk properties and collective behavior of plasmas.

The remainder of this chapter reviews important physical concepts and introduces basic properties of plasmas.

1.1 Speed, energy, and temperature

The kinetic energy of a particle of mass m moving with a speed u is $E = \frac{1}{2}mu^2$. For an assembly of N particles with different kinetic energies, the average energy E_{av} per particle is given as

$$E_{\text{av}} = \frac{1}{2N}\sum_{i=1}^{N} m_i u_i^2.$$

However, there are other ways of measuring the average energy of an assembly of particles. We will see later that for any gas in thermal equilibrium at temperature T, the average energy per particle is $3k_B T/2$, where $k_B = 1.38 \times 10^{-23}\,\text{J}\,\text{K}^{-1}$ is Boltzmann's constant and T is the absolute temperature. A gas in thermal equilibrium has particles with all speeds, and the most probable distribution of these is the so-called Maxwellian (or Maxwell–Boltzmann) distribution, which in a one-dimensional system is given by

$$f(u) = Ae^{-\frac{1}{2}\frac{mu^2}{k_B T}},$$

where $f(u)du$ is the number of particles per unit volume with velocity in the range u to $u + du$. The multiplier A can be determined by noting that the total density N of particles can be found from $f(u)$:

$$N = \int_{-\infty}^{\infty} f(u)du \quad \rightarrow \quad A = N\sqrt{\frac{m}{2\pi k_B T}},$$

where we have used the fact that $\int_{-\infty}^{\infty} e^{-a^2\zeta^2} d\zeta = \sqrt{\pi}/a$. The width of the velocity distribution is characterized by the constant temperature T. The average kinetic energy of the particles can be calculated from $f(u)$ as follows:

$$E_{\text{av}} = \frac{\int_{-\infty}^{\infty} \frac{1}{2} m u^2 f(u) du}{\int_{-\infty}^{\infty} f(u) du}. \tag{1.1}$$

Upon integration of the numerator by parts, we have $E_{\text{av}} = \frac{1}{2} k_{\text{B}} T$. Extending this result to three dimensions, we find that $E_{\text{av}} = \frac{3}{2} k_{\text{B}} T$, or $\frac{1}{2} k_{\text{B}} T$ per degree of freedom.

With $E_{\text{av}} = \frac{3}{2} k_{\text{B}} T$ a gas at 1 K corresponds to an average energy of 2.07×10^{-23} J per particle. Often it is more convenient to measure particle energy in terms of electron volts (eV). If a particle has an electric charge equal in magnitude to the electronic charge $q_e = -1.602 \times 10^{-19}$ C, and is accelerated through a potential difference of 1 V, it has an energy of 1.602×10^{-19} J. This unit of energy is defined as an electron volt, commonly abbreviated as 1 eV. Thus, to express energy in terms of eV, we must divide the kinetic energy in J by $|q_e| = 1.602 \times 10^{-19}$ C. Hence we have

$$E = \frac{mu^2}{2|q_e|} \text{ eV}.$$

The unit of eV is particularly useful in dealing with charged particles, since it directly indicates the potential necessary to produce a singly charged particle of some particular energy.

It is often convenient to express the temperature of a gas in thermodynamic equilibrium in units of energy (eV). Typically, the energy corresponding to $k_{\text{B}} T$ is used to denote the temperature. Using $k_{\text{B}} T = 1 \text{ eV} = 1.6 \times 10^{-19}$ J, the conversion factor is $1 \text{ eV} = 11\,600$ K. Thus, when we refer to a 0.5 eV plasma we mean that $k_{\text{B}} T = 0.5$ eV, or a plasma temperature of $T = 5800$ K, or an average energy (in three dimensions) of $E_{\text{av}} = \frac{3}{2} k_{\text{B}} T = 0.75$ eV. Thus a plasma at a temperature of 300 K has an average energy of 0.0388 eV, and a plasma with 10 keV average energy is at a temperature of $T = 7.75 \times 10^7$ K. With reference to our earlier discussion of the plasma state coming into being at sufficiently high temperature for the material to be ionized, the temperature required to form plasmas from pure substances in thermal equilibrium ranges from ~4000 K for cesium (initially used by Langmuir) to ~20 000 K for elements such as helium which are particularly difficult to ionize.

It should be noted that temperature is an equilibrium concept, and we may not always be faced with equilibrium situations. In such cases, a true temperature cannot always be assigned, although we may still use the term in the sense of average energy. (Note that the average energy can always be calculated for any given distribution using a procedure similar to that given in (1.1).) It should also be noted that by temperature we mean the quantity sometimes called "kinetic temperature," simply the state of energy of the particles. A high value of T does not necessarily mean a lot of heat, since the latter also depends on heat capacity, determined also by the number of particles. As an example, the electron kinetic temperature inside a fluorescent lamp is $\sim$11 000 K but it does not feel nearly as "hot" when one holds the tube while it is lit. The reason is that the free-electron density inside the tube is much less that the number of particles in a gas at atmospheric pressure, so that the total amount of heat transferred to the walls by the impact of electrons is low.

A plasma can have several temperatures at the same time, since often the ions and electrons have separate Maxwellian distributions (of different widths) corresponding to temperatures, respectively, of T_i and T_e. Such equilibria can arise because the collision rate among ions or electrons themselves is larger than that between ions and electrons. Although each species can thus have its own thermal equilibrium, in time the tendency would be for the temperatures to equalize. In a magnetized plasma (i.e., a plasma under the influence of a strong magnetic field), even a single species can have two different temperatures, since the Lorentz forces acting on it along the magnetic field are different than those perpendicular to the field. These different temperatures, typically denoted $T_{e\parallel}$ and $T_{e\perp}$, respectively correspond to Maxwellian distributions of electron velocities along and perpendicular to the magnetic field.

1.2 Quasi-neutrality and plasma oscillations

It was mentioned above that a most fundamental property of a plasma is its tendency to remain electrically neutral and that any small change in local neutrality resulting from charge separation leads to large electric fields which pull electrons back to their original positions. Because of their inertia, the electrons which are pulled back typically oscillate about the initially charged region. However, since this oscillation is typically at a rather high frequency, quasi-neutrality is preserved on a time-average basis.

In this section, we briefly describe the dynamics of this oscillatory behavior of a plasma, which will be studied in detail in later chapters. Consider a steady initial state in which there is a uniform number density $N_e = N_0$ of electrons, neutralized by an equal

Figure 1.2 A simple plasma slab. The positively charged region consists of atoms which have lost one electron, while the negatively charged region has excess electrons.

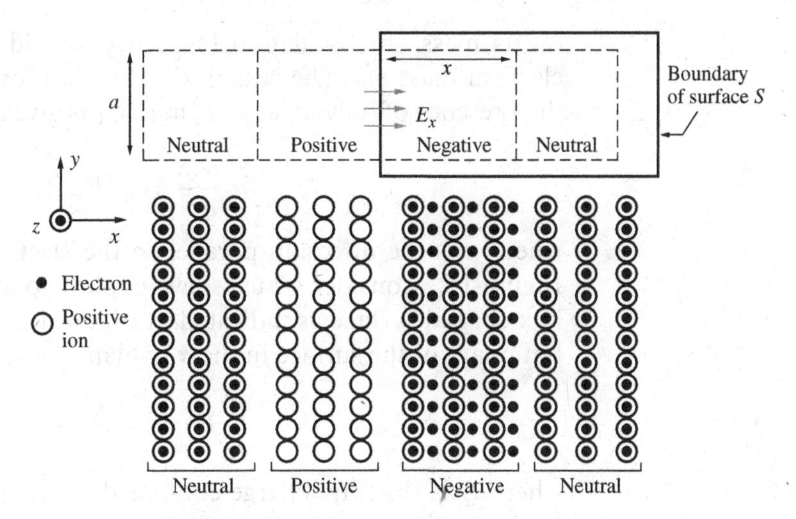

number of ions, i.e., $N_i = N_e = N_0$. We further assume that the plasma is "cold," meaning that the thermal motion of electrons and ions can be neglected. We now perturb this system by transferring a group of electrons (assumed for simplicity to be in a one-dimensional slab) from a given region of space to a nearby region, leaving net positive charge behind (i.e., the ions), as shown in Figure 1.2. This local charge separation gives rise to an electric field **E**, which exerts a force on the electrons and ions. Since the electrons are much lighter than the ions, we make the safe assumption that the electrons move much faster and hence the motion of the ions can be neglected. The electric field **E** acts to reduce the charge separation by pulling the electrons back to their initial locations. The electrons are thus accelerated back to their initial positions. However, as they acquire kinetic energy in this process, their inertia carries them past their neutral positions. The plasma once again becomes non-neutral, and again an electric field is set up (now pointing in a direction opposite to that shown in Figure 1.2) to retard their motion. Now the electrons accelerate to the right and go past their equilibrium positions as a result of their inertia, and once again the charge displacement depicted in Figure 1.2 is set up. In the absence of any damping (due, for example, to collisions of the electrons with ions or other electrons), this oscillatory motion would continue forever. In relatively tenuous (low-density) plasmas, collisional damping can be neglected, so any slight disturbance of the system leads to the oscillation process just described.

We now consider the frequency of this oscillation. Intuitively, we expect that the restoring electric force depends on the amount of charge displaced, i.e., the charge q_e times the density or number of electrons per volume. Since the inertia of a particle depends on its mass, the oscillation frequency should also depend on the electron mass m_e. The equation of motion for a single electron in the presence of only an electric field E_x is given by

$$m_e \frac{d^2 x}{dt^2} = q_e E_x, \tag{1.2}$$

where x is the direction parallel to the electric field, as shown in Figure 1.2. Consider Gauss's law, applied to a closed, rectangular, box-shaped surface as indicated in Figure 1.2, noting that only the boundary of the surface in the x–y plane is shown:

$$\oint_S \mathbf{E} \cdot d\mathbf{s} = \frac{Q}{\epsilon_0},$$

where Q is the total charge contained within the closed surface S. If the equilibrium density of the electrons is N_e, we must have $Q = AxN_e q_e$, where A is the cross-sectional area, and x denotes the displacement of the electrons. Assuming that a is the dimension of the rectangular-box surface in the y direction and that the depth of the box is Δz, we have

$$\oint_S \mathbf{E} \cdot d\mathbf{s} = -a\Delta E_x = \frac{Q}{\epsilon_0} = \frac{ax\Delta z N_e q_e}{\epsilon_0} \quad \rightarrow \quad E_x = -\frac{xN_e q_e}{\epsilon_0}.$$

Substituting back into Equation (1.2), we have

$$m_e \frac{d^2 x}{dt^2} + \frac{N_e q_e^2}{\epsilon_0 m_e} x = 0 \quad \rightarrow \quad \frac{d^2 x}{dt^2} + \omega_p^2 x = 0. \tag{1.3}$$

The solution of this equation is time-harmonic at a frequency $\omega_p = \sqrt{N_e q_e^2/(\epsilon_0 m_e)}$. In other words,

$$x = C_1 \cos(\omega_p t) + C_2 \sin(\omega_p t),$$

where C_1 and C_2 are constants to be determined by initial conditions. Equation (1.3) describes the displacement for free oscillations of the simple plasma slab of Figure 1.2. However, our simplified analysis indicates that any disturbances from equilibrium oscillate at an angular frequency ω_p, which is called the *plasma frequency* or specifically the *electron plasma frequency*:

$$\text{Plasma frequency} \quad \boxed{\omega_p = \sqrt{\frac{N_e q_e^2}{\epsilon_0 m_e}}}. \tag{1.4}$$

It is interesting to note that the plasma oscillations as derived above appear to be entirely local, so that a disturbance does not propagate away. In reality, plasma oscillations do propagate as a result of finite boundaries and thermal effects, as we will see later. One way to think about this is to consider the fact that as thermally agitated electrons stream into adjacent layers of the plasma they carry information about the disturbance (not necessarily through collisions like acoustic waves in a neutral gas but actually via the electric field). The thermal effects can be accounted for by adding an appropriate pressure-gradient term on the right-hand side of (1.2).

In a partially ionized gas where collisions (of electrons and ions with neutrals) is important, plasma oscillations can only develop if the mean free time τ_n between collisions is long enough compared to the oscillation period, or $\omega_p \tau_n > 1$. This condition is sometimes noted as a criterion for an ionized gas to be considered a plasma. However, in some applications, for example at the lower altitudes of the Earth's ionosphere, it is precisely the collisions that lead to some of the most interesting physical effects. Also, there are many other electromagnetic and electrostatic wave modes that can exist in a plasma over a very broad range of frequencies, so that one is not always interested only in plasma oscillations. In general, whether a plasma exhibits the collective wave effects in question depends on a comparison between the frequency of interest (ω) and τ_n.

1.3 Debye shielding

Electron plasma oscillations as discussed above are excited as a result of the effort of the plasma to assert its neutrality in response to a macroscopic perturbation of its essentially neutral equilibrium state. The macroscopic electrical neutrality of the plasma is thus preserved on average over the short time period of these oscillations. Since there are no charge separations, the plasma cannot sustain macroscopic electric fields. On the other hand, if a plasma is deliberately subjected to an external electric field, its free charges redistribute so that the major part of the plasma is shielded from the field. Suppose that the equilibrium state of the plasma is disturbed by the imposition of an electric field due to an external charged particle $+Q$. This electric field may also be that of one of the plasma particles isolated for observation. We now wish to examine the mechanism by which the plasma strives to re-establish its macroscopic electrical neutrality in the presence of this disturbing electric field.

Suppose that we immerse the test particle $+Q$ within an initially uniform plasma at time $t = 0$, such that $N_i = N_e = N_0$. The initial net charge density is thus zero and the electric potential is that due to a single test charge,

$$\Phi(\mathbf{r}) = \frac{1}{4\pi\epsilon_0}\frac{Q}{r},$$

where we have assumed the test charge to be placed at the origin of our spherical coordinate system. Since the ions in the plasma are repelled by the test charge and the electrons are attracted to it, the freely mobile electrons and ions in the plasma in time rearrange themselves in a new equilibrium distribution which takes account of the presence of the test charge. However, the ions move much more slowly than the electrons, so we can assume that their motion on the time scale of our experiment can be neglected. The ion density thus remains the same as before, i.e., $N_i = N_0$. However, the density of electrons near the test charge increases ($N_e > N_0$), so that the new potential distribution $\Phi(\mathbf{r})$ must be evaluated using Poisson's equation:

$$\nabla^2\Phi(\mathbf{r}) = -\frac{\rho}{\epsilon_0} = -\frac{q_e(N_e - N_i)}{\epsilon_0}, \tag{1.5}$$

where we have taken note of the fact that the excess free charge density $\rho = q_e(N_e - N_i)$. We will see later that under equilibrium conditions in the presence of an electrostatic potential $\Phi(\mathbf{r})$ (due to the action of an externally applied conservative electric field $\mathbf{E}$), a non-uniform distribution of particles is established in a plasma, with the number density of the particles given by

$$N(\mathbf{r}) = N_0 e^{-q\Phi(\mathbf{r})/k_B T}.$$

According to this, the electron density distribution in our spherically symmetric system is given by

$$N_e(r) = N_0 e^{-q_e\Phi(r)/k_B T_e}, \tag{1.6}$$

so that, in thermal equilibrium, the electron density is greatest at those locations where the electric potential $\Phi(r)$ is highest. Since T_e is the electron temperature, we note that the density variation is greater when the electron gas is cold than when it is hot. Substituting (1.6) into the spherical coordinate version of (1.5), we have

$$\frac{1}{r^2}\frac{d}{dr}\left(r^2\frac{d\Phi}{dr}\right) = -\frac{q_e N_0}{\epsilon_0}\left[\exp\left(\frac{-q_e\Phi(r)}{k_B T}\right) - 1\right], \quad r > 0. \tag{1.7}$$

This equation is non-linear and must be integrated numerically. However, an approximate solution can be obtained by assuming that the perturbing electrostatic potential is weak, i.e., $|q_e\Phi| \ll k_B T$, in which case we can express the exponential in (1.7) as a power series and retain only the first two terms. We then have

$$\frac{1}{r^2}\frac{d}{dr}\left(r^2\frac{d\Phi}{dr}\right) \simeq \left[\frac{N_0 q_e^2}{\epsilon_0 k_B T_e}\right]\Phi(r) = \frac{1}{\lambda_D^2}\Phi(r),$$

where λ_D, known as the *Debye length* or *Debye shielding length*, is defined as

$$\text{Debye length} \quad \boxed{\lambda_D = \sqrt{\frac{\epsilon_0 k_B T_e}{N_0 q_e^2}}}. \tag{1.8}$$

It is useful to note that $\lambda_D = \omega_p^{-1}\sqrt{k_B T_e/m_e}$.

The solution of the simplified form of (1.7) can be shown to be

$$\Phi(r) = \left[\frac{1}{4\pi\epsilon_0}\frac{Q}{r}\right] e^{-r/\lambda_D}. \tag{1.9}$$

We note that as $r \to 0$ the potential is essentially that of a free charge in free space, whereas for $r \gg \lambda_D$ the potential $\Phi(r) \to 0$ falls much faster than it does for a point charge in free space. While the Coulomb force in vacuum is relatively long-range, in a plasma this force extends only about a Debye length from the source, as a result of the Debye shielding cloud. For a positive test charge, the shielding cloud contains an excess of electrons. Using Gauss's law, it can be shown that the net charge within the shielding cloud is equal and opposite to that of the test charge. The size of the shielding cloud increases as the electron temperature increases, because electrons with greater kinetic energy are better able to overcome the Coulomb attraction associated with the potential. Also, λ_D is smaller for a denser plasma because more electrons (per unit volume) are available to populate the shielding cloud.

Example 1-1 Debye length and plasma frequency
Compute the Debye length for the plasma found in a typical plasma television cell with the following parameters: $N_e = 10^{19}$ m^{-3}, $k_B T = 1$ eV. The cell dimensions are on the order of 100 μm and the plasma is excited using a 250 V signal at 100 kHz.

Solution: To determine the Debye length we need first to find the temperature in Kelvin and then use (1.8). Using the conversion factor of 1 eV = 11 600 K we easily get the plasma temperature and can apply Equation (1.8) as follows:

$$\begin{aligned}\lambda_D &= \sqrt{\frac{\epsilon_0 k_B T_e}{N_0 q_e^2}} \\ &= \sqrt{\frac{(8.85 \times 10^{-12}\,\mathrm{C^2\,m\,J^{-1}})(1.38 \times 10^{-23}\,\mathrm{J\,K^{-1}})(11\,600\,\mathrm{K})}{(10^{19}\,\mathrm{m^{-3}})(1.602 \times 10^{-19}\,\mathrm{C})^2}} \\ &= 2.35 \times 10^{-6}\ \mathrm{m} = 2.35\ \mu\mathrm{m}.\end{aligned}$$

Not surprisingly, the Debye length is much smaller than the physical dimensions of the system. To calculate the plasma frequency we can use Equation (1.4):

$$\omega_p = \sqrt{N_0 q_e^2/(\epsilon_0 m_e)} \tag{1.10}$$

$$= \sqrt{\frac{(10^{19}\,\mathrm{m^{-3}})(1.602 \times 10^{-19}\,\mathrm{C})^2}{(8.85 \times 10^{-12}\,\mathrm{C^2\,m\,J^{-1}})(9.1 \times 10^{-31}\,\mathrm{kg})}} = 179\ \mathrm{GHz} \tag{1.11}$$

Thus the plasma frequency is much higher than the driving frequency of the electronics, which prevents the most basic plasma instabilities from developing. The plasma in a plasma television cell emits UV radiation which when incident on a phosphor coating on the cell is converted to either red, blue, or green light depending on the phosphor.

We can note from (1.9) that the potential very near the test charge is very large (owing to the r^{-1} term) so that the condition $|q_e\Phi| \ll k_B T$ is not likely to be valid. However, this is not a serious limitation since the main purpose of our development is to find the order of magnitude of the distance from the test particle where the potential becomes vanishingly small. A useful numerical expression for λ_D is $\lambda_D \simeq 69\sqrt{T_e/N_0}$, with T_e in units of K and N_0 in units of $\mathrm{m^{-3}}$. In the topside ionosphere of the Earth we have $T_e \simeq 1000$ K and $N_0 \simeq 10^{11}\ \mathrm{m^{-3}}$, so that $\lambda_D \simeq 0.007$ m. Thus, the Debye length is much smaller than the spatial scales relevant to the ionosphere, which is of the order hundreds of kilometers. In the solar wind near 1 AU, we have $T_e \simeq 10^5$ K and $N_0 \simeq 10^7\ \mathrm{m^{-3}}$, so that $\lambda_D \simeq 7$ m, again much smaller than the physical dimensions of the system.

The values of Debye length are generally quite small. If the dimension L of a physical system is much larger than λ_D, then local concentrations of charge that arise for one reason or another, or

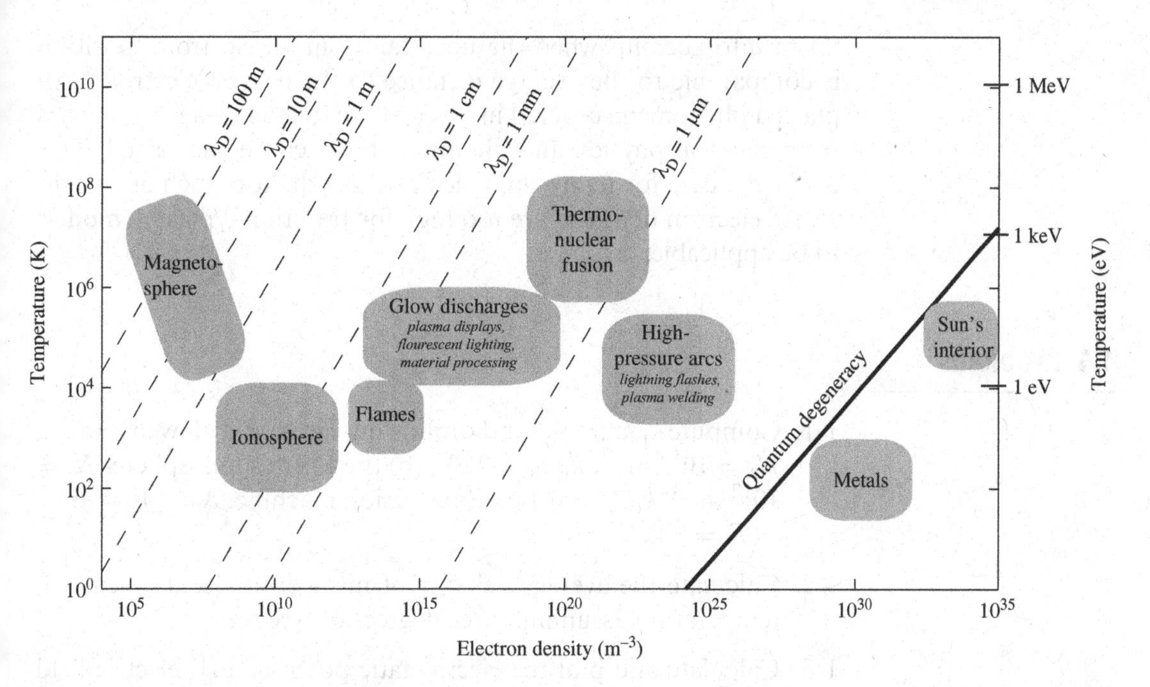

Figure 1.3 Range of temperature, electron density, and Debye length for typical plasma phenomena in nature and in technological applications. Only phenomena to the left of the quantum degeneracy line are considered plasmas and can be treated with formulations from classical physics.

external potentials introduced to the system (e.g., potentials applied on walls), are shielded out in a distance short compared to L, leaving the bulk of the plasma free of large electric potentials or fields. Outside of the sheath on the wall or on an obstacle, $\nabla^2\Phi$ is small and we have $N_e \simeq N_i$, typically to a very high degree of accuracy. The plasma is thus said to be quasi-neutral, so that $N_0 \simeq N_e \simeq N_i$, and N_0 is a common density that can be referred to as the plasma density. An important criterion for an ionized gas to be a plasma is that it be dense enough so that $L \gg \lambda_D$.

Each particle in a plasma can act as a test charge and carry its own shielding cloud. The concept of Debye shielding developed above requires the presence of a sufficiently large number of particles so that "density" can be defined in a statistically meaningful way. A useful parameter in this connection is the number of particles in a Debye sphere, given as

$$N_D = N_0 \left[\frac{4\pi\lambda_D^3}{3}\right] = \frac{1.38 \times 10^6\, T_e^{3/2}}{N_0^{1/2}},$$

where T_e is in K. Thus, a second criterion for an ionized gas to be considered a plasma is that $N_D \gg 1$.

Figure 1.3 shows the ranges of temperature, electron density, and Debye length for typical plasmas found in nature and in technological applications. Also shown in Figure 1.3 is the separation between the quantum and classical regimes. Quantum effects need to be taken into account when the uncertainty in an electron's position is comparable to the average distance to the nearest electron.[4] All plasma phenomena covered in this text are treated using approaches from classical physics, and quantum effects can be neglected. This is not the case for many solid electrical conductors such as metals, where electron densities are too high for traditional plasma models to be applicable.

1.4 Problems

1-1. Compute λ_D and N_D for the following cases: (a) a glow discharge, $N_e = 10^{16}$ m^{-3}, $k_B T_e = 2$ eV; (b) the Earth's ionosphere, $N_e = 10^{12}$ m^{-3}, $k_B T_e = 0.1$ eV; (c) a fusion machine, $N_e = 10^{23}$ m^{-3}, $k_B T_e = 9$ keV.

1-2. Calculate the average velocity of nitrogen molecules at room temperature assuming three degrees of freedom.

1-3. Calculate and plot the electrostatic potential and electric field of a test particle of charge $+Q$ in free space and in a plasma of number density N_0 and temperature T. Label the distance axis of your plot in units of Debye length.

1-4. A metal sphere of radius $r = a$ with charge Q is placed in a neutral plasma with number density N_0 and temperature T. Calculate the effective capacitance of the system. Compare this to the capacitance of the same sphere placed in free space.

1-5. Consider two infinite, parallel plates located at $x = \pm d$, kept at a potential of $\Phi = 0$. The space between the plates is uniformly filled with a gas at density N of particles of charge q. (a) Using Poisson's equation, show that the potential distribution between the plates is $\Phi(x) = [Nq/(2\epsilon_0)](d^2 - x^2)$. (b) Show that for $d > \lambda_D$ the energy needed to transport a particle from one of the plates to the midpoint (i.e., $x = 0$) is greater than the average kinetic energy of the particles. (Assume a Maxwellian distribution of particle speeds.)

[4] The uncertainty in an electron's position is given by the Heisenberg uncertainty principle,

$$\Delta p \Delta x \geq \frac{h}{4\pi},$$

where Δp and Δx are the uncertainties in momentum and position, respectively, and h is Planck's constant, 6.626×10^{-34} J s.

References

[1] I. Langmuir, The interaction of electron and positive ion space charges in cathode sheaths. *Phys. Rev.,* **33** (1929), 954–89. DOI: 10.1103/PhysRev.33.954.

[2] L. Tonks and I. Langmuir, Oscillations in ionized gases. *Phys. Rev.,* **33** (1929), 195–210. DOI: 10.1103/PhysRev.33.195.

[3] O. von Guericke, *Experimenta Nova (ut vocantur) Magdeburgica de Vacuo Spatio* (Amstelodami: Janssonium, 1672).

[4] B. Franklin, *Experiments and Observations on Electricity* (London, 1751).

[5] G. C. Lichtenberg, De nova methodo naturam ac motum fluidi electrici investigandi. *Novi Commentarii Societatis Regiae Scientiarum* (1777).

[6] W. Crookes, On a fourth state of matter. *Proc. R. Soc. London*, **30** (1879), 469–72.

[7] R. Bowers, Plasma in solids. *Sci. Am.*, **209** (1963), 46–53.

CHAPTER

2 Single-particle motion

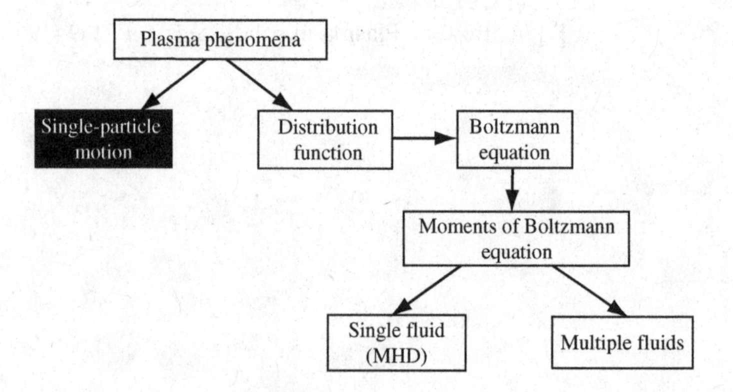

Depending on the density of charged particles, a plasma behaves either as a fluid, with collective effects being dominant, or as a collection of individual particles. In dense plasmas, the electrical forces between particles couple them to each other and to the electromagnetic fields, which affects their motions. In rarefied plasmas, the charged particles do not interact with one another and their motions do not constitute a large enough current to significantly affect the electromagnetic fields; under these conditions, the motion of each particle can be treated independently of any other, by solving the Lorentz force equation for prescribed electric and magnetic fields, a procedure known as the single-particle approach. In magnetized plasmas under the influence of an external static or slowly varying magnetic field the single-particle approach is only applicable if the external magnetic field is quite strong compared to the magnetic field produced by the electric current arising from the particle motions. The single-particle approach is applicable to investigating high-energy particles in the Earth's radiation belts and the solar corona, and also in practical devices such as cathode ray tubes and traveling-wave amplifiers. Although the single-particle approach may only be valid in special circumstances, understanding the individual particle motions is also an important first step in understanding the collective behavior of plasmas. Accordingly, in this chapter we will study single-particle motions, where the fundamental equation of motion for the particles under the influence of the Lorentz force is given by

$$m\frac{d\mathbf{v}}{dt} = q(\mathbf{E} + \mathbf{v} \times \mathbf{B}), \tag{2.1}$$

where m is the particle mass and $\mathbf{v}$ is its velocity. While we only consider non-relativistic motion ($|\mathbf{v}| \ll c$),[1] the above equation is valid for the relativistic case if we simply replace m with $m = m_0\left(1 - v^2/c^2\right)^{-1/2}$, where m_0 is the rest mass and $v = |\mathbf{v}|$. More commonly, the relativistic version of Equation (2.1) is written in terms of the particle momentum $\mathbf{p} = m\mathbf{v}$, rather than velocity $\mathbf{v}$.

2.1 Motion in a uniform B field: gyration

We start by considering the simplest cases of motion in uniform fields. When a particle is under the influence of a static electric field that is uniform in space, the particle simply moves with a constant acceleration along the direction of the field, and this case does not warrant further study. On the other hand, the motion of a charged particle under the influence of a static and uniform magnetic field is of fundamental interest, and is studied in this section. With only a static and uniform magnetic field present, Equation (2.1) reduces to

$$m\frac{d\mathbf{v}}{dt} = q\mathbf{v} \times \mathbf{B}. \tag{2.2}$$

Taking the dot-product of (2.2) with $\mathbf{v}$, we have

$$\mathbf{v} \cdot m\frac{d\mathbf{v}}{dt} = \mathbf{v} \cdot q(\mathbf{v} \times \mathbf{B})$$

$$m\frac{1}{2}\frac{d(\mathbf{v} \cdot \mathbf{v})}{dt} = q[\mathbf{v} \cdot (\mathbf{v} \times \mathbf{B})]$$

$$\frac{d}{dt}\left(\frac{mv^2}{2}\right) = 0,$$

[1] By the same token, we neglect any radiation produced by the acceleration of charged particles. At non-relativistic velocities, such radiation is quite negligible; the radiated electric field at a distance R from the particle is proportional to $q^2a^2/(c^2R^2)$, where q is the charge of the particle, c is the speed of light in free space, and a is the acceleration. For a discussion at an appropriate level, see Chapter 8 of [1].

where $v = |\mathbf{v}|$ is the particle speed and where we have noted that $(\mathbf{v} \times \mathbf{B})$ is perpendicular to $\mathbf{v}$ so that the right-hand side is zero. It is clear that a static magnetic field cannot change the kinetic energy of the particle, since the force is always perpendicular to the direction of motion. Note that this is true even for a spatially non-uniform field, since the derivation above did not use the fact that the field is uniform in space. We first consider the case of a magnetic field configuration consisting of field lines that are straight and parallel, with the magnetic field intensity constant in time and space. Later on, we will allow the magnetic field intensity to vary in the plane perpendicular to the field, while continuing to assume that the field lines are straight and parallel. We can decompose the particle velocity into its components parallel and perpendicular to the magnetic field, i.e.,

$$\mathbf{v} = \mathbf{v}_{\parallel} + \mathbf{v}_{\perp},$$

in which case we can rewrite (2.2) as

$$\frac{d\mathbf{v}_{\parallel}}{dt} + \frac{d\mathbf{v}_{\perp}}{dt} = \frac{q}{m}\,(\mathbf{v}_{\perp} \times \mathbf{B}),$$

since $\mathbf{v}_{\parallel} \times \mathbf{B} = 0$. This equation can be split into two equations in terms of $\mathbf{v}_{\parallel}$ and $\mathbf{v}_{\perp}$, respectively:

$$\frac{d\mathbf{v}_{\parallel}}{dt} = 0 \quad \rightarrow \quad \mathbf{v}_{\parallel} = \text{constant}$$

$$\frac{d\mathbf{v}_{\perp}}{dt} = \frac{q}{m}\,(\mathbf{v}_{\perp} \times \mathbf{B}).$$

It is clear from the above that the magnetic field has no effect on the motion of the particle in the direction along it, and that it only affects the particle velocity in the direction perpendicular to it. To examine the character of the perpendicular motion, consider a static magnetic field oriented along the z axis, i.e., $\mathbf{B} = \hat{\mathbf{z}}B$. We can write (2.2) in component form as

$$m\frac{dv_x}{dt} = qBv_y \tag{2.3a}$$

$$m\frac{dv_y}{dt} = -qBv_x \tag{2.3b}$$

$$m\frac{dv_z}{dt} = 0. \tag{2.3c}$$

The component of the velocity parallel to the magnetic field is often denoted as $v_{\parallel} = v_z$ and is constant since the Lorentz force $q(\mathbf{v} \times \mathbf{B})$ is perpendicular to $\hat{\mathbf{z}}$. To determine the time variations of v_x and v_y,

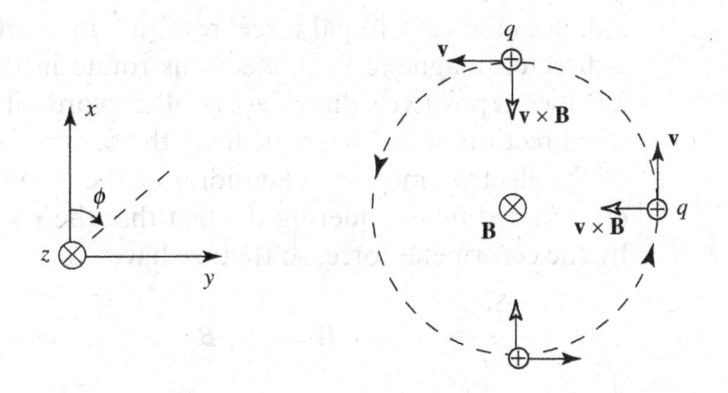

Figure 2.1 Motion of a particle in a magnetic field. A particle with positive charge q and velocity $\mathbf{v}$ experiences a force $q\mathbf{v} \times \mathbf{B}$ in the presence of a magnetic field $\mathbf{B}$.

we can take the second derivatives of (2.3a) and (2.3b) and substitute to find

$$\frac{d^2 v_x}{dt^2} + \omega_c^2 v_x = 0 \tag{2.4a}$$

$$\frac{d^2 v_y}{dt^2} + \omega_c^2 v_y = 0, \tag{2.4b}$$

where $\omega_c = -qB/m$ is the *gyrofrequency* or *cyclotron frequency*.

$$\text{Cyclotron frequency} \quad \boxed{\omega_c = -qB/m}. \tag{2.5}$$

Note that ω_c is an angular frequency (units of rad m^{-1}) and can be positive or negative, depending on the sign of q. A positive value of ω_c in a right-handed coordinate system indicates that the sense of rotation is along the direction of positive ϕ, where ϕ is the cylindrical coordinate azimuthal angle, measured *from* the x axis as shown in Figure 2.1. The solution of (2.4) is in the form of a harmonic motion, given by

$$v_x = v_\perp \cos(\omega_c t + \psi) \tag{2.6a}$$

$$v_y = v_\perp \sin(\omega_c t + \psi) \tag{2.6b}$$

$$v_z = v_\parallel, \tag{2.6c}$$

where ψ is some arbitrary phase angle which defines the orientation of the particle velocity at $t = 0$, and $v_\perp = \sqrt{v_x^2 + v_y^2}$ is the constant speed in the plane perpendicular to $\mathbf{B}$. To appreciate the above result physically, consider the coordinate system and the forces on the particle (assumed to have a positive charge q) as shown in Figure 2.1, at different points along its orbit. It is clear that the particle experiences a $\mathbf{v} \times \mathbf{B}$ force directed inward at all times, which balances the centrifugal force, resulting in a circular motion. For a z-directed magnetic field, electrons rotate in the right-hand sense, i.e., have a positive value of ω_c; in other words, if the thumb points in the direction of the magnetic field, the fingers rotate in the direction of the electron motion. The radius of the circular trajectory can be determined by considering the fact that the $\mathbf{v} \times \mathbf{B}$ force is balanced by the centripetal force, so that we have

$$-\frac{mv_\perp^2}{r} = q\mathbf{v} \times \mathbf{B} = qv_\perp B \quad \rightarrow \quad \boxed{r_c = \frac{-mv_\perp}{qB} = \frac{v_\perp}{\omega_c}},$$

where r_c is called the *gyroradius* or *Larmor radius*.[2] Note that the magnitude of the particle velocity remains constant, since the magnetic field force is at all times perpendicular to the motion. The magnetic field cannot change the kinetic energy of the particle; however, it does change the direction of its momentum. It is important to note that the gyrofrequency ω_c of the charged particle does not depend on its velocity (or kinetic energy) and is only a function of the intensity of the magnetic field. Particles with higher velocities (and thus higher energies) orbit in circles with larger radii but complete one revolution in the same time as particles with lower velocities which orbit in smaller circles. Particles with larger masses also orbit in circles with larger radii, but they complete one revolution in a longer time compared to those with smaller masses. A convenient expression for the gyrofrequency f_{ce} (in Hz) for electrons is

$$f_{ce} = \frac{\omega_c}{2\pi} \simeq 2.8 \times 10^6 B,$$

where B is in units of G (note that 10^4 G = 1 T or Wb m^{-2}). As an example, the Earth's magnetic field at the surface is of the order of $\sim$0.5 G, corresponding to a gyrofrequency of $f_{ce} \simeq 1.4$ MHz.

The particle position as a function of time can be found by integrating (2.6):

$$x = r_c \sin(\omega_c t + \psi) + (x_0 - r_c \sin\psi) \tag{2.7a}$$

$$y = -r_c \cos(\omega_c t + \psi) + (y_0 + r_c \cos\psi) \tag{2.7b}$$

$$z = z_0 + v_\parallel t, \tag{2.7c}$$

where x_0, y_0, and z_0 are the coordinates of the location of the particle at $t = 0$ and ψ is simply the phase with respect to a particular

[2] By the convention we have chosen, r_c can take a negative value. This is a mathematical formalism that allows for writing the expressions for particle trajectories for either positive or negative charges in compact form. The gyroradius should always be interpreted as a real physical distance.

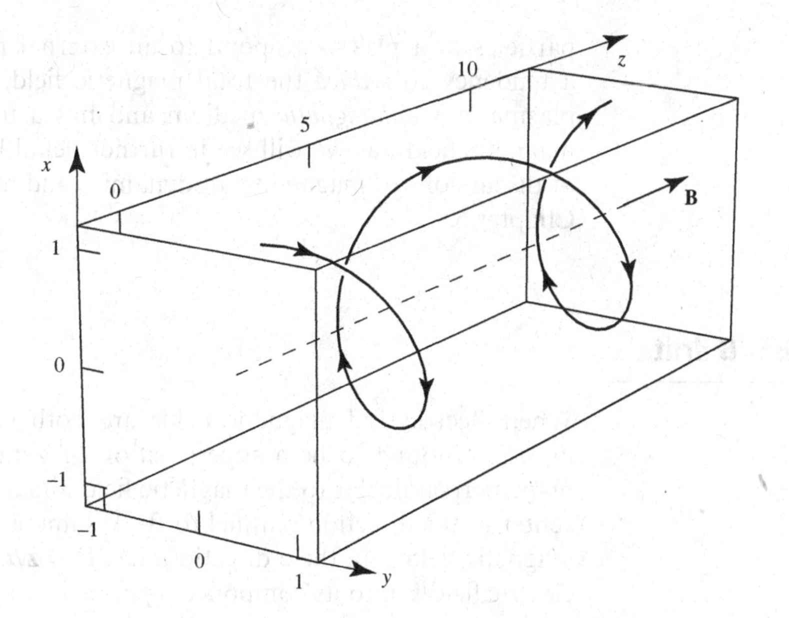

Figure 2.2 Electron guiding-center motion in a magnetic field $\mathbf{B} = \hat{\mathbf{z}}B$.

time of origin. Equations (2.7) show that the particle moves in a circular orbit perpendicular to **B** with an angular frequency ω_c and radius r_c about a *guiding center* $\mathbf{r}_g = \hat{\mathbf{x}}x_0 + \hat{\mathbf{y}}y_0 + \hat{\mathbf{z}}(z_0 + v_\parallel t)$. The concept of a guiding center is useful in considering particle motion in inhomogeneous fields, since the gyration is often much more rapid than the motion of the guiding center. Note from (2.6) that in the present case, the guiding center simply moves linearly along z at a uniform speed $v_\parallel$, although the particle motion itself is helical, as shown in Figure 2.2. The *pitch angle* of the helix is defined as

$$\text{Pitch angle} \quad \boxed{\alpha = \tan^{-1}\left(\frac{v_\perp}{v_\parallel}\right)}. \tag{2.8}$$

It is interesting to note that for both positive and negative charges, the particle gyration constitutes an electric current in the $-\phi$ direction (i.e., opposite to the direction of the fingers of the right hand when the thumb points in the direction of the $+z$ axis). The magnetic moment associated with such a current loop is given by $\mu =$ current $\times$ area or

$$\text{Magnetic moment} \quad \boxed{\mu = \underbrace{\left(\left|\frac{q\omega_c}{2\pi}\right|\right)}_{\text{current}} \underbrace{\left(\pi r_c^2\right)}_{\text{area}} = \frac{mv_\perp^2}{2B}}. \tag{2.9}$$

Note that the direction of the magnetic field generated by the gyration is opposite to that of the external field. Thus, freely mobile particles in a plasma respond to an external magnetic field with a tendency to *reduce* the total magnetic field. In other words, a plasma is a *diamagnetic* medium and has a tendency to exclude magnetic fields, as we will see in further detail later, in the context of discussions of magnetohydrodynamics and magnetic pressure in Chapter 6.

2.2 E × B drift

When electric and magnetic fields are both present, the particle motion is found to be a superposition of gyrating motion in the plane perpendicular to the magnetic field and a drift of the guiding center in the direction parallel to **B**. Assuming once again that the magnetic field is in the z direction, i.e., $\mathbf{B} = \hat{\mathbf{z}}B$, we decompose the electric field **E** into its components parallel and perpendicular to **B**:

$$\mathbf{E} = \mathbf{E}_\perp + \hat{\mathbf{z}}E_\parallel = \hat{\mathbf{x}}E_\perp + \hat{\mathbf{z}}E_\parallel,$$

where we have taken the electric field to be in the x direction, with no loss of generality. Noting that we can also decompose the particle velocity into its two components, i.e., $\mathbf{v}(t) = \mathbf{v}_\perp(t) + \hat{\mathbf{z}}v_z(t)$, the equation of motion can be written as

$$m\frac{d\mathbf{v}_\perp}{dt} = q(\hat{\mathbf{x}}E_\perp + \mathbf{v}_\perp \times \hat{\mathbf{z}}B) \tag{2.10a}$$

$$m\frac{dv_\parallel}{dt} = qE_\parallel. \tag{2.10b}$$

Equation (2.10b) simply indicates constant acceleration along **B**. For the transverse component, we seek a solution of the form

$$\mathbf{v}_\perp(t) = \mathbf{v}_\mathrm{E} + \mathbf{v}_\mathrm{ac}(t), \tag{2.11}$$

where $\mathbf{v}_\mathrm{E}$ is a constant velocity and $\mathbf{v}_\mathrm{ac}$ is the alternating component. Using (2.11) in (2.10a) we have

$$m\frac{d\mathbf{v}_\mathrm{ac}}{dt} = q(\hat{\mathbf{x}}E_\perp + \mathbf{v}_\mathrm{E} \times \hat{\mathbf{z}}B + \mathbf{v}_\mathrm{ac} \times \hat{\mathbf{z}}B). \tag{2.12}$$

We know from the previous section that the left-hand side and the last term on the right-hand side in (2.12) simply describe circular motion (gyration) at a rate $\omega_c = -qB/m$. Thus, if we choose $\mathbf{v}_E$ such that the first two terms on the right-hand side of (2.12) cancel, i.e.,

$$\hat{\mathbf{x}}E_\perp + \mathbf{v}_E \times \hat{\mathbf{z}}B = 0 \quad \rightarrow \quad \boxed{\mathbf{v}_E = \frac{\mathbf{E} \times \mathbf{B}}{B^2}}, \tag{2.13}$$

then (2.12) reduces to the form

$$m\frac{d\mathbf{v}_{ac}}{dt} = q\mathbf{v}_{ac} \times \hat{\mathbf{z}}B, \tag{2.14}$$

which, as mentioned, simply describes rotation at a frequency $\omega_c = -qB/m$. Note that we can use $\mathbf{E}$ rather than $\mathbf{E}_\perp$ in (2.13) since $\hat{\mathbf{z}}E_\parallel \times \mathbf{B} \equiv 0$. Thus, we see that the particle motion in the presence of electric and magnetic fields is given by

$$\mathbf{v}(t) = \hat{\mathbf{z}}v_\parallel(t) + \mathbf{v}_E + \mathbf{v}_{ac}(t), \tag{2.15}$$

consisting of steady acceleration along $\mathbf{B}$, uniform drift velocity $\mathbf{v}_E$ perpendicular to both $\mathbf{B}$ and $\mathbf{E}$, and gyration. Taking the time average of $\mathbf{v}(t)$ over one gyroperiod ($T_c = 2\pi/\omega_c$), we have

$$\langle \mathbf{v} \rangle = \frac{1}{T_c}\int_0^{T_c} \mathbf{v}(t)dt = \hat{\mathbf{z}}v_\parallel + \mathbf{v}_E,$$

showing that $\mathbf{v}_E = (\mathbf{E} \times \mathbf{B})/B^2$ is the average perpendicular velocity. It is interesting to note that the drift velocity $\mathbf{v}_E$ is independent of q, m, and $v_\perp = |\mathbf{v}_\perp|$. The reason can be seen from a physical picture of the drift, as shown in Figure 2.3. As the positively charged particle moves downward (against the electric field) during the first half of its cycle, it loses energy and its r_c decreases. In the second half of its cycle, it regains this energy as it now moves in the direction of the electric field. The acceleration and deceleration of the particle cause

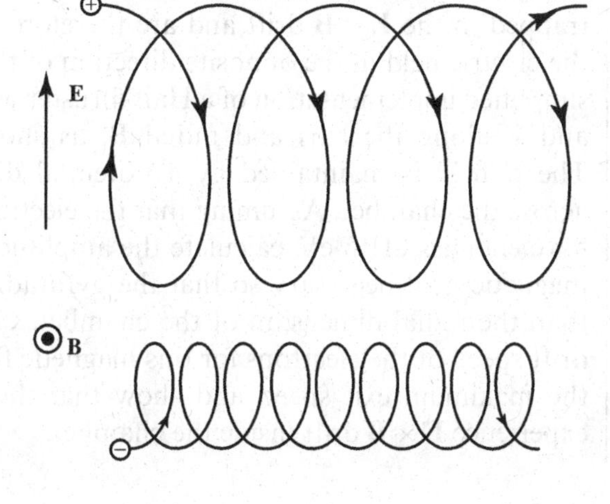

Figure 2.3 Particle drifts in crossed **E** and **B** fields. The negatively charged particle is assumed to have the same velocity ($v_\perp$) as the positively charged one but a smaller mass and therefore a smaller gyroradius. The **E** × **B** drift speed $|\mathbf{v}_E|$ for both particles is the same.

its instantaneous gyroradius to change. This difference in the radius of curvature of its orbit between the top and the bottom of its orbit is the reason for the drift $\mathbf{v}_E$. A negatively charged particle gyrates in the opposite direction but also gains/loses energy in the opposite direction as compared to the positively charged particle. Since we have assumed the negatively charged particle to be lighter, it has a smaller gyroradius r_c. However, its gyrofrequency is larger and the two effects cancel each other out, resulting in the same drift velocity. Two particles of the same mass but different energy (i.e., different $\frac{1}{2}mv_\perp^2$ or $v_\perp$) have the same gyrofrequency ω_c, and although the one with the higher velocity has a higher r_c and hence gains more energy from $\mathbf{E}$ in a half-cycle, the fractional change in r_c for a given change in energy is smaller, so that the two effects cancel out and $\mathbf{v}_E$ is independent of $v_\perp$. The basic source of the $\mathbf{E} \times \mathbf{B}$ drift derived above is the component of electric field perpendicular to $\mathbf{B}$. It is clear from the above procedure that any other constant transverse force $\mathbf{F}_\perp$ acting on a particle gyrating in a constant magnetic field would produce a drift perpendicular to both $\mathbf{F}_\perp$ and $\mathbf{B}$, with the drift velocity given by

$$\boxed{\mathbf{v}_F = \frac{(\mathbf{F}_\perp/q) \times \mathbf{B}}{B^2}}. \tag{2.16}$$

Example 2-1 Hall thruster

A Hall thruster is a spacecraft propulsion device that relies on $\mathbf{E} \times \mathbf{B}$ drift to circulate energetic electrons in a chamber. The electrons are used to ionize heavier atoms, typically Xe (atomic mass $= 2.2 \times 10^{-25}$ kg), which are then accelerated by the electric field and escape the engine, yielding thrust. The efficiency of the Hall thruster relies on the fact that the electrons are largely trapped in the $\mathbf{E} \times \mathbf{B}$ drift and are therefore not accelerated by the electric field in the opposite direction of the ions. Consider a simplified implementation of a Hall thruster with axial symmetry and $\mathbf{E}$ along the axis and radial $\mathbf{B}$, as shown in Figure 2.4. The $\mathbf{E}$ field is maintained by a potential difference of 300 V across the chamber. Assuming that the electrons in the chamber have energies of 15 eV, calculate the amplitude of the minimum magnetic field necessary so that the gyroradius is 10 times less than the radial dimension of the chamber. Calculate the $\mathbf{E} \times \mathbf{B}$ drift speed of the electrons for this magnetic field. Also calculate the maximum exit speed and show that the Xe ions will not experience $\mathbf{E} \times \mathbf{B}$ drift inside the chamber.

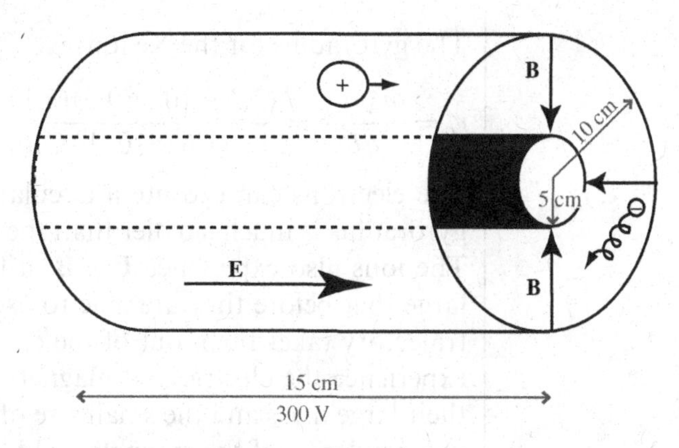

Figure 2.4 Basic geometry of a Hall thruster. Electrons orbit the chamber in a $\mathbf{E} \times \mathbf{B}$ drift, ionizing heavier ions which are expelled, yielding thrust.

Solution: The electron total velocity can be obtained directly from the energy E_e:

$$v = \sqrt{\frac{2E_e}{m_e}} = \sqrt{\frac{2(15\text{ eV})(1.6 \times 10^{-19}\text{ J eV}^{-1})}{9.1 \times 10^{-31}\text{ kg}}} = 2.3 \times 10^6\text{ m s}^{-1}.$$

The magnetic field necessary to make the gyroradius 0.5 cm is then given by

$$B = \frac{m_e v_\perp}{q r_c} = \frac{(9.1 \times 10^{-31}\text{ kg})(2.6 \times 10^6\text{ m s}^{-1})}{(1.6 \times 10^{-19}\text{ C})(0.005\text{ m})} = 2.6\text{ mT},$$

where we have assumed the upper limit of the perpendicular velocity ($v_\perp = v$). The electric field in the chamber is given by $E = (300\text{ V})/(0.15\text{ m}) = 2\text{ kV m}^{-1}$. We can now calculate the $\mathbf{E} \times \mathbf{B}$ drift using Equation (2.13), taking note that the fields are orthogonal everywhere in the chamber:

$$\mathbf{v}_\text{E} = \frac{\mathbf{E} \times \mathbf{B}}{B^2} = \frac{E}{B} = 7.7 \times 10^5\text{ m s}^{-1}.$$

The maximum kinetic energy of the Xe ions, assuming a $+q_e$ charge is accelerated along the entire 300 V potential, is given by $E_{Xe} = (300\text{ V})(q_e) = 300$ eV. The velocity is now obtained as before:

$$v = \sqrt{\frac{2E_{Xe}}{m_{xe}}} = \sqrt{\frac{2(300\text{ eV})(1.6 \times 10^{-19}\text{ J eV}^{-1})}{2.2 \times 10^{-25}\text{ kg}}}$$
$$= 2.1 \times 10^4\text{ m s}^{-1}.$$

The gyroradius for the Xe ions is

$$r_c = \frac{m_{Xe} v_\perp}{qB} = \frac{(2.2 \times 10^{-25}\ \text{kg})(2.1 \times 10^4\ \text{m s}^{-1})}{(1.6 \times 10^{-19}\ \text{C})(0.0026\ \text{T})} = 11\ \text{m} \gg 5\ \text{cm}.$$

The electrons can execute a circular drift motion because their gyroradius is much smaller than the dimensions of the chamber. The ions also experience $\mathbf{E} \times \mathbf{B}$ drift but their gyroradius is so large that before they are able to execute a single gyration their trajectory takes them out of the chamber, where they no longer experience the electric and magnetic fields. In effect, because of their large mass and the small size of the chamber, the ions move independently of the magnetic field.

2.3 Particle motion in non-uniform B fields

Both naturally occurring plasmas and those encountered in many applications may exist in the presence of magnetic fields that do not vary appreciably in time but which vary with one or more coordinates of space, making such fields *non-uniform* or *inhomogeneous*. An important example of a non-uniform magnetic field is the so-called magnetic-mirror configuration which is commonly used to confine plasmas, and is also the mechanism by which energetic particles are trapped in the Earth's radiation belts. Assuming the absence of an electric field, and no temporal variations of the magnetic field,[3] the kinetic energy of the particle must remain zero, since the magnetic force is at all times perpendicular to the motion of the particles, as discussed in Section 2.1. In general, exact analytical solutions for charged-particle motions in a non-uniform magnetic field cannot be found. However, one very important configuration that can be studied analytically is the case in which the gyroradius r_c is much smaller than the spatial scales over which the magnetic field varies. In such cases, the motion of the particle can be decomposed into the fast gyromotion plus some type of relatively slow drift motion. The slow drift is associated with the motion of the guiding center, and the separation of its motion from the rapid gyration is similar to the simplest case analyzed in Section 2.1, where we saw that the guiding center simply moved linearly along the magnetic

[3] Note that any time variation of the magnetic field would lead to an electric field via Faraday's law, i.e., $\nabla \times \mathbf{E} = -\partial \mathbf{B}/\partial t$, which can in turn accelerate the particles.

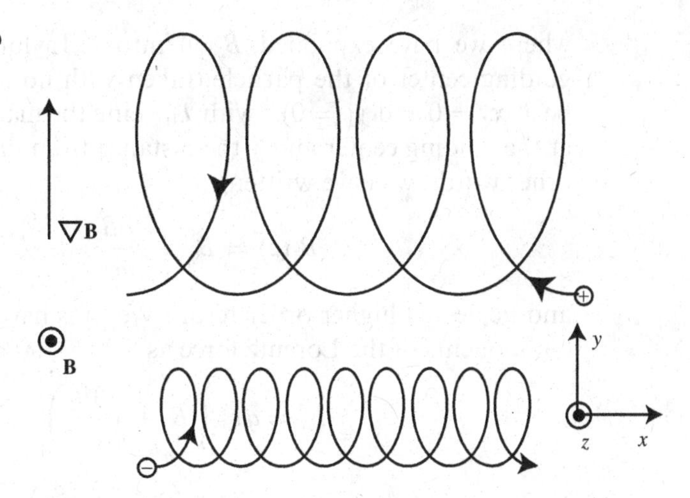

Figure 2.5 Particle drifts due to a magnetic field gradient.

field as the particle executed its complicated gyromotion. We now examine particle motion in different types of non-uniform magnetic fields, assuming the presence of only one type of inhomogeneity in each case.

2.3.1 Gradient drift

We first consider a magnetic field with intensity varying in a direction perpendicular to the magnetic field vector. Without loss of generality, let $\mathbf{B}(y) = \hat{\mathbf{z}}B_z(y)$, as depicted in Figure 2.5. Since the field strength has a non-zero gradient ∇B_z in the y direction, we note that the local gyroradius r_c (i.e., the radius of curvature of the particle orbit) is large in regions where B is small, and vice versa. Thus, on physical grounds alone, we expect a positive charge to drift to the left and a negative charge to drift to the right. To find an expression for the particle drift velocity, we take advantage of expression (2.16), which gives the drift velocity for any force perpendicular to the **B** field. For the field geometry depicted in Figure 2.5, this means a force in either the x or the y direction. Since we started with the premise that the gyration was much more rapid than the relatively slow drift, it is appropriate to determine the net resultant force averaged over one gyroperiod. The force perpendicular to **B** is the Lorentz force given by

$$\mathbf{F} = q(\mathbf{v} \times \mathbf{B}) = \hat{\mathbf{x}}qv_y B_z - \hat{\mathbf{y}}qv_x B_z$$

$$\simeq \hat{\mathbf{x}}qv_y\left(B_0 + y\frac{\partial B_z}{\partial y}\right) - \hat{\mathbf{y}}qv_x\left(B_0 + y\frac{\partial B_z}{\partial y}\right),$$

where we have expanded $B_z(y)$ into a Taylor series around the guiding center of the particle (taken with no loss of generality to be at $x_g = 0$ and $y_g = 0$),[4] with B_0 being the magnetic field intensity at the guiding center and y the distance from the guiding center. In other words, we have written

$$B_z(z) = B_0 + y\frac{\partial B_z}{\partial y} + \cdots$$

and neglected higher-order terms. We thus have the two transverse components of the Lorentz force as

$$F_x = qv_y\left(B_0 + y\frac{\partial B_z}{\partial y}\right) \tag{2.17a}$$

$$F_y = -qv_x\left(B_0 + y\frac{\partial B_z}{\partial y}\right). \tag{2.17b}$$

We wish to determine $\langle F_x \rangle$ and $\langle F_y \rangle$, where the brackets denote averaging over one gyroperiod. To do this, we can assume that the particles by and large follow the orbits for a uniform field, as determined in Section 2.1 (see Equations (2.6) and (2.7)), i.e.,

$$x_c = r_c \sin(\omega_c t + \psi) \tag{2.18a}$$

$$y_c = -r_c \cos(\omega_c t + \psi) \tag{2.18b}$$

$$v_x = v_\perp \cos(\omega_c t + \psi) \tag{2.18c}$$

$$v_y = v_\perp \sin(\omega_c t + \psi), \tag{2.18d}$$

where ω_c has the same sign as q (i.e., is negative for electrons). Substituting in (2.17a) and (2.17b) we have

$$F_x = qv_\perp \sin(\omega_c t + \psi)\left[B_0 - r_c \cos(\omega_c t + \psi)\frac{\partial B_z}{\partial y}\right] \tag{2.19a}$$

$$F_y = -qv_\perp \cos(\omega_c t + \psi)\left[B_0 - r_c \cos(\omega_c t + \psi)\frac{\partial B_z}{\partial y}\right]. \tag{2.19b}$$

The average of F_x over one gyroperiod $(2\pi/\omega_c)$ is zero, since it contains the product of sine and cosine terms. The averaging of F_y

[4] In the more general case this expansion can be written as

$$\mathbf{B} = \mathbf{B}_0 + (\mathbf{r} \cdot \nabla)\mathbf{B}_0 + \cdots,$$

where $\mathbf{B}_0$ is the field at the guiding center and $\mathbf{r}$ is the position vector, with the origin chosen as the guiding center.

has the product of a cosine with a cosine which results in a factor of $\frac{1}{2}$. We thus have

$$\langle F_y \rangle = \frac{q v_\perp r_c}{2} \frac{\partial B_z}{\partial y} = -\frac{m v_\perp^2}{2B} \frac{\partial B_z}{\partial y}. \tag{2.20}$$

Note that the direction of $\langle F_y \rangle$ does not depend on the charge of the particle. The drift velocity is then, from Equation (2.16),

$$\mathbf{v}_\nabla = \frac{(\mathbf{F}_\perp / q) \times \mathbf{B}}{B^2} = \frac{\langle F_y \rangle \hat{\mathbf{y}} \times \hat{\mathbf{z}} B_z}{q B_z^2} = -\frac{m v_\perp^2}{2 q B_z} \frac{\partial B_z}{\partial y} \hat{\mathbf{x}}, \tag{2.21}$$

where the subscript ∇ indicates that the drift velocity is due to the gradient drift. The gradient drift is often written as $\nabla\mathbf{B}$ to emphasize that it arises from a gradient in the magnetic field. Since the magnetic field direction was chosen arbitrarily, we can write (2.21) more generally as

$$\mathbf{v}_\nabla = \frac{m v_\perp^2}{2q} \frac{\mathbf{B} \times \nabla B}{B^3} \tag{2.22}$$

for any magnetic field $\mathbf{B} = \hat{\mathbf{B}} B$. The corresponding, more general expression for the perpendicular gradient force $\mathbf{F}_\nabla$ is

$$\mathbf{F}_\nabla = -\frac{\frac{1}{2} m v_\perp^2}{B} \nabla B = -\frac{W_\perp}{B} \nabla B, \tag{2.23}$$

where $W_\perp$ is the perpendicular kinetic energy of the particle. Equation (2.22) exhibits the dependencies that we expect on a physical basis. Electrons and ions drift in opposite directions and the drift velocity is proportional to the perpendicular energy of the particle, $W_\perp = \frac{1}{2} m v_\perp^2$. Faster particles drift faster, since they have a larger gyroradius and their orbits span a larger range of the field inhomogeneity.

2.3.2 Curvature drift

When particles gyrate rapidly while moving along a magnetic field line which is curved, as depicted in Figure 2.6, they experience a centrifugal force perpendicular to the magnetic field, which produces a drift as defined by (2.16). Assuming once again that the spatial scale of the curvature is much larger than the gyroradius, we can focus our attention on the motion of the guiding center. The outward

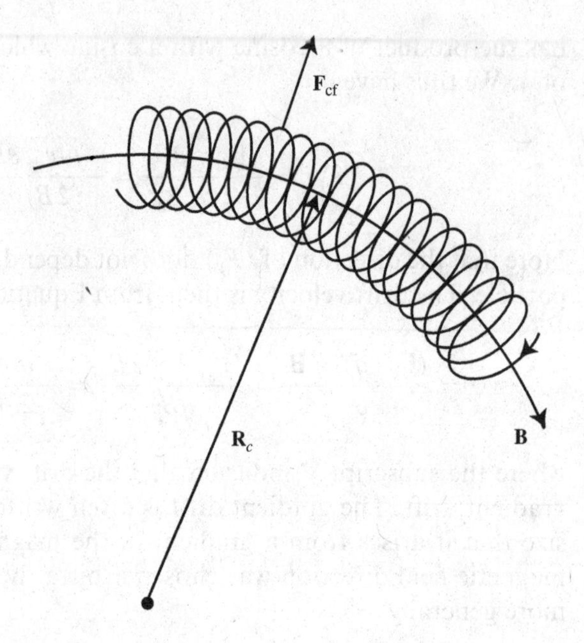

Figure 2.6 Curvature drift. Particle drift in a curved magnetic field.

centrifugal force in the frame of reference moving with the guiding center at a velocity $v_\parallel$ is given by

$$\mathbf{F}_{\mathrm{cf}} = m v_\parallel^2 \frac{\mathbf{R}_c}{R_c^2}, \tag{2.24}$$

where $\mathbf{R}_c$ is the vector pointing radially outward from the center of the circle described by the local curvature of the field and R_c has a magnitude equal to the radius of curvature. Using the force given in (2.24) in (2.16) we find the curvature drift velocity to be

$$\mathbf{v}_{\mathrm{R}} = \frac{(\mathbf{F}_{\mathrm{cf}}/q) \times \mathbf{B}}{B^2} = \frac{m v_\parallel^2}{q} \frac{\mathbf{R}_c \times \mathbf{B}}{R_c^2 B^2}. \tag{2.25}$$

In vacuum, curvature drift cannot by itself be the only drift since the curl of the magnetic field must be zero, i.e., $\nabla \times \mathbf{B} = 0$. In other words, curvature in a magnetic field necessitates a gradient in the magnetic field. Considering cylindrical coordinates with $\mathbf{B} = B_\phi(r)\hat{\phi}$, we must then have

$$(\nabla \times \mathbf{B})_z = \frac{1}{r}\frac{\partial}{\partial r}(r B_\phi) = 0 \quad \rightarrow \quad B_\phi = \frac{A}{r},$$

where A is a constant. The gradient of $\mathbf{B}$ is then $\partial B_\phi / \partial r = -A/r^2 = -B_\phi / r$. More generally, we can write the resulting gradient as

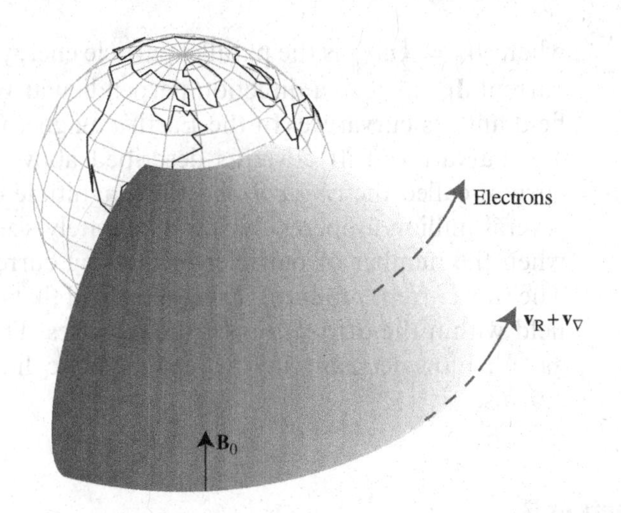

Figure 2.7 Longitudinal drift of radiation belt electrons. Note that the direction of the Earth's magnetic field $\mathbf{B}_0$ is from south to north.

$\nabla = -\left(B/R_c^2\right)\mathbf{R}_c$. Thus, the total drift due to both gradient and curvature effects can be written as

$$\mathbf{v}_{\text{total}} = \mathbf{v}_{\text{R}} + \mathbf{v}_\nabla = \left(v_\parallel^2 + \frac{1}{2}v_\perp^2\right)\frac{\mathbf{B}\times\nabla B}{w_c B^2}. \tag{2.26}$$

An example of gradient plus curvature drift is the *longitudinal drift* of radiation belt electrons around the Earth (see Figure 2.7). Note that the direction of the Earth's magnetic field is from south to north.

It is interesting to note that gradient and curvature drift velocities are both inversely proportional to the charge q, so that electrons and ions drift in opposite directions. The oppositely directed drifts of electrons and ions leads to a transverse current. The gradient-drift current is given by

$$\mathbf{J}_\nabla = N|q_e|[(\mathbf{v}_\nabla)_i - (\mathbf{v}_\nabla)_e]$$
$$= \frac{N}{B^3}[(W_\perp)_i + (W_\perp)_e](\mathbf{B}\times\nabla B),$$

where $N = N_i = N_e$ is the plasma density and $W_\perp = \frac{1}{2}mv_\perp^2$ is the perpendicular particle energy. Note that the gradient-drift current $\mathbf{J}_\nabla$ flows in a direction perpendicular to both the magnetic field and its gradient. Similarly, the different directional curvature drifts of the electrons and ions lead to a curvature drift current given by

$$\mathbf{J}_{\text{R}} = N|q_e|[(\mathbf{v}_{\text{R}})_i - (\mathbf{v}_{\text{R}})_e]$$
$$= \frac{2N}{R_c^2 B^2}[(W_\parallel)_i + (W_\parallel)_e](\mathbf{R}_c \times \mathbf{B}),$$

where $W_{\parallel} = \frac{1}{2}mv_{\parallel}^2$ is the parallel particle energy. The curvature drift current $\mathbf{J}_{\mathrm{R}}$ flows in a direction perpendicular to both the magnetic field and its curvature. In the Earth's magnetosphere, the gradient and curvature drift currents described above create a large-scale current called the *ring current*, the magnitude of which can exceed several million amperes during moderately sized magnetic storms, when the number of particles in the ring current region increases. The ring current produces a magnetic field that decreases the Earth's field within the drift orbits of the particles. This effect is observed as a major decrease in the geomagnetic field during magnetic storms.

2.3.3 Other gradients of B

We have presented particle motion in non-uniform magnetic fields with particular types of inhomogeneities. The various spatial gradients of the magnetic field can be summarized in tensor or dyadic notation as

$$\nabla \mathbf{B} = \begin{bmatrix} \frac{\partial B_x}{\partial x} & \frac{\partial Bx}{\partial y} & \frac{\partial B_x}{\partial z} \\ \frac{\partial B_y}{\partial x} & \frac{\partial B_y}{\partial y} & \frac{\partial B_y}{\partial z} \\ \frac{\partial B_z}{\partial x} & \frac{\partial B_z}{\partial y} & \frac{\partial B_z}{\partial z} \end{bmatrix}.$$

Note that only eight of the nine components of $\nabla\mathbf{B}$ are independent, since the condition $\nabla \cdot \mathbf{B} = 0$ allows us to determine one of the diagonal terms in terms of the other two. In regions where there are no currents ($\mathbf{J} = 0$), we must also have $\nabla \times \mathbf{B} = 0$, imposing additional restrictions on the various components of $\nabla\mathbf{B}$. The diagonal terms are sometimes referred to as the divergence terms and represent gradients along the $\mathbf{B}$ direction, i.e., $\nabla_{\parallel} B$, one of which ($\partial B_z/\partial z$) is responsible for the mirror effect discussed in the following section. The terms $\partial B_z/\partial x$ and $\partial B_z/\partial y$ are known as the gradient terms and represent transverse gradients ($\nabla_{\perp} B$) responsible for the gradient drift studied in Section 2.3.1. The terms $\partial B_x/\partial z$ and $\partial B_y/\partial z$ are known as the curvature terms and represent change of direction of $\mathbf{B}$, i.e., curvature, and were studied in Section 2.3.2. The remaining terms, $\partial B_x/\partial y$ and $\partial B_y/\partial x$, are known as the shear terms and represent twisting of the magnetic field lines; these are not important in particle motion.

2.4 Adiabatic invariance of the magnetic moment

In Section 2.1, we recognized that a gyrating particle constitutes an electric current loop with a magnetic dipole moment given by $\mu = mv_\perp^2/(2B)$. In this section, we demonstrate that this quantity has a remarkable tendency to be conserved (i.e., to be invariant), in spite of spatial or temporal changes in the magnetic field intensity, as long as the changes in B are small over a gyroradius or gyroperiod. This kind of constancy of a variable is termed *adiabatic invariance*, to distinguish such quantities from those that may be absolute invariants, such as total charge, energy, or momentum in a physical system. Consider a particle gyrating in a magnetic field oriented primarily in the z direction but varying in intensity as a function of z, as depicted in Figure 2.8. Assume the field to be azimuthally symmetric, so that there is no ϕ component (i.e., $B_\phi = 0$) and no variation of any of the quantities in ϕ (i.e., $\partial(\cdot)/\partial\phi = 0$). As the particle gyrates around **B** with a perpendicular velocity $v_\perp$ while moving along it at $v_\parallel$, we are primarily concerned with the motion of its guiding center, which moves along the z axis. The force acting on the particle during this motion can be found from (2.23) by noting that the magnetic field has a non-zero gradient in the z direction. We have

$$F_z = -\frac{\frac{1}{2}mv_\perp^2}{B_z}\frac{\partial B_z}{\partial z} = -\frac{W_\perp}{B_z}\frac{\partial B}{\partial z} = -\mu\frac{\partial B}{\partial z}. \tag{2.27}$$

Alternatively, the force F_z can be found directly from the Lorentz force equation. The z component of the Lorentz force results from $q\mathbf{v}_\perp \times \mathbf{B}$ or

$$F_z = q\mathbf{v}_\perp \times \mathbf{B} = qv_\perp B_r, \tag{2.28}$$

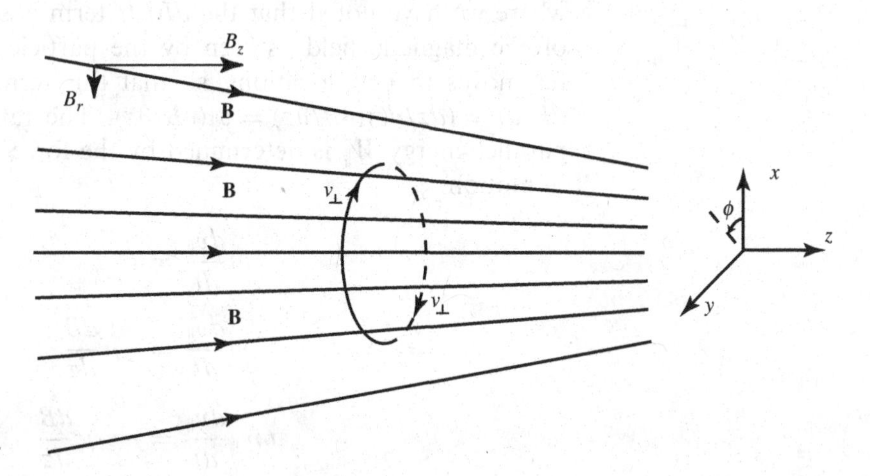

Figure 2.8 Drift of a particle in a magnetic mirror configuration.

where B_r can be found using the fact that we must have $\nabla \cdot \mathbf{B} = 0$, so that

$$\frac{1}{r}\frac{\partial}{\partial r}(r\,B_r) + \frac{\partial B_z}{\partial z} = 0 \quad \rightarrow \quad B_r \simeq -\frac{r}{2}\frac{\partial B_z}{\partial z}, \tag{2.29}$$

assuming that $\partial B_z/\partial z$ does not vary significantly with r. In other words, the total magnetic field in the case of the converging field-line geometry of Figure 2.8 must be given by

$$\mathbf{B} = B_r\hat{\mathbf{r}} + B_z\hat{\mathbf{z}}.$$

Evaluating B_r as given in (2.29) at $r = r_c$ and substituting in (2.28) we find the same expression for F_z as in (2.27). With F_z determined, we can examine the variations of the parallel and perpendicular energies of the particle as its guiding center moves along z. Consider the total energy of the particle,

$$W = W_\perp + W_\|,$$

which must remain constant in the absence of electric fields, so that we have

$$\frac{dW_\perp}{dt} + \frac{dW_\|}{dt} = 0. \tag{2.30}$$

Noting that $W_\perp = \mu B$, the time rate of change of the transverse energy can be written as

$$\frac{dW_\perp}{dt} = \frac{d(\mu B)}{dt} = \mu\frac{dB}{dt} + B\frac{d\mu}{dt} = \mu v_\|\frac{dB}{dz} + B\frac{d\mu}{dt}, \tag{2.31}$$

where we have noted that the dB/dt term is simply the variation of the magnetic field as seen by the particle as its guiding center moves to new locations, so that this term can be written as $dB/dt = (dz/dt)(\partial B/\partial z) = v_\|(\partial B/\partial z)$. The rate of change of the parallel energy $W_\|$ is determined by the force F_z via the equation of motion:

$$m\frac{dv_\|}{dt} = F_z$$

$$m\frac{dv_\|}{dt} = -\mu\frac{dB}{dz}$$

$$mv_\|\frac{dv_\|}{dt} = -v_\|\mu\frac{dB}{dz}$$

$$\frac{d\left(\frac{1}{2}mv_{\|}^{2}\right)}{dt} = -v_{\|}\mu\frac{dB}{dz}$$

$$\frac{dW_{\|}}{dt} = -v_{\|}\mu\frac{dB}{dz}. \tag{2.32}$$

Substituting (2.32) and (2.31) into (2.30) we find

$$\mu v_{\|}\frac{dB}{dz} + B\frac{d\mu}{dt} - v_{\|}\mu\frac{dB}{dz} = 0 \quad \rightarrow \quad \frac{d\mu}{dt} = 0,$$

which indicates that the magnetic moment μ is an invariant of the particle motion. Note from (2.31) and (2.32) that the particle's perpendicular energy increases, while its parallel energy decreases, as it moves toward regions of higher B, so that $dB/dz > 0$. As the particle moves into regions of higher and higher B, its parallel velocity $v_{\|}$ eventually reduces to zero, and it "reflects" back, moving in the other direction. In an asymmetric magnetic field geometry, e.g., a dipole magnetic field like that of the Earth and other magnetized planets, the particle would then encounter a similar convergence of magnetic field lines at the other end of the system, from which it would also reflect, thereby becoming forever trapped in a "magnetic bottle." Until now we have assumed that the magnetic field exhibits no temporal changes, so that $\partial B/\partial t = 0$ and there are no induced electric fields. However, the magnetic moment μ is still conserved when there are time variations, as long as those variations occur slowly in comparison to the gyroperiod of the particles. Noting that temporal variations of the magnetic field would create a spatially varying electric field via Faraday's law, $-\partial\mathbf{B}/\partial t = \nabla \times \mathbf{E}$, let us consider the change in perpendicular energy $W_{\perp}$ due to an electric field. Consider the equation of motion under the influence of the Lorentz force,

$$m\frac{d(\mathbf{v}_{\perp} + \mathbf{v}_{\|})}{dt} = q[\mathbf{E} + (\mathbf{v}_{\perp} + \mathbf{v}_{\|}) \times \mathbf{B}],$$

and take the dot-product of this equation with $\mathbf{v}_{\perp}$ to find

$$\frac{dW_{\perp}}{dt} = q(\mathbf{E} \cdot \mathbf{v}_{\perp}),$$

where we have used the fact that $W_{\perp} = \frac{1}{2}mv_{\perp}^{2}$. The increase in particle energy over one gyration can be found by averaging over a gyroperiod:

$$\Delta W_{\perp} = q\int_{0}^{T_c}(\mathbf{E} \cdot \mathbf{v}_{\perp})dt,$$

where $T_c = 2\pi/\omega_c$. Assuming that the field changes slowly, the particle orbit is not perturbed significantly, and we can replace the integration in time with a line integral over the unperturbed circular orbit. In other words,

$$\Delta W_\perp = q \oint_C \mathbf{E} \cdot d\mathbf{l} = q \int_S (\nabla \times \mathbf{E}) \cdot d\mathbf{s} = -q \int_S \frac{\partial \mathbf{B}}{\partial t} \cdot d\mathbf{s},$$

where $d\mathbf{l}$ is a line element along the closed gyro-orbit C, while $d\mathbf{s}$ is a surface element over the surface S enclosed by the gyro-orbit. For changes much slower than the gyroperiod, we can replace $\partial \mathbf{B}/\partial t$ with $\omega_c \Delta B/(2\pi)$, ΔB being the average change during one gyroperiod. We thus have

$$\Delta W_\perp = \frac{1}{2} q \omega_c r_c^2 \Delta B = \mu \Delta B, \tag{2.33}$$

using previously derived expressions for ω_c, r_c, and μ. However, we know from (2.31) that

$$\Delta W_\perp = \mu \Delta B + B \Delta \mu. \tag{2.34}$$

Comparing (2.33) and (2.34) we find that $\Delta \mu = 0$, indicating that the magnetic moment is invariant even when particles are accelerated in the electric field induced by slow temporal variations in the magnetic field.

Example 2-2 Plasma confinement using magnetic mirrors
Effective confinement of a high-energy plasma is a central issue in achieving controlled thermonuclear fusion. One of the simplest confinement schemes is that of a magnetic mirror. Consider the magnetic mirror configuration shown in Figure 2.9, with axial magnetic field given by

$$B(z) = B_0\left(1 + \delta z^2\right).$$

Calculate the mirror point (reflection point) for an electron with velocity v located initially at $z = 0$ with an initial pitch angle α_0.

Solution: The initial parallel and perpendicular components of the velocity are accessible through the pitch angle:

$$v_{\perp 0} = v \sin(\alpha_0), \quad v_{\parallel 0} = v \cos(\alpha_0).$$

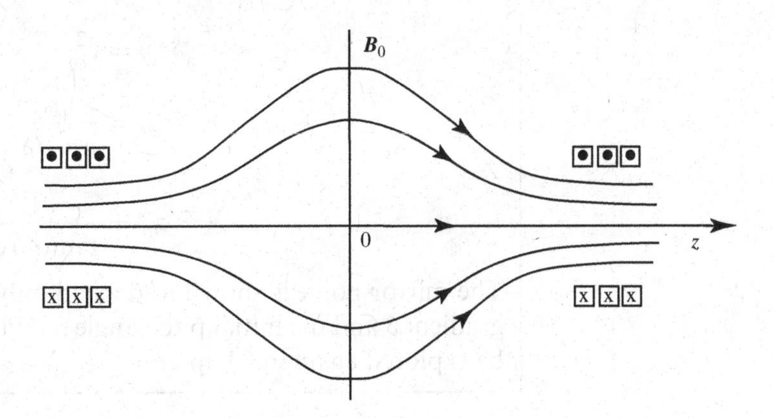

Figure 2.9 Dual magnetic mirror geometry for plasma confinement.

As the electron moves within the magnetic mirror configuration, both the magnetic moment (μ) and total kinetic energy (W) are conserved, making their initial values and their values at the mirror point (z_m) equal:

$$W_0 = W_m, \quad \mu_0 = \mu_m.$$

Working first with the kinetic energy,

$$W_0 = W_m$$

$$\frac{1}{2}m_e\left(v_{\perp_0}^2 + v_{\|_m}^2\right) = \frac{1}{2}m_e\left(v_{\perp_m}^2 + v_{\|_m}^2\right)$$

$$v_{\perp_0}^2 + v_{\|_0}^2 = v_{\perp_m}^2 + v_{\|_m}^2$$

$$v_{\perp_0}^2 + v_{\|_0}^2 = v_{\perp_m}^2,$$

since $v_{\|_m} = 0$ at the mirror point. Turning now to the conservation of the magnetic moment, we have

$$\mu_0 = \mu_m$$

$$\frac{mv_{\perp_0}^2}{2B_0} = \frac{mv_{\perp_m}^2}{2B_0\left(1 + \delta z_m^2\right)}$$

$$mv_{\perp_0}^2 = \frac{m\left(v_{\perp_0}^2 + v_{\|_0}^2\right)}{\left(1 + \delta z_m^2\right)}$$

$$v_{\perp_0}^2\left(1 + \delta z_m^2\right) = v_{\perp_0}^2 + v_{\|_0}^2$$

$$v_{\perp 0}^2 \delta z_m^2 = v_{\parallel 0}^2$$

$$z_m = \frac{v_{\parallel 0}}{v_{\perp 0}\sqrt{\delta}}$$

$$z_m = \frac{1}{\tan(\alpha_0)\sqrt{\delta}}.$$

The mirror point is shown to depend only on the magnetic field gradient δ and the initial pitch angle α_θ. Plasma confinement will be explored again in Chapter 6.

2.5 Particle motion in time-varying electric fields

We have seen the basic dynamics and drifts of particles in non-uniform magnetic fields in the absence of electric fields. In Section 2.2 we showed that particle motion in the presence of static and uniform crossed electric and magnetic fields consists of a gyration superimposed on a drift at velocity $\mathbf{v}_E$. In this section we consider particle motion in a combination of static magnetic and time-varying electric fields. We separately discuss the two limiting cases of slow temporal variations and temporal variations at frequencies comparable to the electron-cyclotron frequency $\omega_c = -q_e|\mathbf{B}|/m_e$. In the former case, we find that the slow temporal variation of $\mathbf{E}$ leads to an additional drift called the "polarization drift." In the latter case, we find that an alternating electric field generates alternating currents not only in its own direction but also in other directions, a fundamental anisotropy of a magnetized plasma.

2.5.1 Polarization drift: slowly varying E field

Consider the fundamental equation of motion (2.1), repeated here:

$$m\frac{d\mathbf{v}}{dt} = q(\mathbf{E} + \mathbf{v} \times \mathbf{B}).$$

We can decompose this equation into its two components in terms of $\mathbf{v}_\perp$ and $\mathbf{v}_\parallel$. We have

$$m\frac{d\mathbf{v}_\perp}{dt} = q(\mathbf{E}_\perp + \mathbf{v}_\perp \times \mathbf{B}) \tag{2.35}$$

$$m\frac{d\mathbf{v}_\parallel}{dt} = q\mathbf{E}_\parallel. \tag{2.36}$$

Once again noting that (2.36) simply denotes acceleration along **B**, we focus our attention on (2.35). Dividing both sides of (2.35) by q and taking the cross-product of both sides with $\mathbf{B}/B^2$, we have

$$\frac{m}{q}\frac{d\mathbf{v}_\perp}{dt} \times \frac{\mathbf{B}}{B^2} = \frac{\mathbf{E}_\perp \times \mathbf{B}}{B^2} + (\mathbf{v}_\perp \times \mathbf{B}) \times \frac{\mathbf{B}}{B^2}$$

$$\frac{m}{q}\frac{d\mathbf{v}_\perp}{dt} \times \frac{\mathbf{B}}{B^2} = \frac{\mathbf{E}_\perp \times \mathbf{B}}{B^2} - \frac{\mathbf{v}_\perp(\mathbf{B}\cdot\mathbf{B})}{B^2} + \frac{\mathbf{B}(\mathbf{v}_\perp \cdot \mathbf{B})}{B^2}$$

$$\frac{m}{q}\frac{d\mathbf{v}_\perp}{dt} \times \frac{\mathbf{B}}{B^2} = \frac{\mathbf{E}_\perp \times \mathbf{B}}{B^2} - \frac{\mathbf{v}_\perp(\mathbf{B}\cdot\mathbf{B})}{B^2} + \frac{\mathbf{B}(\mathbf{v}_\perp \cdot \mathbf{B})}{B^2}$$

$$\frac{m}{q}\frac{d\mathbf{v}_\perp}{dt} \times \frac{\mathbf{B}}{B^2} = \frac{\mathbf{E}_\perp \times \mathbf{B}}{B^2} - \mathbf{v}_\perp + 0, \quad (2.37)$$

where we have used the vector identity

$$(\mathbf{F} \times \mathbf{G}) \times \mathbf{C} \equiv -\mathbf{F}(\mathbf{C}\cdot\mathbf{G}) + \mathbf{G}(\mathbf{C}\cdot\mathbf{F})$$

and noted that $\mathbf{B}\cdot\mathbf{B} = B^2$ and that $\mathbf{v}_\perp \cdot \mathbf{B} = 0$.[5] We now follow a procedure similar to that in Section 2.2 (see Equation (2.11)) and decompose the perpendicular velocity $\mathbf{v}_\perp$ into its various components:

$$\mathbf{v}_\perp = \mathbf{v}_{ac} + \mathbf{v}_E + \mathbf{v}_p, \quad (2.38)$$

where $\mathbf{v}_{ac}$ is the component representing gyromotion, $\mathbf{v}_E$ is the $\mathbf{E} \times \mathbf{B}$ drift velocity found in Section 2.3, and $\mathbf{v}_p$ is the additional drift that results from the temporal variation of **E** considered here. Substituting (2.37) into (2.38) we find

$$\frac{m}{qB^2}\left[\frac{d\mathbf{v}_{ac}}{dt} \times \mathbf{B} + \frac{d\mathbf{v}_E}{dt} \times \mathbf{B} + \frac{d\mathbf{v}_p}{dt} \times \mathbf{B}\right] = \frac{\mathbf{E}_\perp \times \mathbf{B}}{B^2} - \mathbf{v}_{ac} - \mathbf{v}_E - \mathbf{v}_p \quad (2.39a)$$

$$\frac{m}{qB^2}\left[\frac{d\mathbf{v}_{ac}}{dt} \times \mathbf{B} + \frac{d\mathbf{v}_E}{dt} \times \mathbf{B} + \frac{d\mathbf{v}_p}{dt} \times \mathbf{B}\right] = -\mathbf{v}_{ac} - \mathbf{v}_p, \quad (2.39b)$$

where we have noted from Section 2.2 that $\mathbf{v}_E = (\mathbf{E}_\perp \times \mathbf{B})/B^2$. The first term on the left and the first term on the right simply represent gyromotion, just as one would have in the absence of an electric

[5] In fact, what we have found above is intuitively obvious (i.e., without using any vector identity) when we are dealing with the cross-product of two vectors which are orthogonal to one another. The resultant cross-product, which is necessarily orthogonal to both vectors, can be crossed into any one of the original vectors to obtain the other one, with the appropriate polarity (sign) and magnitudes accounted for. In the case in hand, we have $(\mathbf{v}_\perp \times \mathbf{B}) \times (\mathbf{B}/B^2) = -\mathbf{v}_\perp$.

field. The third term on the right is of second order, since it concerns the variation with time of the polarization drift velocity which we are to determine. Thus, we are left with

$$\frac{m}{qB^2}\frac{d\mathbf{v}_\mathrm{E}}{dt} \times \mathbf{B} \simeq -\mathbf{v}_\mathrm{p}$$

$$\frac{m}{qB^2}\frac{d(\mathbf{v}_\mathrm{E} \times \mathbf{B})}{dt} = -\mathbf{v}_\mathrm{p}$$

$$\frac{m}{qB^2}\frac{d(-\mathbf{E}_\perp)}{dt} = -\mathbf{v}_\mathrm{p}$$

$$\frac{m}{qB^2}\frac{d\mathbf{E}_\perp}{dt} = \mathbf{v}_\mathrm{p}, \tag{2.40}$$

where we have used the fact that $\mathbf{v}_\mathrm{E} \times \mathbf{B} = [(\mathbf{E}_\perp \times \mathbf{B})/B^2] \times \mathbf{B} = -\mathbf{E}_\perp$. Thus we see that a slow temporal variation of $\mathbf{E}_\perp$ leads to an additional drift which is superimposed on the $\mathbf{E} \times \mathbf{B}$ drift. This drift is called the *polarization drift* and is qualitatively different in nature from the $\mathbf{E} \times \mathbf{B}$ drift. While the $\mathbf{E} \times \mathbf{B}$ drift velocity $\mathbf{v}_\mathrm{E}$ is independent of the particle mass and charge, the polarization drift velocity is larger for heavier particles, and is in opposite directions for electrons and for ions. When the electric field intensity increases in time ($d\mathbf{E}_\perp/dt > 0$), the ions drift in the direction of the electric field and the electrons in the opposite direction, both picking up energy during the process. (This energy is precisely what is needed to account for the additional drift energy $\frac{1}{2}mv_\mathrm{E}^2$ due to the increased value of the electric field.) Similarly, decreasing the electric field results in an energy loss for the particles. Note that the $\mathbf{E} \times \mathbf{B}$ drift velocity $\mathbf{v}_\mathrm{E}$ is perpendicular to $\mathbf{E}$ and hence does not lead to energy exchange between the field and the particle (when averaged over a gyroperiod). Since electrons and ions drift in opposite directions, the polarization drift creates a polarization current in the plasma, given by

$$\mathbf{J}_\mathrm{p} = N|q_e|\left[(\mathbf{v}_\mathrm{p})_i - (\mathbf{v}_\mathrm{p})_e\right] = \frac{N(m_i + m_e)}{B^2}\frac{d\mathbf{E}_\perp}{dt}. \tag{2.41}$$

Since $m_i \gg m_e$, the polarization current is largely carried by the ions.

2.5.2 Particle motion in static B and arbitrary E fields

We now consider particle motion in the presence of a static and uniform magnetic field. Without loss of generality, we assume the time variation of the electric field to be harmonic with angular frequency ω:

$$\overline{\mathcal{E}}(t) = \mathbf{E}e^{j\omega t}, \tag{2.42}$$

where $\mathbf{E}$ is a complex constant. Note that since the fundamental equation of motion (2.1) is linear, any other arbitrary time variations of the electric field can be synthesized as a summation of terms similar to (2.42), corresponding to all possible values of ω. The equation of motion is simply given by (2.1) with the electric field from (2.42):

$$m\frac{d\mathbf{v}}{dt} = q\left(\mathbf{E}e^{j\omega t} + \mathbf{v} \times \mathbf{B}\right). \tag{2.43}$$

Encouraged by the success of the methodology used in Sections 2.2 and 2.5.1, we proceed by decomposing the velocity vector into two parts:

$$\mathbf{v} = \mathbf{v}_{\mathrm{m}} + \mathbf{v}_{\mathrm{e}}e^{j\omega t}, \tag{2.44}$$

where $\mathbf{v}_{\mathrm{m}}$ contains no time variations with the angular frequency ω and is in fact that part of the motion determined by the static magnetic field $\mathbf{B}$. In other words, $\mathbf{v}_{\mathrm{m}}$ is the same velocity component that we denoted as $\mathbf{v}_{\mathrm{ac}}$ in Sections 2.2 and 2.5.1. The second term represents the velocity component that results from the electric field, and is therefore in general a function of angular frequency ω. Substituting (2.44) into (2.43) we have

$$\frac{d\mathbf{v}_{\mathrm{m}}}{dt} + j\omega\mathbf{v}_{\mathrm{e}}e^{j\omega t} = \frac{q}{m}\left[\mathbf{E}e^{j\omega t} + \mathbf{v}_{\mathrm{m}} \times \mathbf{B} + (\mathbf{v}_{\mathrm{e}}e^{j\omega t}) \times \mathbf{B}\right]. \tag{2.45}$$

This vector equation is in fact a superposition of two equations. The terms that do not contain the periodicity ω are

$$\frac{d\mathbf{v}_{\mathrm{m}}}{dt} = \frac{q}{m}\mathbf{v}_{\mathrm{m}} \times \mathbf{B}$$

and simply define the gyration motion at the cyclotron frequency ω_c. The terms containing the forced oscillation at frequency ω are

$$\left[j\omega\mathbf{v}_{\mathrm{e}} + \frac{q}{m}\mathbf{B} \times \mathbf{v}_{\mathrm{e}}\right] = \frac{q}{m}\mathbf{E}$$

$$\left[j\omega + \frac{q}{m}\mathbf{B}\times\right]\mathbf{v}_{\mathrm{e}} = \frac{q}{m}\mathbf{E}. \tag{2.46}$$

To solve for $\mathbf{v}_{\mathrm{e}}$, we can operate on both sides by $[j\omega - (q/m)\mathbf{B}\times]$ to obtain

$$\left[\frac{q^2}{m^2}B^2 - \omega^2\right]\mathbf{v}_{\mathrm{e}} - \frac{q^2}{m^2}(\mathbf{B}\cdot\mathbf{v}_{\mathrm{e}})\mathbf{B} = \frac{q}{m}\left[j\omega - \frac{q}{m}\mathbf{B}\times\right]\mathbf{E}, \tag{2.47}$$

where we have once again used the vector identity cited in Section 2.5.1:

$$\mathbf{B} \times (\mathbf{B} \times \mathbf{v}_{\mathrm{e}}) \equiv \mathbf{B}(\mathbf{B} \cdot \mathbf{v}_{\mathrm{e}}) - \mathbf{v}_{\mathrm{e}}(\mathbf{B} \cdot \mathbf{B}) = \mathbf{B}(\mathbf{B} \cdot \mathbf{v}_{\mathrm{e}}) - \mathbf{v}_{\mathrm{e}} B^2. \quad (2.48)$$

We can now decompose the vector $\mathbf{v}_{\mathrm{e}}$ into its components parallel and perpendicular to $\mathbf{B}$:

$$\mathbf{v}_{\mathrm{e}} = \mathbf{v}_{\mathrm{e}\perp} + \mathbf{v}_{\mathrm{e}\parallel}. \quad (2.49)$$

Substituting (2.49) into (2.47) yields

$$\mathbf{v}_{\mathrm{e}\parallel} = -\frac{j}{\omega}\frac{q}{m}\mathbf{E}_{\parallel} \quad (2.50\mathrm{a})$$

$$\mathbf{v}_{\mathrm{e}\perp} = \frac{q}{m}\frac{[j\omega - (q/m)\mathbf{B}\times]\mathbf{E}_{\perp}}{\omega_c^2 - \omega^2}, \quad (2.50\mathrm{b})$$

where $\omega_c = -q|\mathbf{B}|/m$ is the cyclotron frequency. The cross-product in (2.50b) indicates that the particle motion occurs in directions other than $\mathbf{E}$. For a Cartesian coordinate system with the static magnetic field oriented in the z direction, i.e., $\mathbf{B} = \hat{\mathbf{z}}B$, (2.50) can be manipulated to express the three velocity components in terms of the three components of the driving electric field. Expressed in matrix form, we can write the result as

$$\begin{bmatrix} v_x \\ v_y \\ v_z \end{bmatrix} = \begin{bmatrix} \dfrac{j\omega(q/m)}{\omega_c^2 - \omega^2} & -\dfrac{\omega_c(q/m)}{\omega_c^2 - \omega^2} & 0 \\ \dfrac{\omega_c(q/m)}{\omega_c^2 - \omega^2} & \dfrac{j\omega(q/m)}{\omega_c^2 - \omega^2} & 0 \\ 0 & 0 & -\dfrac{j(q/m)}{\omega} \end{bmatrix} \begin{bmatrix} E_x \\ E_y \\ E_z \end{bmatrix}. \quad (2.51)$$

We see that in the direction along the magnetic field, the particle motion is 90° out of phase with the forcing electric field, as evidenced by the factor of j. In the plane perpendicular to $\mathbf{B}$, v_x and v_y are each influenced by both E_x and E_y. Likewise, an electric field confined to either the x or y direction drives motion in both x and y directions. This coupling of effects across orthogonal directions is known as anisotropy, and in the context of plasmas is a direct result of the Lorentz force (Equation (2.1)). As we proceed beyond treatment of single particles to discuss collective effects such as currents and waves in later chapters, we will see that anisotropy is a fundamental feature of a magnetized plasma.

An interesting situation occurs when the driving frequency of the electric field is equal to the cyclotron frequency ($\omega = \omega_c$). For this case, Equation (2.51) fails to provide a steady-state solution. This situation is known as cyclotron resonance, and it can be shown that the solution for the perpendicular velocity, including the cyclotron gyration, (see [2]) is

$$\mathbf{v}_{\text{res}\perp} = \mathbf{v}_m + \frac{q}{2m}\left(\mathbf{E}_\perp - j\frac{\omega_c}{|\omega_c|} \times \mathbf{E}_\perp\right) te^{j\omega_c t}. \tag{2.52}$$

Examination of Equation (2.52) shows that under conditions of cyclotron resonance the particle velocity can increase indefinitely with time t, as is illustrated in the example below.

Example 2-3 Cyclotron resonance
For a static magnetic field oriented in the z direction ($\mathbf{B} = B_0\hat{\mathbf{z}}$), find the perpendicular velocity of an electron and an ion under the influence of an electric field $\mathbf{E}_0$ that is right-hand circularly polarized in the $x-y$ plane, with time variation at the cyclotron frequency.

Solution: Since the electric field has time variation at the cyclotron frequency, we need to use Equation (2.52). A right-hand circularly polarized wave implies

$$\overline{\mathcal{E}}(t) = E_0 \cos(\omega_c t)\hat{\mathbf{x}} + E_0 \sin(\omega_c t)\hat{\mathbf{y}},$$

which is written in phasor notation as

$$\overline{\mathcal{E}}(t) = (E_0\hat{\mathbf{x}} - jE_0\hat{\mathbf{y}})e^{jw_c t} = \mathbf{E}_\perp e^{jw_c t}.$$

For the electron, ω_c is positive so $|\omega_c| = \omega$, and plugging this into Equation (2.52) yields

$$\begin{aligned}
\mathbf{v}_{\text{res}\perp} &= \mathbf{v}_m + \frac{q}{2m}[E_0\hat{\mathbf{x}} - jE_0\hat{\mathbf{y}} - j\hat{\mathbf{z}} \times (E_0\hat{\mathbf{x}} - jE_0\hat{\mathbf{y}})]te^{j\omega_c t} \\
&= \mathbf{v}_m + \frac{q}{2m}[E_0\hat{\mathbf{x}} - jE_0\hat{\mathbf{y}} - (j\hat{\mathbf{z}} \times E_0\hat{\mathbf{x}}) + (j\hat{\mathbf{z}} \times jE_0\hat{\mathbf{y}})]te^{j\omega_c t} \\
&= \mathbf{v}_m + \frac{q}{2m}[E_0\hat{\mathbf{x}} - jE_0\hat{\mathbf{y}} - jE_0\hat{\mathbf{y}} + E_0\hat{\mathbf{x}}]te^{j\omega_c t} \\
&= \mathbf{v}_m + \frac{q}{m}[E_0\hat{\mathbf{x}} - jE_0\hat{\mathbf{y}}]te^{j\omega_c t}.
\end{aligned}$$

Taking the real part yields

$$\mathbf{v}_{\text{res}\perp} = \mathbf{v}_m + \frac{E_0}{qm}t\cos(\omega_c t)\hat{\mathbf{x}} + \frac{E_0}{qm}t\sin(\omega_c t)\hat{\mathbf{y}}.$$

We can further evaluate $\mathbf{v}_m$ using Equation (2.6) with $\Psi = 0$:

$$\mathbf{v}_{\text{res}\perp} = \mathbf{v}_m + \left(v_\perp + \frac{E_0}{qm}t\right)\cos(\omega_c t)\hat{\mathbf{x}} + \left(v_\perp + \frac{E_0}{qm}t\right)\sin(\omega_c t)\hat{\mathbf{y}},$$

which is seen to be circular motion at an increasing rate. This phenomenon can be used to increase the speed and hence the kinetic energy of the particles in a plasma. If the applied electric field is that of an electromagnetic wave, the procedure is called radio-frequency heating of a plasma by cyclotron resonance.

It should be noted that, even in the absence of the plasma, a fluctuating electric field such as that given in (2.42) cannot exist without its associated magnetic field, as given by $\nabla \times \mathbf{H} = \epsilon_0 \partial \mathbf{E}/\partial t$, which we implicitly neglected in considering the particle motion. We will see later that this is indeed an excellent assumption, since the magnetic fields associated with electromagnetic waves are typically much smaller than the static magnetic field $\mathbf{B}$ and thus produce only a second-order perturbation of the particle motion.

2.6 Summary

In this chapter we examined the motion of individual charged particles under the influence of electric and magnetic fields. The fundamental equation that governs this behavior is that of the Lorentz force, $\mathbf{F} = q\mathbf{E} + q\mathbf{v} \times \mathbf{B}$. With only a magnetic field present the movement of charged particles is restricted to circular motion known as gyration in a direction perpendicular to the magnetic field plus uninhibited motion along the magnetic field. The center of the circular motion is known as the guiding center, and it is often more convenient to describe the motion of the guiding center than of the actual particle. The addition of a static electric field causes particles with both positive and negative charges to drift in a direction perpendicular to both the magnetic and the electric fields. Similar drifts are created by any force perpendicular to the magnetic field, or even by a gradient in the magnetic field. The fundamental reason for the drifts is that the gyroradius changes over the course of a particle gyration and thus slightly deforms the simple circular motion in the plane perpendicular to the magnetic field.

Although examining the motion of individual charged particles does not fall within the strict definition of plasma physics, such an undertaking is relevant to many applications, including plasma propulsion devices and energetic particles trapped in the

Earth's radiation belts. Moreover, familiarity with the behavior of single particles is a valuable source of intuition for more complicated plasma behavior that will be explored in the remainder of the text.

2.7 Problems

2-1. Compute the gyroradius and cyclotron frequency for the following plasma configurations: (a) A 100 keV electron with pitch angle of 20° in the Earth's radiation belts, which are located at altitudes of 7000–24 000 km. The Earth's magnetic field at these altitudes is in the range of 1 μT. (b) A 2.5 MeV He^{++} particle in a 7 T fusion reactor.

2-2. Consider a particle of mass m and charge q moving in the presence of constant and uniform electromagnetic fields given by $\mathbf{E} = E_0\hat{\mathbf{y}}$ and $\mathbf{B} = B_0\hat{\mathbf{z}}$. Assume that the particle starts from rest and at the origin. Find an expression for the trajectory, $x(t)$ and $y(t)$, of the particle, and plot it.

2-3. A Bainbridge mass spectrometer separates ions according to their mass-to-charge ratio. The ions first enter a velocity selector with perpendicular electric and magnetic fields in which only ions with a particular velocity can pass through in a straight line (not a complex drift). Find the magnitude of the velocity that is "selected," in terms of the applied **E** and **B** fields. Note that this is a special case of the general **E** × **B** drift.

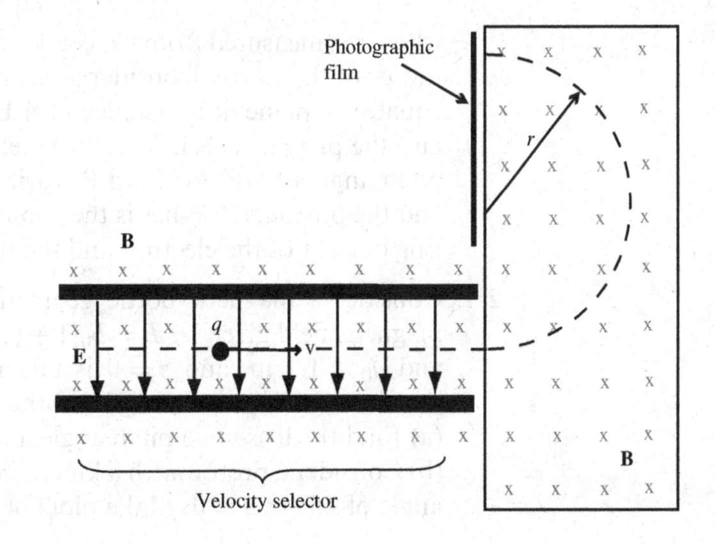

2-4. Consider a particle of charge q and mass m moving in the neighborhood of an infinitely long straight filamentary wire carrying a constant current I and extended along the z axis. At time $t = 0$, the particle is at $z = 0$ and at a radial distance from the wire of $r = r_0$, and has a velocity $v = v_0$ parallel to the wire. (a) Determine and plot the trajectory (i.e., the spatial path followed by the particle as time progresses) of the particle. (b) Assuming the magnitude of the current to be such that $q\mu_0 = m2\pi$, $r_0 = 1$ m, and $v_0 = 0.5$ m s^{-1}, find the location of the particle (i.e., numerical values of its coordinates r and z) at $t = 2$ s.

2-5. For a basic Hall thruster, as discussed in Example 2-1, but with an electric field of 250 V and a magnetic field of 2.5 mT: (a) What is the exit velocity of the xenon ions and the $\mathbf{E} \times \mathbf{B}$ drift velocity for the electrons? (b) If the ionization efficiency of the xenon is 90%, what must the total mass consumption rate from the thruster be in order to achieve a thrust of 0.08 N? (c) What would the equivalent mass flow rate be if a chemical rocket were used instead of the Hall thruster? Assume that the maximum exit velocity for a chemical rocket is 6 km s^{-1}.

2-6. The Earth's magnetic field can be accurately approximated as that of a dipole. The strength of the magnetic field in the equatorial plane is given by the formula (in cylindrical coordinates)

$$B(r) = 3.1 \times 10^{-5} \left(\frac{R_E}{r}\right)^3 \hat{\phi} \text{ (in units of T)},$$

where r is measured from the center of the Earth and R_E is the radius of the Earth. Consider a proton and an electron at the equatorial plane at a distance of 4 Earth radii. The electron and the proton each have total kinetic energy of 10 keV and pitch angle of 90°. (a) Find the drift velocity of the electron and the proton. (b) What is the combined contribution to the ring current of the electron and the proton?

2-7. Consider a magnetic-bottle geometry in a laboratory, with magnetic field given by $B = B_0[1 + (z/a_0)2]$, where $B_0 = 2.5$ T and $a_0 = 1.2$ m, and $z = 0$ is taken to be at the center of the device. The total length of the magnetic bottle is 5 m. (a) Find the loss-cone pitch angle α_{lc} for the magnetic bottle. (b) Consider a proton with a kinetic energy of 1 eV and a pitch angle of 15° at $z = 0$. Make plots of the following quantities versus z: (1) magnetic field versus distance, (2) perpendicular ($v_\perp$) and parallel ($v_\parallel$) velocities, (3) instantaneous pitch angle $\alpha(z)$, (4) kinetic energy, and (5) magnetic moment.

2-8. Consider a magnetic field primarily in the z direction, expressed as $B_z(z) = B_0(1 - k_0 z)$. A particle with charge q and kinetic energy $\frac{1}{2}mv^2$ is gyrating initially at the origin of a cylindrical coordinate system with its guiding center on the z axis and moving in the positive z direction. Find the initial pitch angle (at $z = 0$) that the particle must have in order to be detected by a ring-shaped particle detector located at $z = Z$ and $r = R$ and centered on the z axis, as shown in the diagram below. The magnetic field configuration is such that $|Zk| \ll 1$. Express the answer in terms of Z, R, B_0, k_0, and the fundamental properties of the particle.

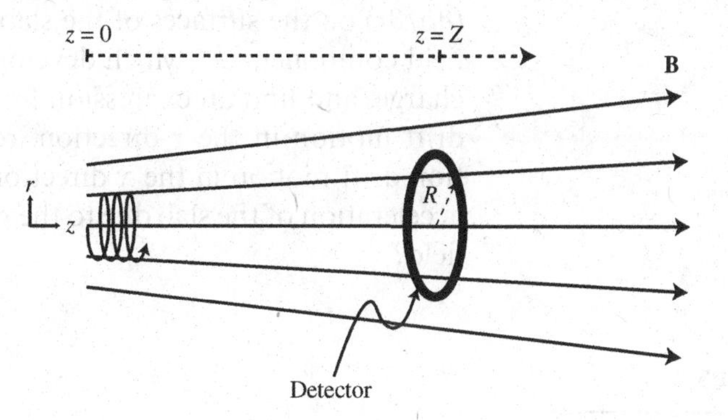

2-9. An electron gyrates in a constant and uniform magnetic field with $\mathbf{B} = 0.1\hat{\mathbf{z}}$ T and $v_\parallel = 0$. A uniform electric field is slowly introduced into the system, with $\mathbf{E}_1 = 0.1t\hat{\mathbf{y}}$ Vm^{-1}. At $t = 0$, the guiding center of the electron is located at the origin. (a) Calculate the displacement of the guiding center of the electron due only to polarization drift at $t = 10$ s. (b) What is the displacement of the guiding center at $t = 10$ s due to $\mathbf{E} \times \mathbf{B}$ drift? (c) At $t = 10$ s, the electric field switches to $\mathbf{E}_2 = \mathbf{E}_1(10) - 0.1t\hat{\mathbf{y}}$ Vm^{-1}. Repeat parts (a) and (b) at $t = 20$ s. (d) At $t = 10$ s, what is the work that has been done by the field on each particle (i.e., $\int q\mathbf{E} \cdot d\mathbf{l}$), and what is the kinetic energy of the $\mathbf{E} \times \mathbf{B}$ drift? This problem is intended to illustrate that the slowly increasing electric field is the agent which imparts energy to the particle, which is then stored as kinetic energy in the $\mathbf{E} \times \mathbf{B}$ drift.

2-10. Given a spatially varying magnetic field of the form

$$\mathbf{B}(x, z) = B_0\left[\alpha z\hat{\mathbf{x}} + (1 + \alpha x)\hat{\mathbf{z}}\right],$$

where α is a positive constant: (a) Determine and make a plot of the locus of points in the x–z plane where a charged particle with $\mathbf{v} = \mathbf{v}_{\|} + v_{\perp}$ can exist without experiencing any drift. (b) Given that the magnetic field lines will follow the equation $\frac{dx}{dz} = \frac{B_x}{B_z}$, plot the magnetic field lines and explain the result of part (a).

2-11. Consider a rectangular coordinate system with z vertically upward. A slab of charge-neutral plasma is infinite in the y and z directions but is bounded in the x direction between $x = \pm a$. A gravitational field $\mathbf{g} = -g\hat{\mathbf{z}}$ and a magnetic field $\mathbf{B} = \hat{\mathbf{y}}B$ are present. Assume that $\mathbf{B}$ is sufficiently strong that all particle motion can be analyzed in terms of drift motions. (a) Describe the motions of electrons and ions in the x direction. Determine the rate of accumulation of surface charge $(\partial\rho/\partial t)$ on the surfaces of the slab. (b) Determine the electric field component E_x which develops as a result of the surface charge, and find an expression for the time-dependent $\mathbf{E} \times \mathbf{B}$ drift motion in the z direction. (c) Determine the polarization drift motion in the x direction. (d) What is the resultant acceleration of the slab due to the presence of the gravitational field?

References

[1] M. A. Heald and J. B. Marion, *Classical Electromagnetic Radiation*, 3rd edn (Philadelphia: Saunders College Publishing, 1995), 274–6.
[2] J. A. Bittencourt, *Fundamentals of Plasma Physics*, 3rd edn (New York: Springer-Verlag, 2004), 106–8.

CHAPTER

3 Kinetic theory of plasmas

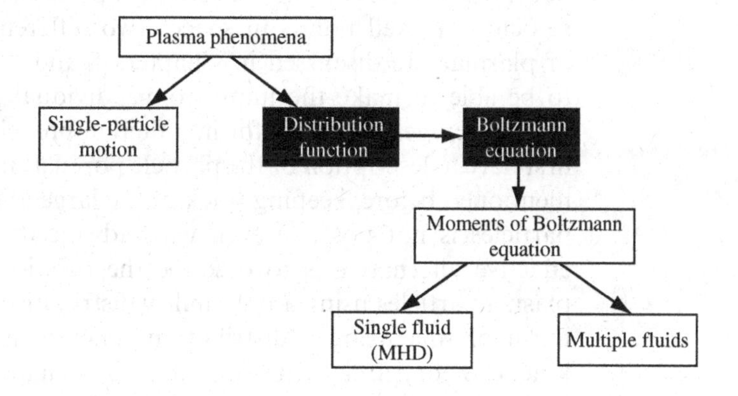

3.1 Introduction

Our examination of single-particle behavior in the previous chapter provided our first insight into plasma behavior. We have seen that particles execute gyrating motions along magnetic fields while drifting across the magnetic field, owing to a variety of mechanisms. In these analyses, we have implicitly assumed the electric and magnetic fields to be separately specified, without discussing the specific sources for the fields. Such an assumption exemplifies the limitations of the single-particle approach. While a static magnetic is typically imposed for confinement or heating of plasmas, or is naturally present as in a magnetized planetary body, an external electric field in general cannot be sustained in a plasma, as a result of the Debye shielding discussed in Chapter 1. Non-zero electric fields in a plasma generally arise self-consistently as a result of the collective motion of many plasma particles. Furthermore, the parameters modeled in single-particle analyses (e.g., particle position and velocity) are in general not measurable and cannot be related to observations. The measurable quantities, such as the bulk plasma velocity and particle density, cannot easily be derived from the single-particle parameters, the dependencies on which are rather complicated. There is thus a practical need to describe the behavior of large quantities of particles.

Most plasma physics problems can be treated in a useful manner using meaningful statistics and averages over a large number of individual particles. The most common scheme for describing such bulk or macroscopic behavior is the fluid approach, whereby the the plasma is described in terms of fluid equations, appropriately modified to account for electromagnetic forces. In using a fluid treatment, one is not concerned with the motions of individual particles but rather focuses on the collective properties (e.g., density, average velocity) at fixed points in space. Two different fluid descriptions of plasmas are discussed in Chapters 5 and 6. However, in order to be able to make the jump from individual particle motions to useful averages (as called for in a fluid approach), it is necessary to first have a description of the particle population. Since, as has been mentioned before, keeping track of the large numbers of individual particles is not possible even with advanced supercomputers, an effective alternative is to describe the positions and velocities of plasma particles using a probability distribution function. Describing a plasma using a distribution function is known as plasma kinetic theory and is the subject of this chapter. As is illustrated in the hierarchical chart above, kinetic theory forms the foundation of the majority of approaches to plasma phenomena.

Plasma kinetic theory is based on concepts originally developed to describe the dynamical behavior of neutral gases, namely a velocity distribution function describing the number of particles in any given point in space and having a particular set of component velocities, and an integro-differential equation, known as the *Boltzmann equation*, which describes the time variation of this distribution function. It is the most fundamental description of the plasma state, and is in fact the only means at our disposal for analysis of non-equilibrium plasmas. For the vast majority of applications involving plasmas under quasi-equilibrium conditions, simplified equations (i.e., fluid equations) derived from the Boltzmann equation can be used, and measurable quantities (e.g., density, velocity, temperature) can be obtained as suitable averages of the velocity distribution function. We start with a brief review of properties of gases and continue by introducing the concept of the velocity distribution function as a means for statistical characterization of plasma dynamics. We then discuss methods by which average values of particle properties can be derived from this distribution function. After deriving the Boltzmann equation from fundamental principles, we introduce the particular distribution function known as the Maxwell–Boltzmann distribution, which any gas in thermal equilibrium attains given enough time, in the absence of external forces.

3.2 Comparison of properties of gases and plasmas

Kinetic theory of plasmas makes maximum use of the analogy between a plasma and an ordinary neutral gas. The two parameters which characterize the physical state of a gas are the density of molecules N_0 and the temperature T. The mean kinetic energy of the particles depends only on the temperature (see Section 1.1), while the pressure is proportional to both the density and the temperature. For gases in thermal equilibrium, the pressure p is related to N_0 and T by

$$p = N_0 k_B T, \tag{3.1}$$

where k_B is Boltzmann's constant.[1] When a gas is in thermal equilibrium at a certain temperature T, the individual gas molecules do not all have the same energy. On the contrary, it can be shown theoretically, and measured experimentally, that the molecules have widely different velocities, ranging from zero to very large values. In the mid-1890s, J. C. Maxwell showed that the number of gas molecules which have speeds lying in the range between v and $v + dv$ is given by

$$N(v) = Av^2 e^{-mv^2/(2k_B T)} \Delta v, \tag{3.2}$$

where the coefficient A can be determined on the basis that the number of molecules per unit volume is equal to the density N_0. Summing over all possible particle velocities, which is in effect an integration, we find

$$N_0 = \int_0^\infty Av^2 e^{-mv^2/(2k_B T)} dv \quad \rightarrow \quad A = N_0 \left[\frac{2}{\pi} \left(\frac{m}{k_B T} \right)^3 \right]^{1/2}.$$

[1] Equation (3.1) describes only one of many possible thermodynamic equations of state, known in this case as the *isothermal* equation of state. We will see other types of state equations in the following chapter.

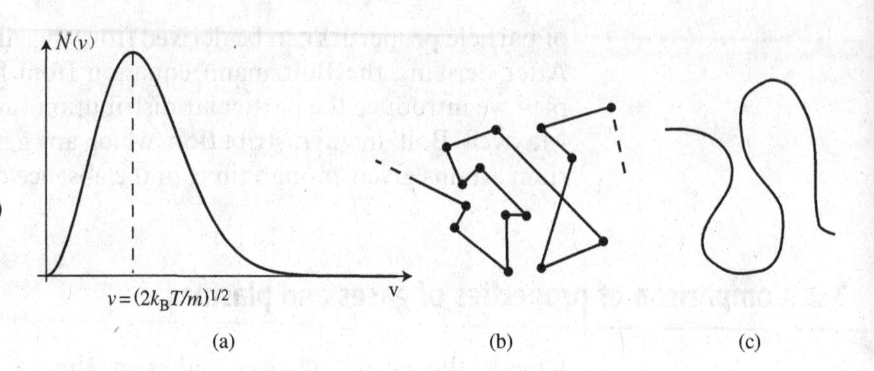

Figure 3.1 A gas in thermal equilibrium. (a) Maxwellian distribution of particle speeds in a gas in thermal equilibrium. (b) Random motion of a neutral gas molecule under the influence of elastic collisions. (c) Trajectory of a charged particle in a plasma under the influence of Coulomb collisions.

This speed distribution function is shown in Figure 3.1a. The maximum of this function occurs at $v = \sqrt{k_B T/m}$, indicating that this is the most probable speed for any given molecule. Another feature is the very rapid decrease in the number of molecules at low velocities and very high velocities. Figure 3.1b shows the random thermal motion of a particular gas molecule. Each break in this curve is a result of an elastic collision of the molecule with some other molecule of the gas. The duration of each collision is very small compared to the time between collisions. The elementary kinetic theory of gases is based on the assumption that after a collision the particle can, with equal probability, continue in any direction, regardless of its initial momentum. The mean (or average) length of the straight-line sections making up the zig-zag path of the molecules in a gas is known as the *mean free path*, generally denoted by λ. The average time taken by a molecule to traverse a straight-line section is known as the mean time between collisions and is denoted by τ. The mean collision frequency, or the average number of collisions experienced by the particle per second, is then $\nu = \tau^{-1}$.

Since plasmas are basically ionized gases, we can expect many of the concepts discussed above to apply when we consider a collection of charged (rather than neutral) particles. This expectation is borne out to some degree, but important differences exist, especially in the nature of "collisions" between particles. In gases, the length scales of the system are often much larger than the mean free path and the time scales involved are much longer than the mean time between collisions. Under such conditions, the gas attains thermal equilibrium on a much finer scale of length and time than the spatial and temporal variations of macroscopic variables (e.g., pressure, temperature) so that these variables can be accurately described by a fluid treatment. The fundamental reason for this is the very short range of action of the forces between neutral molecules (i.e., the particles have to essentially touch one another, at atomic distances of $\sim 10^{-8}$ cm).

In plasmas, the interaction between particles is via Coulomb forces, which are long-range and relatively weak, compared to direct-impact collisions. Therefore, the path of a charged particle in a plasma is quite different from the zig-zag path of a neutral gas molecule shown in Figure 3.1b. The path of the charged particles is smoother and cannot readily be resolved into straight-line segments which begin and end at points at which collisions take place. Each individual particle moves under the influence of the average electric field due to all of the other particles and its motion is subject to continuous fluctuations in magnitude and direction. This average field is generally weak, so that the particle motion resembles the random walk of a person in a desert at night, as shown in Figure 3.1c. Another way to look at the difference between the motions described in Figures 3.1b and 3.1c is to note that the long-range Coulomb forces actually allow the particles to avoid collisions. A useful analogy can be made by observing large numbers of students on bicycles rapidly streaming through a busy plaza, skillfully avoiding collisions with all but an occasional few of the many other cyclists. As a result of the reduced number of collisions, establishment of thermal equilibrium is slower in plasmas than in neutral gases, and many physical processes can occur over time scales which are faster than the relaxation time to equilibrium. As a result, non-equilibrium distributions are more often encountered in plasmas, and must be analyzed with a kinetic treatment.

3.3 Velocity distribution function

We start by assuming that the location of each particle is documented by a position vector $\mathbf{r}$ drawn from the origin to the physical point at which the particle resides. In other words, we have

$$\mathbf{r} = \hat{\mathbf{x}}x + \hat{\mathbf{y}}y + \hat{\mathbf{z}}z. \tag{3.3}$$

We consider a small elemental volume $d\mathbf{r} = dxdydz$, also denoted as d^3r. Note that $d\mathbf{r}$ is not a vector, but simply represents a three-dimensional volume element. Note that the volume element $d\mathbf{r}$ must be large enough to contain a great number of particles, but small enough so that macroscopic quantities such as pressure, temperature, and velocity vary only slightly within this element. As an example, a cube 0.01 mm on each side of a gas at standard temperature and pressure contains $\sim 10^{10}$ molecules, and knowledge of macroscopic quantities with a spatial resolution of 0.01 mm is certainly sufficient in most applications. Let the linear velocity of the particle be

$$\mathbf{v} = \hat{\mathbf{x}}v_x + \hat{\mathbf{y}}v_y + \hat{\mathbf{z}}v_z \tag{3.4}$$

so that the particle speed is $|\mathbf{v}| = v = \sqrt{v_x^2 + v_y^2 + v_z^2}$. In analogy with configuration space, we think of the components v_x, v_y, and v_z as being coordinates in *velocity space*. Thus the velocity vector $\mathbf{v}$ documents the location of the particle in this velocity space. For the sake of compactness, it is often convenient to introduce the concept of *phase space*, defined by the six coordinates x, y, z, v_x, v_y, and v_z. Thus, the position $\mathbf{r}$ and the velocity $\mathbf{v}$ of a particle at any given time can be represented as a point in this six-dimensional space. Now consider a time interval dt, centered around the time of observation t, which is long enough compared to the mean time the particle takes to traverse the volume element $d\mathbf{r}$ (if it is not scattered by collisions) but short enough compared to the time scales of variations of the macroscopic parameters. As an example, at a mean molecular velocity of hundreds of m s^{-1} for a gas at standard temperature and pressure, a particle travels 0.01 mm in less than 10^{-7} s, and variations over such time scales are rarely of interest in gas dynamics. The number of particles in the elemental volume $d\mathbf{r}$, when averaged over the time interval dt, is then simply given by $N(\mathbf{r}, t)d\mathbf{r}$, where $N(\mathbf{r}, t)$ is the number density of the particles, regardless of the shape of the volume element $d\mathbf{r}$. To each of the $N(\mathbf{r}, t)d\mathbf{r}$ particles contained in the volume $d\mathbf{r}$ corresponds a point in velocity space, which is denoted by the velocity point $\mathbf{v}$ and which represents the velocity of this particle in the volume $d\mathbf{r}$ at the considered instant of time (i.e., in the time interval dt centered around the observation time t). The distribution of these velocity points varies with time as a result of collisions and of the particle flux through $d\mathbf{r}$ (i.e., particles entering or leaving the elemental volume $d\mathbf{r}$). In the same manner that we defined a particle density in configuration space, we may statistically define a density of the $Nd\mathbf{r}$ points in velocity space. This density, proportional to $d\mathbf{r}$ and a function of $\mathbf{r}$ and t as well as $\mathbf{v}$, is denoted $f(\mathbf{r}, \mathbf{v}, t)d\mathbf{r}$. With this definition, the probable number of particles found at time t in the element $d\mathbf{r}$, possessing velocities between $\mathbf{v}$ and $\mathbf{v} + d\mathbf{v}$, is $f(\mathbf{r}, \mathbf{v}, t)d\mathbf{r}d\mathbf{v}$. The function $f(\mathbf{r}, \mathbf{v}, t)$ is called the *velocity distribution function* and is the probable density of representative points in the six-dimensional phase space. Note that if the number of velocity points $f(\mathbf{r}, \mathbf{v}, t)d\mathbf{r}d\mathbf{v}$ contained in the elemental volume $d\mathbf{v}$ is summed up over all possible velocities, we obtain the total number $N(\mathbf{r}, t)d\mathbf{r}$ of velocity points in the entire velocity space. Thus, it follows that

$$N(\mathbf{r}, t) = \int_{-\infty}^{\infty} f(\mathbf{r}, \mathbf{v}, t)d\mathbf{v} = \iiint_{-\infty}^{\infty} f(\mathbf{r}, \mathbf{v}, t)dv_x dv_y dv_z. \tag{3.5}$$

It is obvious that the distribution function $f(\mathbf{r}, \mathbf{v}, t)$ must be finite, continuous, and positive for all values of t and that it must approach zero as the speed v tends to infinity. The velocity distribution function representation of the plasma is quite general. Inhomogeneous plasmas, in which physical quantities depend on location, are described by distribution functions $f(\mathbf{r}, \mathbf{v}, t)$ which are explicit functions of $\mathbf{r}$, whereas distribution functions for homogeneous plasmas do not depend on $\mathbf{r}$ and are only functions of $\mathbf{v}$ and t. In anisotropic plasmas, $f(\mathbf{r}, \mathbf{v}, t)$ is dependent on the orientation of the velocity vector $\mathbf{v}$, while in isotropic plasmas, $f(\mathbf{r}, \mathbf{v}, t)$ is not dependent on the direction of $\mathbf{v}$. The velocity distribution function representation of a plasma retains the full statistical information on all of the particles and hence is a *microscopic* description. However, often the most important use of the velocity distribution function $f(\mathbf{r}, \mathbf{v}, t)$ is that *macroscopic* (i.e., ensemble average) values of various plasma parameters (e.g., density, flux, current) can be obtained from $f(\mathbf{r}, \mathbf{v}, t)$. Consider any property $g(\mathbf{r}, \mathbf{v}, t)$ of a particle. The value of this quantity averaged over all velocities is then given by

$$g_{\text{av}}(\mathbf{r}, t) \equiv \langle g(\mathbf{r}, \mathbf{v}, t) \rangle = \frac{1}{N(\mathbf{r}, t)} \int g(\mathbf{r}, \mathbf{v}, t)\, f(\mathbf{r}, \mathbf{v}, t)\, d\mathbf{v}. \tag{3.6}$$

Example 3-1 Phase-space distribution function

Consider a system of particles confined to a cube with side length s_0 centered at the origin. The particles are uniformly distributed in the cube with velocities given by the distribution function

$$f(\mathbf{r}, \mathbf{v}, t) = C_0^2 - v_i^2, \quad |v_i| < C_0, \quad |i| < \frac{s_0}{2} \quad (i = x, y, z)$$

$$= 0 \quad \text{otherwise.}$$

Find the number density N_0 of the particles and of the average particle kinetic energy, and plot the distribution function in spatial and velocity coordinates.

Solution: Using Equation (3.5) we can find the particle density as

$$N_0(\mathbf{r}, t) = \iiint_{-\infty}^{\infty} f(\mathbf{r}, \mathbf{v}, t) dv_x dv_y dv_z$$

$$= \iiint_{-C_0}^{C_0} \left(C_0^2 - v_x^2\right)\left(C_0^2 - v_y^2\right)\left(C_0^2 - v_z^2\right) dv_x dv_y dv_z$$

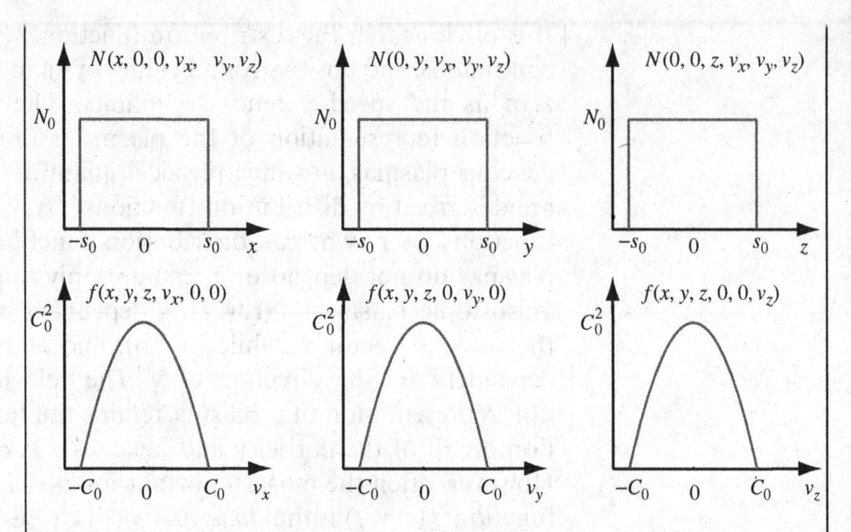

Figure 3.2 Plots of distribution function discussed in Example 3-1.

$$= \left[C_0^2 v_x - \frac{v_x^3}{3}\right]_{-C_0}^{C_0} \left[C_0^2 v_y - \frac{v_y^3}{3}\right]_{-C_0}^{C_0} \left[C_0^2 v_z - \frac{v_z^3}{3}\right]_{-C_0}^{C_0}$$

$$= \left(\frac{4}{3}C_0^3\right)\left(\frac{4}{3}C_0^3\right)\left(\frac{4}{3}C_0^3\right) = \frac{64}{27}C_0^9.$$

The kinetic energy of a single particle is $\frac{1}{2}mv^2 = \frac{1}{2}m(v_x^2 + v_y^2 + v_z^2)$ and by applying Equation (3.6) we can find the average kinetic energy of all the particles:

$$\left\langle \frac{1}{2}mv^2 \right\rangle = \frac{1}{N_0} \iiint_{-C_0}^{C_0} \frac{1}{2}m\left(v_x^2 + v_y^2 + v_z^2\right)\left(-v_x^2 + C_0^2\right) \times \left(-v_y^2 + C_0^2\right)\left(-v_z^2 + C_0^2\right) dv_x dv_y dv_z$$

$$= \frac{m}{2N_0}\left(\frac{4C_0^5}{15}\right)^3 = \frac{mC_0^6}{250}.$$

The distribution function is shown in the six plots in Figure 3.2, corresponding to the six dimensions of phase space.

3.4 The Boltzmann equation

How the velocity distribution function $f(\mathbf{r}, \mathbf{v}, t)$ evolves in time is described by the Boltzmann equation. The positions of the particles will vary in time because of their non-zero velocities. The velocities

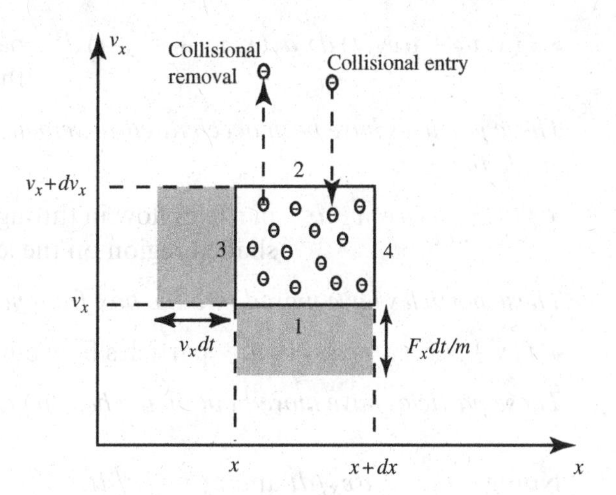

Figure 3.3 Phase space for a one-dimensional velocity distribution function. The particles are shown as negatively charged electrons as an example; the same process is valid for every distinct particle population.

will in turn vary in the presence of acceleration which is the result of any forces acting on the particles. In this straightforward manner, the representative points in phase space move as a function of time. We start by deducing the Boltzmann equation for a one-dimensional velocity distribution function, which depends on the position coordinate x, the velocity coordinate v_x, and time t. The corresponding two-dimensional phase space is shown in Figure 3.3.

Consider a time interval dt which is long compared to the average time interval of interaction between any two particles, so that most interactions which begin in the interval dt are completed in the same interval. On the other hand, we must have dt short enough compared to the average time between interactions so that each particle interacts at most once with another particle in the interval dt. Under these conditions, the trajectory of a particle may be composed of segments where only the external forces act, joined by very short trajectories during which there is interaction between particles. It is only under these conditions that the Boltzmann equation is valid.

If the external force acting on a particle is $\mathbf{F} = \hat{\mathbf{x}}F_x$, its acceleration is $\mathbf{a} = \hat{\mathbf{x}}F/m$. As particles drift in phase space under the action of the macroscopic force F_x, they flow into and out of the fixed two-dimensional "volume" $dxdv_x$ shown in Figure 3.3. The distribution function can be derived on the basis of conservation of particles, considering the net flow of particles into and out of the volume through each of its four faces in a time interval dt:

- $f(x, v_x, t)\, dx\, a_x(x, v_x, t)\, dt$ particles flow in through side 1 (from the shaded region below).

These particles have been accelerated so that they have velocities within the box.

- $f(x, v_x + dv_x, t)\, dx\, a_x(x, v_x + dv_x, t)\, dt$ particles flow out through side 2.

These particles have been accelerated to velocities greater than those inside the box.

- $f(x, v_x, t)\, dv_x\, v_x\, dt$ particles flow in through side 3 (from the shaded region on the left).

These particles have moved into the box from positions on the left.

- $f(x + dx, v_x, t)\, dv_x\, v_x\, dt$ particles flow out through side 4.

These particles have moved out of the box to positions on the right.

Note that $a_x \equiv dv_x/dt$ and $v_x \equiv dx/dt$.

So far we have accounted for changes in the distribution function arising from intrinsic motion of the particles or from external forces. In addition, in the time interval dt, the particles in the range dx at x interact or collide at most once with other particles and have their velocities changed. Velocities of some of the particles which were in the range dv_x at v_x change to other values so they leave the two-dimensional volume, while velocities of some other particles outside the range dv_x at v_x now fall within the range. Note that such collisions can nearly instantaneously change the velocity of the particle, but not its position. Until we are ready to be specific about the nature of the collisions and their effectiveness in moving particles in phase space, we schematically represent the resultant net gain or loss of particles due to this process by a collision term given by

$\left(\frac{\partial f}{\partial t}\right)_{\text{coll}} dx dv_x dt$ particles flow in as a result of collisions.

The total number of particles in the velocity space volume element $dx\, dv_x$ is given by $f(x, v_x, t)\, dx\, dv_x$. The time rate of change of the number of particles is determined by the difference between the number of particles flowing into and out of the two-dimensional volume:

$$\begin{aligned}\frac{\partial}{\partial t}&[f(x, v_x, t)\, dx\, dv_x] \\ &= [f(x, v_x, t)\, a_x(x, v_x, t) - f(x, v_x + dv_x, t) a_x(x, v_x + dv_x, t)]\, dx \\ &\quad + [f(x, v_x, t)\, v_x - f(x + dx, v_x, t)\, v_x]\, dv_x + \left(\frac{\partial f}{\partial t}\right)_{\text{coll}} dx dv_x.\end{aligned}$$

Dividing by $dx\,dv_x$ we have

$$\frac{\partial}{\partial t} f(x, v_x, t) = -\frac{\partial}{\partial x}(f\,v_x) - \frac{\partial}{\partial v_x}(f\,a_x) + \left(\frac{\partial f}{\partial t}\right)_{\text{coll}}. \tag{3.7}$$

Noting that v_x and x are independent variables, and assuming that the force F_x is independent of the velocity v_x so that a_x is independent of v_x, we can write

$$\frac{\partial f}{\partial t} + v_x \frac{\partial f}{\partial x} + \frac{F_x}{m} \frac{\partial f}{\partial v_x} = \left(\frac{\partial f}{\partial t}\right)_{\text{coll}}, \tag{3.8}$$

where we have substituted $a_x = F_x/m$. Note that the assumption of F_x being independent of v_x is not restrictive, especially in view of the fact that the external force in the case of plasmas is almost always the Lorentz force, $\mathbf{F} = q[\mathbf{E} + \mathbf{v} \times \mathbf{B}]$, so that F_x is in fact not dependent on v_x (see Equation (3.12b)). It is clear from the above discussion that the Boltzmann equation is a statement of the conservation of the number of representative points in phase space. By following the same procedure as above for an elemental volume $d\mathbf{r}d\mathbf{v} = dxdydzdv_xdv_ydv_z$ in six-dimensional phase space, we can derive the three-dimensional Boltzmann equation:

Boltzmann equation

$$\boxed{\frac{\partial}{\partial t} f(\mathbf{r}, \mathbf{v}, t) + (\mathbf{v} \cdot \nabla_{\mathbf{r}})\, f(\mathbf{r}, \mathbf{v}, t) + \left[\left(\frac{\mathbf{F}}{m}\right) \cdot \nabla_{\mathbf{v}}\right] f(\mathbf{r}, \mathbf{v}, t) = \left(\frac{\partial f}{\partial t}\right)_{\text{coll}}}, \tag{3.9}$$

where

$$\nabla_{\mathbf{r}} \equiv \hat{\mathbf{x}} \frac{\partial}{\partial x} + \hat{\mathbf{y}} \frac{\partial}{\partial y} + \hat{\mathbf{z}} \frac{\partial}{\partial z} \tag{3.10a}$$

$$\nabla_{\mathbf{v}} \equiv \hat{\mathbf{x}} \frac{\partial}{\partial v_x} + \hat{\mathbf{y}} \frac{\partial}{\partial v_y} + \hat{\mathbf{z}} \frac{\partial}{\partial v_z}, \tag{3.10b}$$

so that the second and third terms in (3.9) are

$$(\mathbf{v} \cdot \nabla_{\mathbf{r}})\, f(\mathbf{r}, \mathbf{v}, t) = v_x \frac{\partial f}{\partial x} + v_y \frac{\partial f}{\partial y} + v_z \frac{\partial f}{\partial z} \tag{3.11a}$$

$$\left[\left(\frac{\mathbf{F}}{m}\right) \cdot \nabla_{\mathbf{v}}\right] f(\mathbf{r}, \mathbf{v}, t) = \frac{F_x}{m} \frac{\partial f}{\partial v_x} + \frac{F_y}{m} \frac{\partial f}{\partial v_y} + \frac{F_z}{m} \frac{\partial f}{\partial v_z}. \tag{3.11b}$$

Note that in the plasma context, the force acting on the particles is the Lorentz force, given by

$$\mathbf{F} = q[\mathbf{E} + \mathbf{v} \times \mathbf{B}] \tag{3.12a}$$

$$F_x = qE_x + q\, v_y\, B_z - q\, v_z\, B_y \tag{3.12b}$$

$$F_y = qE_y + q\, v_z\, B_x - q\, v_x\, B_z \tag{3.12c}$$

$$F_z = qE_z + q\, v_x\, B_y - q\, v_y\, B_x, \tag{3.12d}$$

where the electric field **E** and the magnetic field **B** are continuous macroscopic fields obtained by averaging over an elemental volume whose dimensions are large enough to contain a large number of particles but small enough compared to the spatial scale of variation of the physical quantities of interest. Note that since each component of **F** is independent of the corresponding component of particle velocity **v**, the transition we made from (3.7) to (3.8) is valid. Although we still are far from having an explicit expression for the collision term on the right-hand side, we can note that it is generally possible to express $(\partial f/\partial t)_{\text{coll}}$ as an appropriately weighted multiple integral of $f(\mathbf{r}, \mathbf{v}, t)$. Thus, the Boltzmann equation (3.9) is in general an integro-differential equation. In a typical plasma there are at least three different species of particles, for example, electrons, ions, and neutrals. The Boltzmann equation (3.9) is separately valid for each species, each being characterized by a different distribution function $f_i(\mathbf{r}, \mathbf{v}, t)$. Since only electromagnetic forces are considered, the Boltzmann equation for the neutrals does not have the force term. The three Boltzmann equations for electrons, ions, and neutrals are coupled through the collision term. For example, the collision term in the Boltzmann equation for electrons contains the velocity distribution functions of the ions and neutrals as well as that for electrons.

3.5 The Maxwell–Boltzmann distribution

In the previous sections we introduced the concepts of velocity space, the six-dimensional phase space, and the distribution function as a means of specifying the probable locations and velocity components of an assembly of particles. We have also derived the Boltzmann equation as a mathematical statement of the conservation of particles. The Boltzmann equation is the very foundation of kinetic theory. It is not specific to plasmas but is valid for any gas under equilibrium or non-equilibrium conditions. The logical next step is to investigate possible solutions to the Boltzmann equation.

Given enough time, any collection of freely mobile charged or neutral particles will reach an equilibrium state. Consider a gas of particles suddenly introduced into a bounded volume (e.g., a container with walls). Depending on how the particles were introduced, they may initially be located in specific positions (e.g., the corner from which they were introduced) and may have velocities predominantly in one direction, so that the probability of finding particles at other locations and with velocities in other directions is low. In other words, the distribution function f is initially inhomogeneous (a function of $\mathbf{r}$) and anisotropic (a function of the direction of $\mathbf{v}$). As time progresses, and as a result of collisional interactions between particles and between particles and the walls, the particle distribution function evolves toward a homogeneous one, so that it is equally probable to find particles anywhere in the container. Furthermore, the velocities of particles passing through any point in the volume become uniformly distributed in all directions, so that the distribution function f becomes isotropic. If the container is maintained at a constant temperature, the gas eventually attains thermal equilibrium with its temperature equal to that of the container. Under thermal equilibrium conditions, a homogeneous gas with no external forces attains a specific homogeneous and isotropic distribution known as the Maxwell–Boltzmann distribution. An example of a homogeneous but anisotropic distribution function representing a gas of N_0 electrons all moving in a given direction with a speed v_0 is the so-called beam-like distribution given by

$$f_{\text{beam}}(\mathbf{r}, \mathbf{v}, t) = f_{\text{beam}}(\mathbf{v}) = N_0\,\delta(v_x - v_0)\,\delta(v_y)\,\delta(v_z), \tag{3.13}$$

where $\delta(x)$ is the Dirac delta function. It can easily be shown by substitution that (3.13) is a solution of the Boltzmann equation. However, although a distribution function $f(\mathbf{r}, \mathbf{v}, t) = f_{\text{beam}}(\mathbf{v})$ satisfies the Boltzmann equation at any given time, the distribution f does not stay the same in time but instead evolves toward a Maxwell–Boltzmann distribution. An example of a homogeneous and isotropic distribution function which also cannot remain in effect as a function of time is the so-called shell distribution, representing a uniform gas of particles moving in all directions with equal probability (i.e., isotropic) but all with the same speed v_0, i.e.,

$$f_{\text{shell}}(\mathbf{r}, \mathbf{v}, t) = f_{\text{shell}}(\mathbf{v}) = A\,\delta(v - v_0), \tag{3.14}$$

where $v = |\mathbf{v}| = \sqrt{v_x^2 + v_y^2 + v_z^2}$. Equation (3.14) is solution of the Boltzmann equation at a given time, but if the distribution function of a gas $f(\mathbf{r}, \mathbf{v}, t)$ is equal to $f_{\text{shell}}(\mathbf{v})$ at any given time, it cannot remain equal to it and must evolve in time toward a

Maxwell–Boltzmann distribution. We stated above that a homogeneous gas with no external forces attains the Maxwell–Boltzmann distribution under thermal equilibrium (or steady-state) conditions. Mathematically, these conditions can be represented by $\mathbf{v} \cdot \nabla_{\mathbf{r}} = 0$ (homogeneous), $(\mathbf{F}/m) \cdot \nabla_{\mathbf{v}} = 0$ (no external forces), and $\partial f/\partial t = 0$ (steady-state), reducing the Boltzmann equation (3.8) to

$$0 + 0 + 0 = \left(\frac{\partial f}{\partial t}\right)_{\text{coll}} \quad \rightarrow \quad \left(\frac{\partial f}{\partial t}\right)_{\text{coll}} = 0. \tag{3.15}$$

In words, (3.15) states that, under equilibrium conditions, in any given time interval dt as many particles enter an elemental volume in phase space as a result of collisions as leave. By considering the mechanics of two-particle collisions in which total energy is conserved under conditions specified by (3.15), it can be shown [1, 2] that the particle distribution function must necessarily be the Maxwell–Boltzmann distribution. The Maxwell–Boltzmann distribution has the general form

$$f(\mathbf{r}, \mathbf{v}, t) = f(v) = Ce^{-av^2}, \tag{3.16}$$

where C and a are positive constants (independent of $\mathbf{r}$ and $\mathbf{v}$) and $v = \sqrt{v_x^2 + v_y^2 + v_z^2}$ is the particle speed. The physical parameters of the gas, such as the particle number density N_0 and temperature T, can be derived from (3.16), as we show below. Note that both N_0 and T are constants at all points in space (i.e., they are independent of $\mathbf{r}$).

3.5.1 Number density

The total number of particles in the system can be obtained by integrating (3.16) over the entire phase space. Since f does not depend on $\mathbf{r}$, this means simply that we integrate (3.16) over velocity space. Using spherical coordinates as appropriate here, the elemental volume $d\mathbf{v}$ in velocity space is $d\mathbf{v} = v^2 \sin\theta\, dv d\theta d\phi$, and we have

$$N_0 = \int_0^{2\pi} \int_0^{\pi} \int_0^{\infty} Ce^{-av^2} v^2 dv \sin\theta\, d\theta d\phi$$

$$= \int_0^{\infty} 4\pi C v^2 e^{-av^2}\, dv$$

$$N_0 = C\left(\frac{\pi}{a}\right)^{\frac{3}{2}} \quad \rightarrow \quad C = N_0\left(\frac{a}{\pi}\right)^{\frac{3}{2}},$$

where we have used the fact that

$$\int_0^\infty \zeta^2 e^{-a\zeta^2} d\zeta = \sqrt{\frac{\pi}{16a^3}}. \tag{3.17}$$

Having determined C, we can rewrite (3.16) as

$$f(v) = N_0 \left(\frac{a}{\pi}\right)^{\frac{3}{2}} e^{-a v^2}.$$

Note that the constant a is yet to be determined. It represents the width of the distribution function, and is thus related to the average kinetic energy of particles, or temperature.

3.5.2 Temperature

The temperature of a gas in thermal equilibrium was defined in Chapter 1 in terms of the average kinetic energy of the particles constituting it. The distribution function representation of a plasma allows us to readily calculate the average values of any physical property of the individual particles. Let us first consider the average value of the component of the particle kinetic energy due to its motion in the x direction, i.e., $\frac{1}{2}mv_x^2$. The number of velocity points in the elemental volume $d\mathbf{v}$ around the position vector $\mathbf{v}$ in velocity space is $f(\mathbf{r}, \mathbf{v}, t)d\mathbf{r}d\mathbf{v}$. Thus, the value of $\frac{1}{2}mv_x^2$ summed over all the velocity points in the volume element $d\mathbf{v}$ is $\frac{1}{2}mv_x^2 f(\mathbf{r}, \mathbf{v}, t)d\mathbf{r}d\mathbf{v}$. The total value of this quantity summed over all velocity points in velocity space can be found by integrating over all possible velocities:

$$d\mathbf{r} \int_{-\infty}^{\infty} \frac{1}{2}mv_x^2 f(\mathbf{r}, \mathbf{v}, t)d\mathbf{v}.$$

To find the average value of the quantity $\frac{1}{2}mv_x^2$, we simply divide by the total number of velocity points, $N(\mathbf{r}, t)d\mathbf{r}$, with $N(\mathbf{r}, t)$ related to $f(\mathbf{r}, \mathbf{v}, t)$ as in (3.5):

$$\left\langle \frac{1}{2}mv_x^2 \right\rangle = \frac{1}{N(\mathbf{r}, t)} \int_{-\infty}^{\infty} \frac{1}{2}mv_x^2 f(\mathbf{r}, \mathbf{v}, t)d\mathbf{v}. \tag{3.18}$$

For the specific distribution function (3.16) we have

$$\left\langle \frac{1}{2}mv_x^2 \right\rangle = \frac{1}{N_0} \int_{-\infty}^{\infty} \int_{-\infty}^{\infty} \int_{-\infty}^{\infty} \frac{1}{2}mv_x^2 N_0 \left(\frac{a}{\pi}\right)^{\frac{3}{2}} \times v_x^2 e^{-a\left(v_x^2+v_y^2+v_z^2\right)} dv_x dv_y dv_z$$

$$= \frac{m}{2}\left(\frac{a}{\pi}\right)^{\frac{3}{2}} \int_{-\infty}^{\infty} v_x^2 e^{-av_x^2} dv_x \int_{-\infty}^{\infty} e^{-av_y^2} dv_y \int_{-\infty}^{\infty} e^{-av_z^2} dv_z$$

$$\frac{1}{2} m \left\langle v_x^2 \right\rangle = \frac{m}{4a},$$

where we have used (3.17) and the fact that

$$\int_0^{\infty} e^{-a\zeta^2} d\zeta = \left(\frac{\pi}{4a}\right)^{\frac{1}{2}}. \tag{3.19}$$

Since the distribution (3.16) is isotropic, we must have

$$\frac{1}{2} m \left\langle v_x^2 \right\rangle = \frac{1}{2} m \left\langle v_y^2 \right\rangle = \frac{1}{2} m \left\langle v_z^2 \right\rangle = \frac{m}{4a}.$$

The average total kinetic energy of the particles is then given by

$$\frac{1}{2} m \left\langle v^2 \right\rangle = \frac{1}{2} m \left\langle v_x^2 + v_y^2 + v_z^2 \right\rangle = \frac{3m}{4a}.$$

As mentioned in Chapter 1, the temperature T of a gas in thermal equilibrium is defined by the relation

$$\text{Average energy} = \frac{3}{2} k_B T \quad \rightarrow \quad \frac{1}{2} m \langle v^2 \rangle = \frac{3}{2} k_B T \quad \rightarrow \quad a = \frac{m}{2 k_B T}.$$

Substituting into (3.16) gives the Maxwell–Boltzmann distribution, written in terms of the physical parameters N_0 and T:

Maxwell–Boltzmann distribution

$$\boxed{f(v) = N_0 \left(\frac{m}{2\pi k_B T}\right)^{\frac{3}{2}} e^{-mv^2/(2k_B T)}}. \tag{3.20}$$

The distribution function described by (3.20) is the only steady-state distribution possible for a homogeneous (constant density N_0 everywhere) gas in thermal equilibrium (fixed temperature T) in the absence of external forces. All other distributions approach that given in (3.16) as time evolves. The distribution function (3.20) is referred to either as the Maxwell–Boltzmann distribution or as the Maxwellian distribution.

3.5.3 Velocity in one dimension and speed

Having defined the Maxwellian distribution in Equation (3.20), we can find the distribution for one component of the velocity by integrating over the remaining components. For example, the distribution of velocities in the x direction is given by

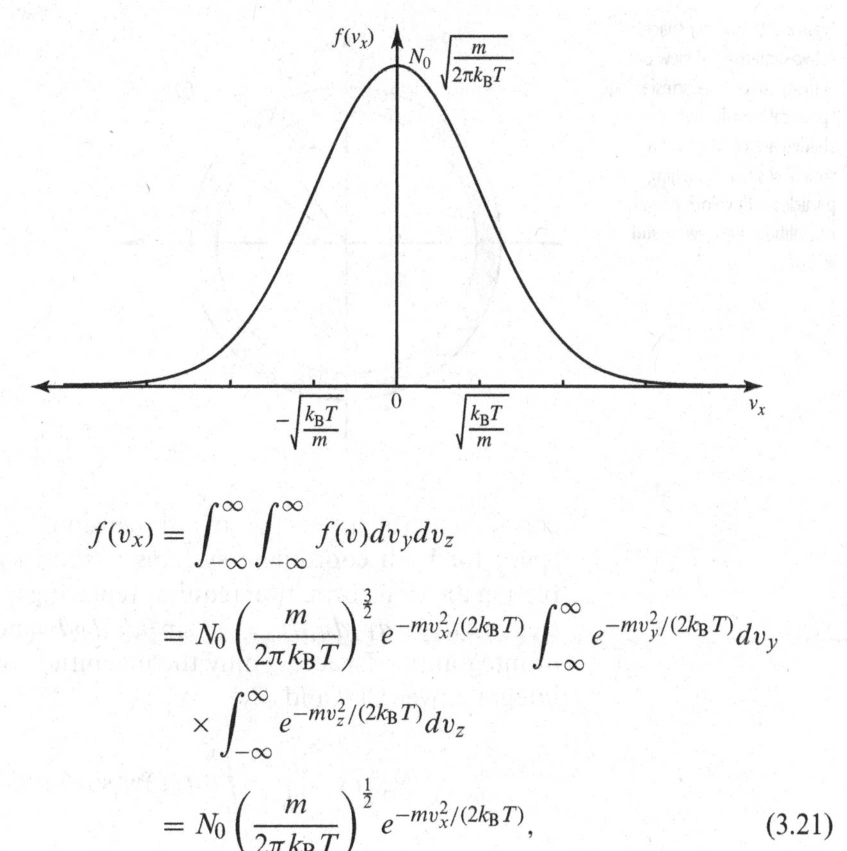

Figure 3.4 Distribution of velocity in one direction. The distribution function for particle velocity in one direction in a Maxwell–Boltzmann distribution is the well-known Gaussian distribution, with a standard deviation of $\sqrt{k_B T/m}$.

$$
\begin{aligned}
f(v_x) &= \int_{-\infty}^{\infty}\int_{-\infty}^{\infty} f(v)dv_y dv_z \\
&= N_0\left(\frac{m}{2\pi k_B T}\right)^{\frac{3}{2}} e^{-mv_x^2/(2k_B T)} \int_{-\infty}^{\infty} e^{-mv_y^2/(2k_B T)}dv_y \\
&\quad \times \int_{-\infty}^{\infty} e^{-mv_z^2/(2k_B T)}dv_z \\
&= N_0\left(\frac{m}{2\pi k_B T}\right)^{\frac{1}{2}} e^{-mv_x^2/(2k_B T)}, \qquad (3.21)
\end{aligned}
$$

where we have used the fact that each of the integrals is equal to $(2\pi k_B T/m)^{1/2}$. By symmetry, the expression in Equation (3.21) applies to any of the velocity components and can be readily identified as a *Gaussian* or *normal* distribution, with zero mean and a standard deviation of $\sqrt{k_B T/m}$. Equation (3.21) is plotted in Figure 3.4 in units of $\sqrt{k_B T/m}$. It is worth noting that the Gaussian distribution is the most important distribution in statistics; according to the central limit theorem the distribution of the sum of a large number of random events will always approach such a distribution. In this connection, it is to be expected that each of the velocity components in a Maxwell–Boltzmann distribution will have a Gaussian distribution since the average velocity of particles in a gas or plasma is the sum of many random motions and interactions. It is this statistical convergence that underpins the fact that the Maxwell–Boltzmann distribution is the most probable distribution for a gas at equilibrium.

We can also find useful expressions for the distribution of particle speeds $v = |\mathbf{v}|$. For this purpose it is useful to once again define velocity space in polar coordinates (v, θ, ϕ) instead of Cartesian

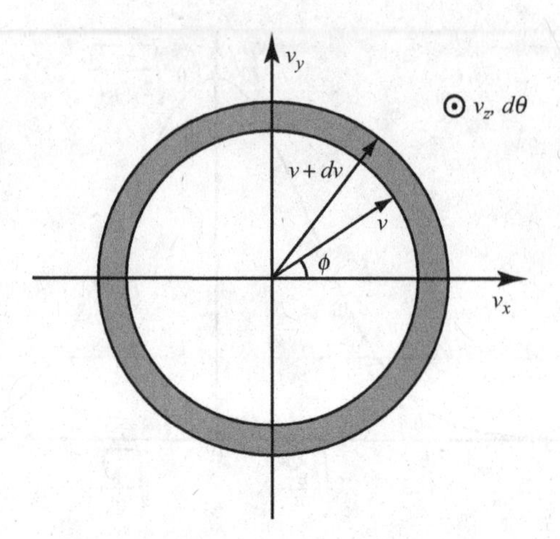

Figure 3.5 Velocity space. A two-dimensional view of velocity space for Cartesian and spherical coordinates. The shaded area represents a spherical shell containing particles with velocities with magnitudes between v and $v + dv$.

coordinates (v_x, v_y, v_z). A two-dimensional slice through velocity space for both coordinate systems is shown in Figure 3.5. Completing the transformation requires replacing the differential volume element $d\mathbf{v} = dv_x dv_y dv_z = v^2 \sin\theta d\theta d\phi dv$ and changing the limits of integration. To access only the magnitude of velocity we need to integrate over all θ and ϕ:

$$N_{\rm sp}(v) = \int_0^{2\pi}\int_0^{\pi} f(v)v^2 \sin\theta d\theta d\phi$$

$$= N_0 4\pi \left(\frac{m}{2k_{\rm B}T\pi}\right)^{\frac{3}{2}} v^2 e^{-mv^2/(2k_{\rm B}T)}, \tag{3.22}$$

which is identical to (3.2) and is plotted in Figure 3.1a. The most probable speed of the particles is given by $v_{\rm max} = \sqrt{2k_{\rm B}T/m}$. Although it was written specifically for the kinetic energy $\frac{1}{2}mv_x^2$, Equation (3.18) also shows how we can find the average value of any other physical quantity. For example, we can find the average velocity $\langle\mathbf{v}\rangle$ of particles by simply replacing $\frac{1}{2}mv_x^2$ with $\mathbf{v}$ under the integrand in (3.18):

$$\langle\mathbf{v}\rangle = \left(\frac{m}{2\pi k_{\rm B}T}\right)^{\frac{3}{2}} \int_{-\infty}^{\infty} (\hat{\mathbf{x}}v_x + \hat{\mathbf{y}}v_y + \hat{\mathbf{z}}v_z)\, e^{-mv^2/(2k_{\rm B}T)}\, dv_x dv_y dv_z = 0,$$

since the integrands are clearly odd functions. The fact that the average value of velocities in a Maxwellian distribution is zero simply indicates that there are no preferred directions for particle motion. The average speed of the particles can be found using the speed distribution (3.22) and determining an average similar to that in (3.18):

$$\langle|\mathbf{v}|\rangle = \frac{1}{N_0}\int_{-\infty}^{\infty} v N_{\rm sp}(v)dv$$

$$= 4\pi\left(\frac{m}{2k_{\rm B}T\pi}\right)^{\frac{3}{2}}\int_{-\infty}^{\infty} v v^2 e^{-mv^2/(2k_{\rm B}T)}dv = \left(\frac{8k_{\rm B}T}{\pi m}\right)^{\frac{1}{2}},$$

where we have used the fact that

$$\int_0^{\infty}\zeta^3 e^{-a\zeta^2}d\zeta = \frac{1}{2a^2}. \tag{3.23}$$

Similarly, it can be shown that the root-mean-square (rms) velocity of particles in a Maxwellian distribution is given by

$$v_{\rm rms} = \sqrt{\langle v^2\rangle} = \left(\frac{3k_{\rm B}T}{m}\right)^{\frac{1}{2}}.$$

3.5.4 Degree of ionization: the Saha equation

Under the equilibrium conditions of the Maxwell–Boltzmann distribution it is also possible to estimate the degree of ionization of a gas at a given average temperature. It was noted in Chapter 1 that the ionization of individual atoms and molecules typically requires thermal energies of greater than 20 000 K. However, when a gas is in thermal equilibrium a considerable degree of ionization still occurs, even if the mean temperature of the gas is far below the ionization energy. This is because even a low-temperature gas at equilibrium contains a small but finite number of particles with very high velocities, in the "tail" of the Maxwellian distribution. The degree of ionization is determined by the balance between ionization by collision with high-energy particles and recombination. The exact solution of the problem, which takes into account quantum mechanical aspects, is known as the Saha equation:

Saha equation
$$\boxed{\frac{N_i}{N_n} = 2.405\times 10^{21}\,\frac{T^{3/2}}{N_i}\,e^{-U/(k_{\rm B}T)}}, \tag{3.24}$$

where N_i is the density of ionized atoms, N_n is the density of neutrals, and U is the ionization energy. The action of recombination is manifested in the presence of N_i in the denominator on the right-hand side. An ion–electron pair quickly recombines when in close proximity, and because of quasi-neutrality their densities can be approximated as being equal ($N_e \simeq N_i$). Thus the higher the density of ions, the smaller the equilibrium degree of ionization, making it much easier to sustain a plasma state in gases at low densities.

Nevertheless, gases in equilibrium at temperatures of only a few thousand degrees can have a significant degree of ionization, as illustrated in Example 3-2.

Example 3-2 Ionization fraction of air
Estimate the degree of ionization of air at 1 atm pressure and room temperature (300 K). Repeat the calculation for a temperature of 8000 K. The main component of air is nitrogen, with an ionization potential of 14.5 eV.

Solution: We first find the total density of particles, N_{tot}, comprising neutral atoms N_n, ions N_i, and electrons N_e. Since the number of ions and electrons will be equal, $N_{tot} = 2N_i + N_n$. Given the pressure and temperature, the total density can be found using the ideal gas law:

$$p = N_{tot} k_B T$$

$$N_{tot} = \frac{p}{k_B T}$$

$$N_{tot}(T = 300\ \text{K}) = \frac{(1\ \text{atm})(101\,325\ \text{Pa/atm})}{(1.38 \times 10^{-23}\ \text{J/K})(300\ \text{K})} = 2.4 \times 10^{25}\ \text{m}^{-3}$$

$$N_{tot}(T = 8000\ \text{K}) = \frac{(1\ \text{atm})(101\,325\ \text{Pa/atm})}{(1.38 \times 10^{-23}\ \text{J/K})(8000\ \text{K})} = 9.2 \times 10^{23}\ \text{m}^{-3}.$$

We can now use the Saha equation, noting that $N_n = N_{tot} - 2N_i$:

$$\frac{N_i}{N_{tot} - 2N_i} = 2.4 \times 10^{21} \frac{T^{3/2}}{N_i} e^{-U/(k_B T)}$$

$$\frac{N_i^2}{N_{tot} - 2N_i} = 2.4 \times 10^{21}\ T^{3/2} e^{-U/(k_B T)} = S_t$$

$$N_i^2 + 2S_t N_i - S_t N_{tot} = 0.$$

This is a simple quadratic in N_i, with positive solution given by

$$N_i = -S_t + \sqrt{S_t^2 + S_t N_{tot}}.$$

For $T = 300$ K we have

$$S_t = 2.4 \times 10^{21}\ T^{3/2} e^{-U/(k_B T)}$$

$$= 2.4 \times 10^{21}\ (300\ \text{K})^{3/2}$$

$$\times e^{-[(14.5\ \text{eV})(1.6 \times 10^{-19}\ \text{J/eV})]/[(1.38 \times 10^{-23}\ \text{J/K})(300\ \text{K})]}$$

$$= 5.28 \times 10^{-219},$$

yielding

$$\frac{N_i}{N_i + N_n} \simeq 10^{-122},$$

which is a fraction so incredibly low as to be negligible, with no observable plasma effects. The higher temperature case, on the other hand, yields

$$S_t = 1.28 \times 10^{18}$$

$$\frac{N_i}{N_i + N_n} = 0.0012,$$

which shows that a significant portion of the atoms are ionized even though the average temperature is much less than 1 eV.

3.5.5 Shifted Maxwellian distribution

A simple generalization of the equilibrium distribution function allows us to represent a plasma which is in thermal equilibrium but which is streaming at a velocity $\mathbf{v}_0 = \hat{\mathbf{x}}v_{x0} + \hat{\mathbf{y}}v_{y0} + \hat{\mathbf{z}}v_{z0}$. For such a gas with a non-zero average velocity the distribution function is the so-called shifted or drifting Maxwellian distribution, given by

$$f(v_x, v_y, v_z) = N_0 \left(\frac{m}{2\pi k_B T}\right)^{\frac{3}{2}} e^{-(m/(2k_B T)[(v_x - v_{x0})^2 + (v_y - v_{y0})^2 + (v_z - v_{z0})^2]}.$$

It can be readily shown that the average values of the component velocities are given by the components of $\mathbf{v}_0$. In other words, we have

$$\langle v_x \rangle = v_{x0} \quad \text{or, in general,} \quad \langle \mathbf{v} \rangle = \mathbf{v}_0.$$

3.6 The Vlasov equation

Now that we are more used to conversing about six-dimensional phase space, it is useful to reconsider the derivation of the Boltzmann equation. In this connection, it is important to note that the position and velocity variables are completely independent, i.e., the instantaneous velocity of a particle is not a function of its position. Thus, the set of six numbers x, y, z, v_x, v_y, v_z merely represent a six-dimensional coordinate space. For the sake of the following discussion, let us consider an assembly of non-interacting particles (i.e., no collisions). Consider a small volume element in phase space, written compactly as $d\mathbf{r}d\mathbf{v}$. Assume that the surface area of this volume in configuration space is $d\mathbf{s}_r$, whereas that in velocity space is $d\mathbf{s}_v$. We now impose conservation of particles: the rate of change of the number of particles in the volume element is equal to the net flux of particles into the volume. In configuration space the flux is

$$f\frac{d\mathbf{r}}{dt}\cdot d\mathbf{s}_r = f\mathbf{v}\cdot d\mathbf{s}_r.$$

In velocity space the flux is

$$f\frac{d\mathbf{v}}{dt}\cdot d\mathbf{s}_v = f\mathbf{a}\cdot d\mathbf{s}_v.$$

Hence we have

$$\frac{\partial}{\partial t}fd\mathbf{r}d\mathbf{v} = -f\mathbf{v}\cdot d\mathbf{s}_r d\mathbf{v} - f\mathbf{a}\cdot d\mathbf{s}_v d\mathbf{r}.$$

Using the divergence theorem, $(\nabla\cdot\mathbf{G})dv = \mathbf{G}\cdot d\mathbf{s}$, we can write

$$\frac{\partial}{\partial t}fd\mathbf{r}d\mathbf{v} = -\nabla_{\mathbf{r}}\cdot(f\mathbf{v})d\mathbf{r}d\mathbf{v} - \nabla_v\cdot(f\mathbf{a})d\mathbf{v}d\mathbf{r}.$$

Further assuming that the acceleration $\mathbf{a}$ (and thus the force $\mathbf{F} = m\mathbf{a}$) is not a function of $\mathbf{v}$, we find the collisionless Boltzmann equation

$$\frac{\partial}{\partial t}f + (\mathbf{v}\cdot\nabla_{\mathbf{r}})f + \left(\frac{\mathbf{F}}{m}\cdot\nabla_{\mathbf{v}}\right)f = 0. \tag{3.25}$$

As written above in terms of a general force $\mathbf{F}$, the Boltzmann equation is quite general and valid for both neutral gases and plasmas. For plasmas, the operative force of interest is the Lorentz force, which when substituted gives

Vlasov equation

$$\boxed{\frac{\partial f}{\partial t} + (\mathbf{v}\cdot\nabla_{\mathbf{r}})f + \frac{q}{m}[(\mathbf{E}+\mathbf{v}\times\mathbf{B})\cdot\nabla_{\mathbf{v}}]f = 0}. \tag{3.26}$$

Equation (3.26), the collisionless Boltzmann equation with $\mathbf{F}$ replaced by the Lorentz force, is known as the Vlasov equation, and is one of the most important equations of plasma physics. Maxwell's equations describing the fields $\mathbf{E}(\mathbf{r}, t)$ and $\mathbf{B}(\mathbf{r}, t)$ and the Vlasov equation represent a complete set of self-consistent equations. In this connection, it should be noted that from the point of view of an individual particle, the fields $\mathbf{B}$ and $\mathbf{E}$ in (3.26) represent the fields due to the rest of the plasma plus any externally applied fields. The current and charge density terms in Maxwell's equations can of course be expressed in terms of the particle velocity as follows:

$$\mathbf{J}(\mathbf{r}, t) = \sum_i q_i N_i(\mathbf{r}, t)\mathbf{u}_i(\mathbf{r}, t) \tag{3.27a}$$

$$\rho(\mathbf{r}, t) = \sum_i q_i N_i(\mathbf{r}, t), \tag{3.27b}$$

where $\mathbf{u}_i = \langle \mathbf{v}_i \rangle$ is the fluid velocity and the summation is to be carried over all particle species. In some specialized cases, e.g., in very-high-temperature plasmas, particle velocities may be high enough that relativistic effects may need to be taken into account. In such cases, (3.26) is valid if we replace the velocity $\mathbf{v}$ with the momentum $\mathbf{p}$. Note that the momentum $\mathbf{p}$ of a particle is given by

$$\mathbf{p} = m\mathbf{v} = \frac{m_0 \mathbf{v}}{\sqrt{1 - |\mathbf{v}|^2/c^2}},$$

where m_0 is the rest mass and c is the speed of light. The Vlasov equation can then be written as

$$\frac{\partial f}{\partial t} + \left(\frac{\mathbf{p}}{m} \cdot \nabla_{\mathbf{r}}\right) f + \frac{q}{m}\left[\left(\mathbf{E} + \frac{\mathbf{p}}{m} \times \mathbf{B}\right) \cdot \nabla_{\mathbf{p}}\right] f = 0. \tag{3.28}$$

3.6.1 The convective derivative in physical space and in phase space

A concept that sets the stage for describing a plasma as a fluid (to be explored in later chapters) and also provides another angle for analysis of the Vlasov equation is the *convective derivative*. The equation of motion for a single particle is simply given by Newton's second law of motion,

$$m\frac{d\mathbf{v}}{dt} = \mathbf{F}, \tag{3.29}$$

where $\mathbf{F}$ is any general force. In Equation (3.29) the derivative is taken at the position of the particle, and changes are measured with respect to a stationary reference frame. If we now want to consider a collection of particles, under the assumptions that all particles are the same and that the net effects of collisions are minimal, we can simply multiply both sides of Equation (3.29) by the particle density, yielding

$$Nm\frac{d\mathbf{u}}{dt} = N\mathbf{F}, \tag{3.30}$$

where $\mathbf{u}$ is the average velocity of the particles that make up a finite fluid element. Equation (3.30) is still in the stationary reference frame and must be evaluated at the position of the moving particles (or fluid elements), which makes it inconvenient to use. It would be much more useful to have an expression in which the derivatives could be evaluated at a fixed position but could apply directly to all the particles within a larger volume of interest. This can be accomplished by shifting to a reference frame that moves along with the fluid elements. Since such a derivative expresses changes as experienced within a moving coordinate system, it is known as a convective derivative or total time derivative. The key to the utility of the convective derivative is that representing a collection of individual particles as a fluid element means that all the particles previously defined only at single points are now represented by a continuous function (extending in all spatial dimensions) at a specific set of coordinates. From a reference frame moving with the fluid, observation of a change in any property associated with the mobile fluid element can result either from the property changing in time or from the fluid element moving into a region where the property is different. A useful analogy is to imagine an automobile moving on a crowded motorway with the property of interest being the density of traffic. The traffic density observed by the driver can change either because of more cars entering or leaving the motorway at his location or because the driver arrives at a more or less crowded section of the motorway. Cars entering the motorway represent a local change in traffic, while the driver driving into traffic is change experienced by moving against a traffic gradient. If we first consider movement in only one dimension and take $\mathbf{A}$ to be any property of a fluid element (such as traffic density in the example above) we can write the total change experienced by the fluid element as

$$\frac{d\mathbf{A}(x,t)}{dt} = \frac{\partial \mathbf{A}}{\partial t} + \frac{\partial \mathbf{A}}{\partial x}\frac{dx}{dt} = \frac{\partial \mathbf{A}}{\partial t} + v_x \frac{\partial \mathbf{A}}{\partial x},$$

where the last term on the right-hand side represents changes in $\mathbf{A}$ experienced by the fluid element as a result of movement into spatial regions where $\mathbf{A}$ is different. Generalizing the expression to three dimensions, we can express the convective derivative as

$$\text{Convective derivative} \quad \boxed{\frac{d\mathbf{A}}{dt} = \frac{\partial \mathbf{A}}{\partial t} + (\mathbf{v}\cdot\nabla)\mathbf{A}}. \tag{3.31}$$

The total time derivative of any physical property of a fluid in motion can be decomposed into two such components. For example, consider the time derivative of the number density $N(x, t)$ in a fluid flowing with velocity $\mathbf{v}$:

$$\frac{dN}{dt} = \frac{\partial N}{\partial t} + \frac{\partial N}{\partial x}\frac{\partial x}{\partial t} + \frac{\partial N}{\partial y}\frac{\partial y}{\partial t} + \frac{\partial N}{\partial z}\frac{\partial z}{\partial t}$$
$$\frac{dN}{dt} = \frac{\partial N}{\partial t} + (\mathbf{v} \cdot \nabla_{\mathbf{r}})N. \quad (3.32)$$

The first term, $\partial N/\partial t$, is the explicit variation of density with time, while the second term, $(\mathbf{v} \cdot \nabla_{\mathbf{r}})N$, is the variation due to the motion of the fluid. At a point moving with the fluid, the density can be varying simply because of the existing spatial variation of density as the fluid elements move into new regions of physical space. In an analogous manner, we can interpret the collisionless Boltzmann equation (3.25) as the total derivative of the distribution function in phase space. In this connection, we should note that the phase space is simply a six-dimensional coordinate system of independent variables, so that we can write

$$\frac{df}{dt} = \frac{\partial f}{\partial t} + \frac{\partial f}{\partial x}\frac{\partial x}{\partial t} + \frac{\partial f}{\partial y}\frac{\partial y}{\partial t} + \frac{\partial f}{\partial z}\frac{\partial z}{\partial t} + \frac{\partial f}{\partial v_x}\frac{\partial v_x}{\partial t} + \frac{\partial f}{\partial v_y}\frac{\partial v_y}{\partial t} + \frac{\partial f}{\partial v_z}\frac{\partial v_z}{\partial t}$$
$$= \frac{\partial}{\partial t} f + (\mathbf{v} \cdot \nabla_{\mathbf{r}})f + \left(\frac{d\mathbf{v}}{dt} \cdot \nabla_{\mathbf{v}}\right) f.$$

Thus the collisionless Boltzmann equation (3.25) can be simply stated as

$$\frac{df}{dt} = 0, \quad (3.33)$$

i.e., the total derivative of the distribution function f is always zero for a collisionless assembly of particles. In other words, as a particle moves around in phase space, it sees a constant f in its local frame. This fundamental result is known as Liouville's theorem.

3.7 Equivalence of the particle equations of motion and the Vlasov equation

We noted above that the particle equation of motion and the Vlasov equation are fundamentally equivalent. One result of this equivalence is that any function of any physical quantity that remains constant during the particle's motion is a solution of the Vlasov equation. This result is known as Jeans's theorem, first stated by Jeans in connection with stellar dynamics.[2] We can demonstrate

[2] See [3]. In this connection, note that the kinetic description of galaxies as a "gas" consisting of stars instead of molecules has similarities with the kinetic theory of plasmas. Since collisions between stars in galaxies are rather rare, the evolution of the distribution function of stars in phase space can be described by a continuity equation similar to the

this very general result by considering the Vlasov equation (3.26) and its simple form given in (3.33). Suppose that the macroscopic electromagnetic fields **E** and **B** are known. The position and velocity of each particle in the system are then completely determined at all times as a solution of (2.1) for a given initial position $\mathbf{r}_0 = \hat{\mathbf{x}}\zeta_1 + \hat{\mathbf{y}}\zeta_2 + \hat{\mathbf{z}}\zeta_3$ and initial velocity $\mathbf{v}_0 = \hat{\mathbf{x}}\zeta_3 + \hat{\mathbf{y}}\zeta_4 + \hat{\mathbf{z}}\zeta_6$ of the particle, where ζ_i are constants. Thus, the position $\mathbf{r}(t)$ and the velocity $\mathbf{v}(t)$ of the particle can be written as functions of these six constants ζ_i:

$$\mathbf{r} = \mathbf{r}(\zeta_1, \ldots, \zeta_6, t) \tag{3.34a}$$

$$\mathbf{v} = \mathbf{v}(\zeta_1, \ldots, \zeta_6, t). \tag{3.34b}$$

Note that the actual functional forms of **r** and **v** depend on $\mathbf{E}(\mathbf{r}, t)$ and $\mathbf{B}(\mathbf{r}, t)$, which in turn must be self-consistently determined via a simultaneous solution of Maxwell's equations (with the current density and charge terms given by (3.27)) and the Vlasov equation. Equation (3.34) represents a relationship between the six constants ζ_i and six components of the combined vectors **r** and **v** at any given t. In principle, we can invert this relationship and write the six constants ζ_i as functions of **r**, **v**, and t:

$$\zeta_1 = \zeta_1(\mathbf{r}, \mathbf{v}, t) \tag{3.35a}$$

$$\zeta_2 = \zeta_2(\mathbf{r}, \mathbf{v}, t) \tag{3.35b}$$

$$\ldots \tag{...}$$

$$\zeta_6 = \zeta_6(\mathbf{r}, \mathbf{v}, t). \tag{3.35f}$$

Note that as the values of **r** and **v** change with time during the particle motion, the values of ζ_i remain constant. Now consider an arbitrary distribution function which depends only on **r**, **v**, and t, in terms of functional combinations specified in (3.35). In other words, consider a distribution function of the form

$$f(\mathbf{r}, \mathbf{v}, t) = g(\zeta_1, \ldots, \zeta_6), \tag{3.36}$$

where g is an arbitrary function. Substituting (3.36) into (3.33) and using the chain rule of differentiation, we have

$$\frac{dg}{dt} = \sum_{i=1}^{6} \frac{\partial g}{\partial \zeta_i}\frac{d\zeta_i}{dt} = 0, \tag{3.37}$$

one we derive in this chapter as the zeroth-order moment of the Boltzmann equation (see Chapter 4 for more on moments of the Boltzmann equation). Note that in the case of stars, each star interacts with the rest of the stars in the galaxy via the local gravitational potential established as a result of the collective gravitational effects of all the stars.

showing that the Vlasov equation is satisfied for arbitrary g, since the ζ_i are constants and their total derivative is by definition zero. Thus, (3.36) is a solution of Vlasov's equation regardless of the nature of the $\partial g/\partial\zeta_i$ terms. In other words, *any* function of the constants of motion of a particle is a solution of the Vlasov equation, a result which is known as Jeans's theorem. As an example, consider a plasma in the absence of an electric field, i.e., $\mathbf{E} = 0$. In this case, there can be no acceleration of any particle and its energy remains constant, so $\frac{1}{2}mv^2$ is a constant of motion. Thus, any function of $\frac{1}{2}mv^2$ is a solution of the Vlasov equation, e.g., the Maxwell–Boltzmann distribution, $f(v) = e^{-mv^2/(2k_B T)}$.

Expanding on this concept, we can find the equilibrium distribution function for a gas in the presence of an external conservative force. A conservative force is one that can be written as a gradient of a scalar potential. This is another way of saying that the action of such a force is reversible and leads directly to a change in potential energy. Gravity and the electric and magnetic forces of the Lorentz force are conservative forces. Frictional and drag forces, on the other hand, are not conservative since they lead to a loss of kinetic energy that cannot be retrieved. Under the action of a conservative force, the sum of the potential and kinetic energies of particles should therefore be a constant of motion. If the force is specified in terms of a potential energy $U(r)$ by

$$F(\mathbf{r}) = -\nabla U(\mathbf{r})$$

then the equilibrium distribution function of a gas in the presence of this force is given by

$$f(v) = N_0 \left(\frac{m}{2\pi k_B T}\right)^{\frac{3}{2}} e^{\left(-\frac{1}{2}mv^2 - U\right)/(k_B T)}, \tag{3.38}$$

which is the unperturbed Maxwell–Boltzmann distribution of Equation (3.20), multiplied by a correction term $e^{-U(\mathbf{r})/k_B T}$ known as the Boltzmann factor. We can now consider the problem of a plasma in the presence of an electrostatic field. If there is an electrostatic field $\mathbf{E}(\mathbf{r})$ present it can be specified by an electrostatic potential $\Phi(\mathbf{r})$. The Boltzmann factor in this case is simply $e^{-q\Phi(\mathbf{r})/k_B T}$, making the distribution function

$$f(v) = N_0 \left(\frac{m}{2\pi k_B T}\right)^{\frac{3}{2}} e^{\left(-\frac{1}{2}mv^2 - q\Phi\right)/(k_B T)}.$$

The number density for this situation, obtained by integrating over velocity space, is

$$N(\mathbf{r}) = N_0 e^{(-q\Phi(\mathbf{r}))/(k_B T)}, \tag{3.39}$$

which is the expression that was used in Chapter 1 (Equation (1.6)) to derive the Debye length.

3.8 Summary

In this chapter we introduced kinetic theory, the description of a plasma using a probability distribution function $f(\mathbf{r}, \mathbf{v}, t)$ describing the likelihood of finding particles at a given position, at a given velocity, and at a given time. Employing a probability distribution function is a logical way to proceed from discussing single-particle motions, as was done in Chapter 2, to the description of the collective behavior of the particles that make up a plasma. In kinetic theory a plasma is described in a six-dimensional phase-space coordinate system with three spatial dimensions (x, y, z) and three velocity dimensions (v_x, v_y, v_z). The way a particle distribution changes under the influence of external forces is described by the Boltzmann equation. Derivation of the Boltzmann equation follows from the fact that velocity changes will originate from the Lorentz force and position changes can be derived from velocities. If we neglect collisions, the Boltzmann equation reduces to the Vlasov equation. In a limiting case, the Vlasov equation can be interpreted as stating that, without external forces or collisions, the distribution function (as measured by a total derivative) in fact does not change, which is an intuitively pleasing result.

In the second half of the chapter we discussed the most probable distribution function that a plasma will assume under equilibrium. This is the Maxwell–Boltzmann distribution, which all plasmas and gases will approach, given enough time, if there is no net energy flow into or out of the system. The Maxwell–Boltzmann distribution is prevalent in many applications of plasma physics and has remarkable properties. At the same time there are many applications where a Maxwell–Boltzmann distribution cannot be assumed nor be observed. It is important to emphasize that kinetic theory, including the Boltzmann and Vlasov equations, is general and can be applied to all plasma distributions. Plasma fluid theories, discussed in the following chapters, follow directly from kinetic theory but are not as general and are applicable only to situations where equilibrium or quasi-equilibrium conditions can be assumed.

3.9 Problems

3-1. Write down the distribution function for the following cases: (a) two infinite particle beams each with density N_0 moving in opposite directions along the x axis at a speed of v; (b) an infinite particle population with all speeds less than the maximum speed v_{max} being equally probable.

3-2. Calculate the average number density $N(\mathbf{x}, t)$ and the average kinetic energy $\langle \frac{1}{2} m \mathbf{v}^2 \rangle$ for the following distribution functions: (a)$f(\mathbf{r}, \mathbf{v}, t) = K_0 \delta(v_x)\delta(v_y - v_0)\delta(v_z)$; (b)$f(\mathbf{r}, \mathbf{v}, t) = (A_0^2 - v_x^2)(A_0^2 - v_y^2)(A_0^2 - v_z^2)$ for $|v_i| < A_0$ $(i = x, y, z)$.

3-3. The electrons inside a system of two coaxial magnetic mirrors can be described by the so-called loss-cone distribution function,

$$f(\vec{v}) = \left(\frac{4}{\pi}\right)^{\frac{3}{2}} \frac{1}{\alpha_\perp^2 \alpha_\parallel} \left(\frac{v_\perp^2}{\alpha_\perp^2}\right) \exp\left[-\frac{v_\perp^2}{\alpha_\perp^2} - \frac{v_\parallel^2}{\alpha_\parallel^2}\right],$$

where $v_\perp$ and $v_\parallel$ denote the electron velocities in the directions perpendicular and parallel to the magnetic bottle axis, respectively, and where $\alpha_\perp^2 = 2k_B T_\perp / m_e$ and $\alpha_\parallel^2 = 2k_B T_\parallel / m_e$. (a) Determine the number density of electrons N_0 in the magnetic bottle. (b) Determine the average perpendicular and parallel energies.

3-4. Given a particle distribution function of the form

$$f(\mathbf{r}; \mathbf{v}) = \begin{cases} Ae^{-x/a}\left[1 + \cos\left(\dfrac{\pi v^3}{v_0^3}\right)\right] & v \le v_0 \\ 0 & v \ge v_0 \end{cases},$$

where v_0 is a constant and $v = |\mathbf{v}|$: (a) Determine the constant A, given that the particle number density at $x = 0$ is N_0. (b) If an application requires that the spatial dependence of f remain constant, how could the spatial distribution be maintained? Give a quantitative description of the proposed scheme.

3-5. Consider an equilibrium (Maxwellian) plasma with electron temperature T_e and electron density N_e immersed in a constant uniform magnetic field $\mathbf{B}$. (a) Find an expression for $\langle \mu \rangle$, the average value of the magnetic moment of the electron due to its gyration around $\mathbf{B}$, where $\mu = m_e v_\perp^2/(2B)$.

(b) For $B = 5 \times 10^{-5}$ T and $T_e = 300$ K, compare the magnitude of $\langle \mu \rangle$ to the spin magnetic moment μ_B of the electron, which is called the *Bohr magneton* and is given by

$$\mu_B = \frac{|q_e|h}{4\pi m_e},$$

where $h = 6.626 \times 10^{-34}$ J s is Planck's constant.

3-6. In some plasmas confined in tokamak machines, distinct populations of energetic ions are created which do not have a Maxwellian distribution function. An example is the so-called "slowing-down" distribution, given by

$$f_{ion} = \begin{cases} \dfrac{A}{v^3 + a^3} & v \leq v_0 \\ 0 & v > v_0 \end{cases},$$

where v_0 is the velocity at which the energetic ions are created (by fusion reactions) and a is a constant determined by the rate of collisions. (a) Determine the constant A if the average ion density is N_i. (b) Determine v_{max}, the most probable value of v for this distribution.

3-7. Use direct substitution to show that a distribution function of the form

$$f = f\left(mv^2/2 + q\phi\right)$$

is a steady-state solution of the Boltzmann equation, where ϕ is an electric potential. Assume only one dimension and neglect collisions.

3-8. Use the Saha equation to show that the solar wind, which is an electron and proton plasma, must be almost completely ionized. The electron density of the solar wind is 3.5×10^6 m^{-3} and the temperature is 10^5 K. The ionization potential of hydrogen is 13.6 eV.

3-9. A candle flame has a maximum temperature of ~1600 K. Find the degree of ionization of the flame and discuss whether any plasma properties can be observed. Also find the Debye length.

3-10. A cylindrical hot-water heater, initally OFF, has a vertical temperature gradient of 50°C m^{-1} increasing from bottom to top. The heat source surrounding the cylinder is turned ON, increasing the temperature at a rate of 1°C s^{-1}. What will be the temperature change experienced by a rock, dropped into the water at the top of the cylinder, that falls to the bottom at a speed of 0.5 m s^{-1}? The rock is dropped in at the same time the heater is turned ON.

References

[1] E. H. Holt and R. E. Haskel, *Foundations of Plasma Dynamics.* (New York: Macmillan, 1965), Section 5.4.
[2] J. A. Bittencourt, *Fundamentals of Plasma Physics*, 3rd edn (New York: Springer-Verlag, 2004), 589–607.
[3] S. Chandrasekhar, *Principles of Stellar Dynamics* (Chicago: University of Chicago Press, 1942).

CHAPTER

4 Moments of the Boltzmann equation

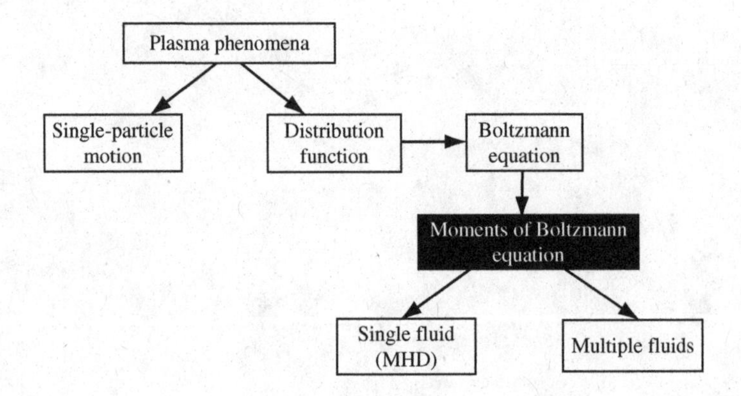

4.1 Introduction

In deriving the Boltzmann equation in Chapter 3, we did not use any physical principle other than the equation of motion relating the particle acceleration to the Lorentz force. The rest of the derivation was simply a matter of keeping a tally of the number of particles into and out of a volume element in phase space. We also saw how useful macroscopic information could be obtained by integrating over the distribution function in velocity space. For example, the number density is

$$N(\mathbf{r}, t) = \int f(\mathbf{r}, \mathbf{v}, t) d\mathbf{v}. \tag{4.1}$$

The mean plasma velocity or "fluid" velocity is

$$\mathbf{u}(\mathbf{r}, t) \equiv \langle \mathbf{v} \rangle = \frac{1}{N(\mathbf{r}, t)} \int \mathbf{v} f(\mathbf{r}, \mathbf{v}, t) d\mathbf{v}. \tag{4.2}$$

In general, the average value of any quantity $\mathbf{Q}(\mathbf{r}, \mathbf{v}, t)$ is given by

$$\mathbf{Q}(\mathbf{r}, t) \equiv \langle \mathbf{Q} \rangle = \frac{1}{N(\mathbf{r}, t)} \int \mathbf{Q} \, f(\mathbf{r}, \mathbf{v}, t) d\mathbf{v}. \tag{4.3}$$

All integrals are to be taken over the entire velocity space. It is clear that knowing the distribution function gives one access to nearly all relevant information in a plasma, even though the exact trajectories of the constituent particles are not known. Under long-term equilibrium conditions, the distribution function will become the Maxwell–Boltzmann distribution. In principle the distribution for non-equilibrium situations can be obtained by solving the Boltzmann equation directly. Unfortunately, solving the Boltzmann equation is usually not straightforward. Fortunately, however, we are often not interested in the details of the particle distribution function but simply need to know the macroscopic quantities (e.g., number density of particles, mean velocity, etc.) in physical (or configuration) space. In other words, we seek the distribution only in order to integrate over it and obtain the desired macroscopic values. In this chapter we will show that under certain assumptions it is not necessary to obtain the actual distribution function if one is only interested in the macroscopic values. Instead of first solving the Boltzmann (or Vlasov) equation for the distribution function and then integrating, it is possible to first take appropriate integrals over the Boltzmann equation and then solve for the quantities of interest. This approach is referred to as "taking the moments of the Boltzmann equation." The resulting equations are known as the macroscopic transport equations, and form the foundation of plasma fluid theory.

The basic procedure for deriving macroscopic equations from the Boltzmann equation involves multiplying it by powers of the velocity vector $\mathbf{v}$ and integrating over velocity space. It is important to realize that in performing such an integration we intrinsically lose information on the details of the velocity distribution. This is a further simplification of the type we have been making since considering single-particle motions. In Chapter 3 we gave up knowledge of individual particle trajectories and contented ourselves with a description of likely positions and velocities. Here we will cede information on the velocity distribution in order to obtain single-value macroscopic quantities.

To carry out the necessary integrations, we need to know something about the behavior of the distribution function at large values of $|\mathbf{v}|$. Since no particle can have infinite velocity, it is physically reasonable to assume that f falls off rapidly as $|\mathbf{v}| \to \infty$. This assumption ensures that the various surface integrals containing

f in the integrand vanish when extended over a sphere of radius $|\mathbf{v}| \to \infty$ in velocity space.

4.2 The zeroth-order moment: continuity equation

The first of these transport equations is the continuity equation, which in fact is a statement of conservation of charge and mass. To evaluate the zeroth-order moment, we multiply (3.9) by $v^0 = 1$ and integrate to find

$$\int \frac{\partial f}{\partial t} d\mathbf{v} + \int (\mathbf{v} \cdot \nabla_{\mathbf{r}}) f \, d\mathbf{v} + \frac{q}{m} \int [(\mathbf{E} + \mathbf{v} \times \mathbf{B}) \cdot \nabla_{\mathbf{v}}] f \, d\mathbf{v} = \int \left(\frac{\partial f}{\partial t}\right)_{\text{coll}} d\mathbf{v}, \tag{4.4}$$

where we note that $d\mathbf{v} = dv_x dv_y dv_z$. We separately examine below each term of (4.4). Recalling (3.5), we can see that the first term is

$$\frac{\partial}{\partial t} \int f(\mathbf{r}, \mathbf{v}, t) d\mathbf{v} = \frac{\partial}{\partial t} N(\mathbf{r}, t), \tag{4.5}$$

whereas the second term is

$$\begin{aligned}
\int (\mathbf{v} \cdot \nabla_{\mathbf{r}}) f \, d\mathbf{v} &= \int v_x \frac{\partial}{\partial x} f \, d\mathbf{v} + \int v_y \frac{\partial}{\partial y} f \, d\mathbf{v} + \int v_z \frac{\partial}{\partial z} f \, d\mathbf{v} \\
&= \frac{\partial}{\partial x} \int v_x f \, d\mathbf{v} + \frac{\partial}{\partial y} \int v_y f \, d\mathbf{v} + \frac{\partial}{\partial z} \int v_z f \, d\mathbf{v} \\
&= \nabla_{\mathbf{r}} \cdot [N(\mathbf{r}, t) \langle \mathbf{v} \rangle] \\
&= \nabla_{\mathbf{r}} \cdot [N(\mathbf{r}, t) \mathbf{u}(\mathbf{r}, t)],
\end{aligned} \tag{4.6}$$

where $\mathbf{u}(\mathbf{r}, t)$ is the average plasma velocity or "fluid" velocity, and where we have used the fact that v_x and x are independent variables. Note that in deriving the above we have used the fact that $\mathbf{v}$ and $\mathbf{r}$ are independent variables, so that the operator $\nabla_{\mathbf{r}}$ can be taken out of the integral on the variables $d\mathbf{v}$. For the third term in (4.4), let us consider the $\mathbf{E}$ and $\mathbf{B}$ terms separately. We have

$$\int (\mathbf{E} \cdot \nabla_{\mathbf{v}}) f d\mathbf{v} = \int \nabla_{\mathbf{v}} \cdot (f\mathbf{E}) d\mathbf{v} = \oint_{S_v} f \mathbf{E} \cdot d\mathbf{s}_v = 0, \tag{4.7}$$

where we have used the divergence theorem in velocity space (i.e., S_v is the area enclosing the velocity space volume over which we integrate). The surface integral in (4.7) vanishes because when we take the surface to infinity its area increases as v^2, while any physical distribution function approaches zero much more quickly (e.g., a Maxwellian distribution goes as e^{-v^2}). The **B** term is

$$\int [(\mathbf{v}\times\mathbf{B})\cdot\nabla_{\mathbf{v}}] f \, d\mathbf{v} = \int \nabla_{\mathbf{v}}\cdot[f(\mathbf{v}\times\mathbf{B})]\, d\mathbf{v} - \int f \nabla_{\mathbf{v}}\cdot(\mathbf{v}\times\mathbf{B})\, d\mathbf{v}$$
$$= \oint_{S_v} f(\mathbf{v}\times\mathbf{B})\cdot d\mathbf{s}_v - \int f\nabla_{\mathbf{v}}\cdot(\mathbf{v}\times\mathbf{B})\, d\mathbf{v} = 0, \tag{4.8}$$

where we have used the fact that

$$\nabla(a\mathbf{A}) \equiv \mathbf{A}\cdot\nabla a + a\nabla\cdot\mathbf{A}$$

for any scalar a and vector **A**. The first term in (4.8) vanishes for the same reason as (4.7), while the second term vanishes because $(\mathbf{v}\times\mathbf{B})$ is always perpendicular to $\nabla_{\mathbf{v}}$, i.e., any given component of the magnetic force is independent of the velocity component in the same direction. The collision term in (4.4) also vanishes, i.e.,

$$\int \left(\frac{\partial f}{\partial t}\right)_{\text{coll}} d\mathbf{v} = \left[\frac{\partial}{\partial t}\int f\, d\mathbf{v}\right] = 0 \tag{4.9}$$

if we assume that the total number of particles of the species considered must remain constant as collisions proceed. In other words, the coordinates of the particles in configuration space are unaltered by collisions, or equivalently, the number density of particles in an element of volume in configuration space cannot be changed by collisions.[1] Collisions displace particles in velocity space but do not alter their density in configuration space. Combining our results in (4.5)–(4.9), we arrive at the continuity equation,

Continuity equation for mass or charge transport

$$\boxed{\frac{\partial}{\partial t}N(\mathbf{r},t) + \nabla_{\mathbf{r}}\cdot[N(\mathbf{r},t)\mathbf{u}(\mathbf{r},t)] = 0}, \tag{4.10}$$

which is simply a statement of the conservation of particles. Equation (4.10) can be simply deduced from basic fluid-dynamical principles by considering the average flow of particles in and out

[1] We are for now implicitly neglecting recombination or ionization events, which can result in the removal of particles from the distribution or the creation of new particles. For example, electrons can recombine with ions, leading to the removal of one electron and one ion from their respective distributions, while creating a new neutral molecule, thus changing the distribution function of the neutrals. We will consider these processes in Section 4.2.1.

of an enclosed volume. In this context, the first term represents the rate of change of particle concentration within the volume, while the second term represents the divergence of particles or the flow of particles out of the volume. The continuity equation simply states that these two processes must balance under the stated assumption that no new particles are created or destroyed. By multiplying (4.10) by m or q, we can obtain the equations describing conservation of mass or charge, respectively. The similarity of the continuity equations for mass and charge so obtained implies that the quantity $\Gamma \equiv mN\mathbf{u}$ is analogous to electric current density $\mathbf{J} = qN\mathbf{u}$. Indeed, $\Gamma = mN\mathbf{u}$ is known as the particle current, more commonly referred to as "flux."

4.2.1 Closer consideration of collisions and conservation of particles

In the previous section it was assumed that the total number of of particles of any type does not change. Under this assumption collisions do not affect the continuity equation. It is important to note, however, that the number of charged particles is not always conserved in an ionized gas. In a partially ionized gas consisting of electrons, ions, and neutrals, various mechanisms can lead to loss or gain of particles. These processes are naturally described within the paradigm of collisions since such particle–particle interactions require a collision of some sort.

Collision phenomena can be broadly divided into two categories, elastic and inelastic. In elastic collisions, mass, momentum, and energy are conserved and there is neither creation nor annihilation of particles. An elastic collision is one between hard spheres, analogous to that between two steel balls; they do not change shape or gain mass, but simply exchange momentum and energy. This is the type of collision that was considered in Equation (4.9).

In an inelastic collision between two particles, either or both of the particles may have their internal states changed and the total number of particles may increase or decrease. The collision between a bullet and a wall into which it is embedded on impact is an example of an inelastic collision. In inelastic collisions, an electron may *recombine* with an ion to form a neutral atom, or it may *attach* itself to a neutral particle to form a much heavier (than an electron) negative ion. Collisions may also raise the energy state of electrons in an atom so that new electrons are stripped off, in a process referred to as *ionization*. If only one species of positive ions is present in a plasma, then the rate of recombination is proportional to numbers of both electrons and ions, which are equal because of macroscopic neutrality of the plasma. Thus, the loss rate due to recombination processes is given by $-\alpha N^2$, where α is the recombination rate, usually determined experimentally. The rate of electron attachment is proportional to the number densities of electrons and neutral particles. However, in a weakly ionized gas the neutral density can be considered to be constant, so that the loss rate due to attachment processes is given by $-\nu_a N$, where ν_a is the attachment rate. The rate at which electrons are added to the system as a result of ionization is given by $+\nu_i N$, where ν_i is the ionization rate. Thus, a more general version of the continuity equation (4.10) is

$$\frac{\partial N}{\partial t} + \nabla \cdot [N\mathbf{u}] = -\alpha N^2 - \nu_a N + \nu_i N. \tag{4.11}$$

Example 4-1 Electron density in the ionosphere: day versus night
Above an altitude of about 80 km the upper atmosphere contains significant numbers of free electrons and ions because of ionization by the Sun and cosmic rays. This lowest region of ionization is known as the ionospheric D-region, and its properties change remarkably between day and night. At night the ionization rate, which is due only to cosmic rays, is $\nu_{i_{cosmic}} = 0.007\,\text{s}^{-1}$. During the day the Sun's rays are responsible for photoionization at a rate of $\nu_{i_{photo}} = 0.44\,\text{s}^{-1}$. The electron–ion recombination rate is $\alpha = 10^{-10}\,\text{m}^3\,\text{s}^{-1}$ and the attachment rate is $\nu_a = 0.0018\,\text{s}^{-1}$ for both day and night. Ignoring any flow, find the ambient (steady-state) electron densities for night and day.

Solution: We can use the steady-state version of Equation (4.11) with zero fluid velocity, in which all $\frac{\partial}{\partial t} = 0$ and $\mathbf{u} = 0$:

$$0 = -\alpha N_e^2 - \nu_a N_e + \nu_i N_e = N_e(-\alpha N_e - \nu_a + \nu_i)$$

$$N_e = \frac{\nu_i - \nu_a}{\alpha}.$$

The ionization rate during the day is $\nu_{i_{day}} = \nu_{i_{photo}} + \nu_{i_{cosmic}} = 0.447\,\text{s}^{-1}$ and during the night it is $\nu_{i_{night}} = \nu_{i_{cosmic}} = 0.007\,\text{s}^{-1}$, making the ambient electron densities

$$N_e(day) = \frac{(0.447 - 0.002)\,\text{s}^{-1}}{10^{-10}\,\text{m}^3\,\text{s}^{-1}} = 4.45 \times 10^9\,\text{m}^{-3}$$

$$N_e(night) = \frac{(0.007 - 0.002)\,\text{s}^{-1}}{10^{-10}\,\text{m}^3\,\text{s}^{-1}} = 5.00 \times 10^7\,\text{m}^{-3}.$$

4.3 The first-order moment: momentum transport equation

The first-order moment of the Boltzmann equation is obtained by multiplying (3.9) by $m\mathbf{v}$ and integrating to find

$$m\int \mathbf{v}\frac{\partial f}{\partial t}\,d\mathbf{v} + m\int \mathbf{v}\,(\mathbf{v}\cdot\nabla_{\mathbf{r}})\,f\,d\mathbf{v} + q\int \mathbf{v}\,[(\mathbf{E}+\mathbf{v}\times\mathbf{B})\cdot\nabla_{\mathbf{v}}]\,f\,d\mathbf{v}$$
$$= \int m\mathbf{v}\left(\frac{\partial f}{\partial t}\right)_{\text{coll}} d\mathbf{v}. \tag{4.12}$$

The first term gives

$$m\frac{\partial}{\partial t}\int \mathbf{v}\,f(\mathbf{r},\mathbf{v},t)\,d\mathbf{v} = m\frac{\partial}{\partial t}[N(\mathbf{r},t)\,\mathbf{u}(\mathbf{r},t)], \tag{4.13}$$

where we have used (4.2). Next consider the third term, substituting $\mathbf{G} = (\mathbf{E}+\mathbf{v}\times\mathbf{B})$:

$$q\int \mathbf{v}\,[\mathbf{G}\cdot\nabla_{\mathbf{v}}]\,f\,d\mathbf{v} = q\int\left[\mathbf{v}G_x\frac{\partial}{\partial v_x}f + \mathbf{v}G_y\frac{\partial}{\partial v_y}f + \mathbf{v}G_z\frac{\partial}{\partial v_z}f\right]d\mathbf{v}. \tag{4.14}$$

We can examine each term of (4.14) separately; integrating by parts, we find

$$\begin{aligned}
q\int \mathbf{v}G_x\frac{\partial f}{\partial v_x}d\mathbf{v} &= q\int G_x\,dv_y\,dv_z\int \mathbf{v}\frac{\partial f}{\partial v_x}\,dv_x \\
&= q\int G_x\,dv_y\,dv_z\left[\mathbf{v}f\Big|_{-\infty}^{\infty} - \int f\frac{\partial}{\partial v_x}\mathbf{v}\,dv_x\right] \\
&= -q\int G_x\,f\frac{\partial}{\partial v_x}\mathbf{v}\,dv_x\,dv_y\,dv_z \\
&= -qN(\mathbf{r},t)\left\langle\frac{\partial}{\partial v_x}[G_x\mathbf{v}]\right\rangle,
\end{aligned} \tag{4.15}$$

where we have once again observed that the distribution f would vanish much faster than $|\mathbf{v}|$ as $v_x \to \infty$. Noting that the other terms of (4.14) will reduce to similar expressions, we can write

$$q\int \mathbf{v}\,[\mathbf{G}\cdot\nabla_{\mathbf{v}}]\,f\,d\mathbf{v} = -qN(\mathbf{r},t)\,\langle\nabla_{\mathbf{v}}\cdot(\mathbf{G}\mathbf{v})\rangle, \tag{4.16}$$

where the term $(\mathbf{Gv})$ is a tensor product or dyad.[2] Using the property

$$\nabla \cdot (\mathbf{Gv}) = \mathbf{v}\,(\nabla \cdot \mathbf{G}) + (\mathbf{G} \cdot \nabla)\,\mathbf{v},$$

we have

$$\nabla_{\mathbf{v}} \cdot (\mathbf{Gv}) = \mathbf{v}\underbrace{\left[\frac{\partial G_x}{\partial v_x} + \frac{\partial G_y}{\partial v_y} + \frac{\partial G_z}{\partial v_z}\right]}_{=0 \text{ since } G_i \text{ is independent of } v_i} + \left[G_x\frac{\partial}{\partial v_x} + G_y\frac{\partial}{\partial v_y} + G_z\frac{\partial}{\partial v_z}\right]\mathbf{v}$$

$$= \hat{\mathbf{x}}\left[G_x\frac{\partial v_x}{\partial v_x} + \underbrace{G_y\frac{\partial v_x}{\partial v_y} + G_z\frac{\partial v_x}{\partial v_z}}_{=0 \text{ since } v_i \text{ is independent of } v_j}\right]$$

$$+\,\hat{\mathbf{y}}\left[\underbrace{G_x\frac{\partial v_y}{\partial v_x}}_{=0} + G_y\frac{\partial v_y}{\partial v_y} + \underbrace{G_z\frac{\partial v_y}{\partial v_z}}_{=0}\right]$$

[2] The tensor product or dyad $\mathbf{AB}$ of two vectors is defined as

$$\mathbf{AB} \equiv \begin{bmatrix} A_x B_x & A_x B_y & A_x B_z \\ A_y B_x & A_y B_y & A_y B_z \\ A_z B_x & A_z B_y & A_z B_z \end{bmatrix}.$$

The *tensor dot-product* is itself a vector, defined as

$$(\mathbf{AB}) \cdot \mathbf{C} \equiv \begin{bmatrix} A_x B_x & A_x B_y & A_x B_z \\ A_y B_x & A_y B_y & A_y B_z \\ A_z B_x & A_z B_y & A_z B_z \end{bmatrix}\begin{bmatrix} C_x \\ C_y \\ C_z \end{bmatrix};$$

$$\mathbf{C} \cdot (\mathbf{AB}) \equiv \begin{bmatrix} C_x & C_y & C_z \end{bmatrix}\begin{bmatrix} A_x B_x & A_x B_y & A_x B_z \\ A_y B_x & A_y B_y & A_y B_z \\ A_z B_x & A_z B_y & A_z B_z \end{bmatrix}.$$

With these definitions, it can be shown that

$$(\mathbf{AB}) \cdot \mathbf{C} = \mathbf{A}(\mathbf{B} \cdot \mathbf{C}) = (\mathbf{C} \cdot \mathbf{B})\mathbf{A}$$

$$\mathbf{C} \cdot (\mathbf{AB}) = (\mathbf{C} \cdot \mathbf{A})\mathbf{B}$$

$$\nabla \cdot (\mathbf{AB}) = \mathbf{B}(\nabla \cdot \mathbf{A}) + (\mathbf{A} \cdot \nabla)\mathbf{B}.$$

$$+\hat{\mathbf{z}}\left[\underbrace{G_x\frac{\partial v_z}{\partial v_x}+G_y\frac{\partial v_z}{\partial v_y}}_{=0}+G_z\frac{\partial v_z}{\partial v_z}\right]$$

$$=\hat{\mathbf{x}}\,G_x\frac{\partial v_x}{\partial v_x}+\hat{\mathbf{y}}\,G_y\frac{\partial v_y}{\partial v_y}+\hat{\mathbf{z}}\,G_z\frac{\partial v_z}{\partial v_z}=\mathbf{G}, \qquad (4.17)$$

where the derivatives of the components of $\mathbf{G}$ are zero because G_i is independent of V_i, and all cross-derivatives are zero because all v_i and v_i are independent of v_j for $i \neq j$. We can therefore rewrite (4.16) as

$$-qN(\mathbf{r},t)\left\langle\nabla_{\mathbf{v}}\cdot(\mathbf{G}\,\mathbf{v})\right\rangle=-qN(\mathbf{r},t)\,\langle\mathbf{G}\rangle=-qN(\mathbf{r},t)\,\langle(\mathbf{E}+\mathbf{v}\times\mathbf{B})\rangle$$
$$=-qN(\mathbf{r},t)\,(\mathbf{E}+\mathbf{u}\times\mathbf{B}). \qquad (4.18)$$

This is the final form of the third term of (4.12). We are now ready to consider the second term of (4.12). Noting that $\mathbf{r}$ and $\mathbf{v}$ are independent variables, we have

$$m\int\mathbf{v}\,(\mathbf{v}\cdot\nabla_{\mathbf{r}})\,f\,d\mathbf{v}$$
$$=m\int\nabla_{\mathbf{r}}\cdot(f\,\mathbf{v}\,\mathbf{v})\,d\mathbf{v}=m\,\nabla_{\mathbf{r}}\cdot\int f\,\mathbf{v}\,\mathbf{v}\,d\mathbf{v}=m\,\nabla\cdot[N\langle\mathbf{v}\,\mathbf{v}\rangle],$$

where $\mathbf{v}\,\mathbf{v}$ is once again a tensor product. At this point, it is useful to separate the velocity $\mathbf{v}$ into an average (i.e., fluid) velocity $\mathbf{u}$ and a random (thermal) velocity $\mathbf{w}$, i.e., $\mathbf{v}=\mathbf{u}+\mathbf{w}$. We then have

$$m\,\nabla_{\mathbf{r}}\cdot[N\langle\mathbf{v}\,\mathbf{v}\rangle]=m\,\nabla_{\mathbf{r}}\cdot(N\mathbf{u}\mathbf{u})+m\,\nabla_{\mathbf{r}}\cdot[N\langle\mathbf{w}\,\mathbf{w}\rangle]$$
$$+m\,\nabla_{\mathbf{r}}\cdot N[\mathbf{u}\langle\mathbf{w}\rangle+\langle\mathbf{w}\rangle\mathbf{u}]. \qquad (4.19)$$

The final term on the right-hand side of (4.19) is zero, since $\langle\mathbf{w}\rangle\equiv 0$. The first term in (4.19) can be written as

$$m\,\nabla_{\mathbf{r}}\cdot(N\mathbf{u}\mathbf{u})=m\,\mathbf{u}\nabla_{\mathbf{r}}\cdot[N\mathbf{u}]+m\,N[\mathbf{u}\cdot\nabla_{\mathbf{r}}]\mathbf{u}. \qquad (4.20)$$

The second term on the right-hand side of (4.19) contains the quantity $mN\langle\mathbf{w}\,\mathbf{w}\rangle$, which has dimensions of energy density (J m^{-3}), or force per unit area, or pressure. This quantity is defined as the pressure tensor or dyad and is denoted by Ψ:

$$\Psi=m\,N\,\langle\mathbf{w}\,\mathbf{w}\rangle=\begin{bmatrix}p_{xx}&p_{xy}&p_{xz}\\p_{yx}&p_{yy}&p_{yz}\\p_{zx}&p_{zy}&p_{zz}\end{bmatrix}s.$$

Note that Ψ is in fact a measure of the thermal motion in a fluid. If all particles moved with the same steady velocity $\mathbf{v}$, we would have $\mathbf{w} = 0$ and thus $\Psi = 0$ (i.e., a "cold" plasma). The components of Ψ represent transport of momentum. For example, the top row of Ψ, namely p_{xx}, p_{xy}, and p_{xz}, represents the three velocity components of the x component of momentum. The three rows of Ψ together constitute a set of three flux vectors associated with momentum transfer. Noting that $p_{xy} = mN\langle w_x w_y\rangle$ and $p_{yx} = mN\langle w_y w_x\rangle = p_{xy}$, it is clear that Ψ is a symmetric tensor, with only six independent components. The diagonal components of this tensor have the meaning of normal pressure, e.g., p_{xx} is the force per unit area in the x direction exerted on a plane surface in the gas normal to the x axis. The off-diagonal elements represent the shearing stresses, e.g., p_{yx} is the force per unit area in the x direction exerted on a plane surface perpendicular to the y axis. In an ordinary fluid, the off-diagonal elements are associated with viscosity arising from the interaction of the fluid with the walls of pipes or other confining boundaries. In plasmas similar effects can occur because of the gyrating motion of the particles, transverse to the bulk fluid motion. To appreciate the significance of the components of the pressure tensor, we can consider the pressure across three orthogonal planes at some point A in the gas, as shown in Figure 4.1a. In this context, it is useful to remember that pressure is simply force per unit area, which is also the rate of transfer of momentum. In other words, pressure in a fundamental sense is momentum flux and can as such be defined at any point without actually having a physical surface.

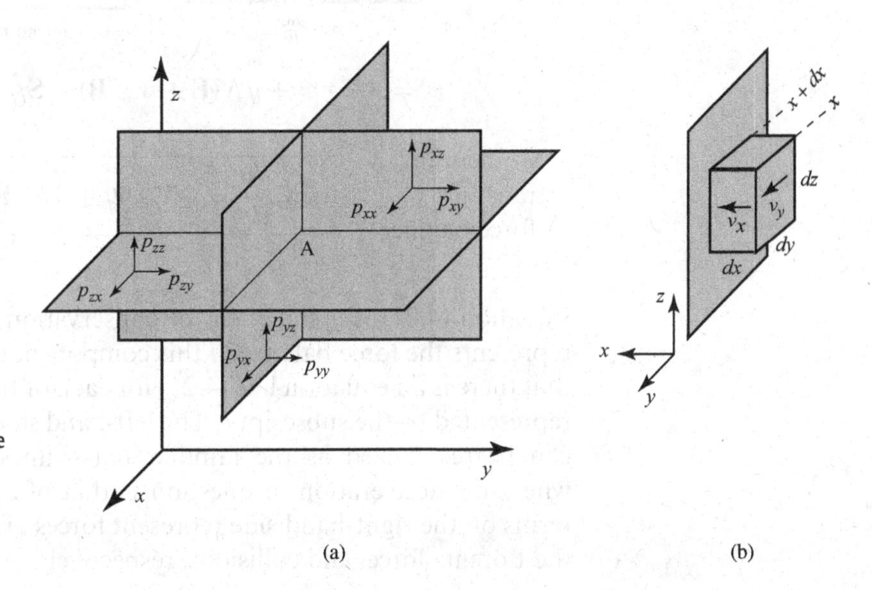

Figure 4.1 Pressure tensor. (a) Components of the pressure tensor on three orthogonal planes. (b) Differential element of volume to illustrate momentum flux.

Collecting all the terms of (4.12) together (i.e., (4.13), (4.18), and (4.20)) we have

$$m\frac{\partial}{\partial t}[N\mathbf{u}] + m\,\mathbf{u}\nabla_{\mathbf{r}} \cdot [N\mathbf{u}]$$
$$+ m\,N[\mathbf{u} \cdot \nabla_{\mathbf{r}}]\mathbf{u} + \nabla \cdot \Psi - qN(\mathbf{E} + \mathbf{u} \times \mathbf{B}) = \underbrace{\int m\mathbf{v}\left(\frac{\partial f}{\partial t}\right)_{\text{coll}} d\mathbf{v}}_{\mathbf{S}_{ij}}.$$

The collision term will be denoted by $\mathbf{S}_{ij}$ and represents the rate of change of momentum density due to collisions between different plasma species i and j. Note that collisions between like particles cannot produce a net momentum change of that species since the total average momentum of the species will remain unchanged. Upon further manipulation, and using (4.10), we can derive the final version of the momentum transport equation:

$$m\frac{\partial}{\partial t}[N\mathbf{u}] + m\,\mathbf{u}\nabla_{\mathbf{r}} \cdot [N\mathbf{u}] + m\,N[\mathbf{u} \cdot \nabla_{\mathbf{r}}]\mathbf{u}$$
$$= -\nabla \cdot \Psi + qN(\mathbf{E} + \mathbf{u} \times \mathbf{B}) + \mathbf{S}_{ij}$$
$$mN\frac{\partial \mathbf{u}}{\partial t} + m\mathbf{u}\frac{\partial N}{\partial t} + m\mathbf{u}\nabla_{\mathbf{r}} \cdot (N\mathbf{u}) + m\,N[\mathbf{u} \cdot \nabla_{\mathbf{r}}]\mathbf{u}$$
$$= -\nabla \cdot \Psi + qN(\mathbf{E} + \mathbf{u} \times \mathbf{B}) + \mathbf{S}_{ij}$$
$$\underbrace{mN\frac{\partial \mathbf{u}}{\partial t} + m\,N[\mathbf{u} \cdot \nabla_{\mathbf{r}}]\mathbf{u}}_{mN\frac{d\mathbf{u}}{dt}} + \mathbf{u}\underbrace{\left[\frac{\partial N}{\partial t} + m\nabla_{\mathbf{r}} \cdot (N\mathbf{u})\right]}_{=0\ \text{(continuity equation)}}$$
$$= -\nabla \cdot \Psi + qN(\mathbf{E} + \mathbf{u} \times \mathbf{B}) + \mathbf{S}_{ij}$$

Momentum transport (force balance)

$$\boxed{mN_i\frac{d\mathbf{u}}{dt} = -\nabla \cdot \Psi_i + qN_i(\mathbf{E} + \mathbf{u} \times \mathbf{B}) + \mathbf{S}_{ij}}. \tag{4.21}$$

Equation (4.21) is a statement of conservation of momentum and represents the force balance in this component of the plasma. Note that there is an equation like (4.21) for each of the plasma species as represented by the subscript i. The left-hand side of Equation (4.21) can be recognized as the familiar mass-times-acceleration term, where the acceleration in question is that of a fluid element. The terms on the right-hand side represent forces arising from pressure, the Lorentz force, and collisions, respectively.

4.3.1 The pressure and collision terms

Fundamentally, the momentum density (i.e., $m\mathbf{u}$) of a fluid element can change not only because of external forces acting on the particles that constitute the fluid but also because of motion in and out of the element of particles which carry momentum with them. Consider (with the help of Figure 4.1b) the flux in the x direction of y-directed momentum. This flux is given by the number of particles per unit area per second passing through the surface of constant x, times the momentum in the y direction, mv_y, carried by each particle. The differential number of particles in the phase space element $d\mathbf{v}d\mathbf{r}$ is simply $f(\mathbf{r}, \mathbf{v})d\mathbf{v}d\mathbf{r}$. This element of phase space "empties out" in the x direction in a time interval $dt = dx/v_x$. The differential number of particles carried per second across the surface of constant x by this element of phase space is $fd\mathbf{v}d\mathbf{r}/dt = v_x f d\mathbf{v}dydz$. These specific particles carry y-directed momentum mv_y, so that the differential amount of momentum flux in this direction carried per second across a surface of constant x by this element of phase space is $mv_y v_x f d\mathbf{v}dydz$. To obtain the total momentum flux, i.e., the total quantity of y-directed momentum crossing a surface of constant x per unit area per second, we divide out the differential area $dydz$ and integrate over velocity space. The total flux of y-directed momentum in the x direction becomes $\int mv_y v_x f d\mathbf{v}$, which is simply the ensemble average, $mN\langle v_y v_x\rangle$. The rate of change of y-directed momentum, averaged over all the particles, is then given as the divergence of fluxes of momentum across the various surfaces of the volume element. In other words,

$$\frac{\partial(mNu_y)}{\partial t} = -\frac{\partial}{\partial x}\left(mN\langle v_y v_x\rangle\right) - \frac{\partial}{\partial y}\left(mN\langle v_y v_y\rangle\right) - \frac{\partial}{\partial z}\left(mN\langle v_y v_z\rangle\right),$$

which is simply the y component of the $-\nabla\cdot\Psi$ term in (4.21). In general, Equation (4.21) is only useful in those cases where the distribution of random velocities is sufficiently well behaved that the pressure tensor Ψ can be represented in a relatively simple way. In the simplest case, the distribution of random velocities $\mathbf{w}$ is isotropic, so that the diagonal entries of Ψ vanish, and the three diagonal entries are all equal to one another. In such a case we have

$$\nabla\cdot\Psi = \nabla p,$$

where p is the scalar pressure.[3] In another simplification of the pressure tensor, the off-diagonal entries vanish but the diagonal entries are all different. This situation may arise in the absence of collisions, when compression of the gas increases the root-mean-square random velocity in that direction without affecting the mean velocities in the other directions. Similar effects can also occur in magnetized plasmas. Neglecting collisions and assuming isotropic conditions, the momentum transport equation can be written in its simplest form as

$$mN\left[\frac{\partial \mathbf{u}}{\partial t} + (\mathbf{u}\cdot\nabla_{\mathbf{r}})\mathbf{u}\right] = -\nabla p + qN(\mathbf{E} + \mathbf{u}\times\mathbf{B}), \qquad (4.22)$$

which is reminiscent of the *Navier–Stokes equation* of ordinary fluid mechanics,

$$\rho_{\mathrm{m}}\left[\frac{\partial \mathbf{u}}{\partial t} + (\mathbf{u}\cdot\nabla_{\mathbf{r}})\mathbf{u}\right] = -\nabla p + \rho_{\mathrm{m}}\,\nu_{\mathrm{v}}\,\nabla^2\mathbf{u}, \qquad (4.23)$$

where ρ_{m} is the mass density and ν_{v} is the kinematic viscosity coefficient. The similarity of (4.22) and (4.23) is the basis for the fluid treatments of plasmas, which we will study in the next few chapters.

The collision term $\mathbf{S}_{ij}$ represents the total momentum transferred (gained) by species i via its collisions with species j. If there is only one species this term will by definition be zero since, as stated before, the total momentum change per unit volume for a given particle species will not be changed by collisions between same-species particles. However, a plasma contains ions and electrons and can also contain large populations of neutral molecules. If we consider only elastic collisions, we can approximate the collision

[3] Note that the general form of $\nabla\cdot\Psi$ is

$$\nabla\cdot\Psi \equiv \left[\frac{\partial}{\partial x}\;\frac{\partial}{\partial y}\;\frac{\partial}{\partial z}\right]\begin{bmatrix} p_{xx} & p_{xy} & p_{xz} \\ p_{yx} & p_{yy} & p_{yz} \\ p_{zx} & p_{zy} & p_{zz}\end{bmatrix} = \begin{bmatrix} \frac{\partial p_{xx}}{\partial x} + \frac{\partial p_{yx}}{\partial y} + \frac{\partial p_{zx}}{\partial z} \\ \frac{\partial p_{xy}}{\partial x} + \frac{\partial p_{yy}}{\partial y} + \frac{\partial p_{zy}}{\partial z} \\ \frac{\partial p_{xz}}{\partial x} + \frac{\partial p_{yz}}{\partial y} + \frac{\partial p_{zz}}{\partial z}\end{bmatrix}.$$

Since $p = p_{xx} = p_{yy} = p_{zz}$, the above expression reduces to

$$\nabla\cdot\Psi = \begin{bmatrix} \frac{\partial p}{\partial x} \\ \frac{\partial p}{\partial y} \\ \frac{\partial p}{\partial z}\end{bmatrix} = \begin{bmatrix} \frac{\partial}{\partial x} \\ \frac{\partial}{\partial y} \\ \frac{\partial}{\partial z}\end{bmatrix} p = \nabla p.$$

term as proportional to the difference between the particle species' mean velocities:

$$\mathbf{S}_{ij} = -\sum_j mN_i \nu_{ij}(\mathbf{u}_i - \mathbf{u}_j). \tag{4.24}$$

The ν_{ij} term is called the collision frequency between particles of type i and j, and has units of s^{-1}. Equation (4.24) can be used when the difference in mean velocities is not too great and when each particle species has a Maxwell–Boltzmann velocity distribution. A further simplification can be made if the dominant collision process is between electrons and neutral molecules. This is the case in low-pressure gas discharges and in the lower heights of the ionosphere. The neutral particles are assumed to be stationary on average ($\mathbf{u}_j = 0$), and if we consider only one neutral species the momentum transport equation can be reduced to

$$mN\frac{d\mathbf{u}}{dt} = qN\mathbf{E} - \nabla p - mN\nu\,\mathbf{u}, \tag{4.25}$$

where the background magnetic field is assumed to be negligible, as is the case in highly collisional plasmas (see Problem 4-3).

Example 4-2 Fluorescent lamp

Common fluorescent lamps contain a low-pressure plasma called a glow discharge. As shown in Figure 4.2, the gas in the tube is partially ionized by the application of a voltage across the electrodes. The free electrons then impact a small population of mercury atoms, causing ultraviolet (UV) radiation. When the UV

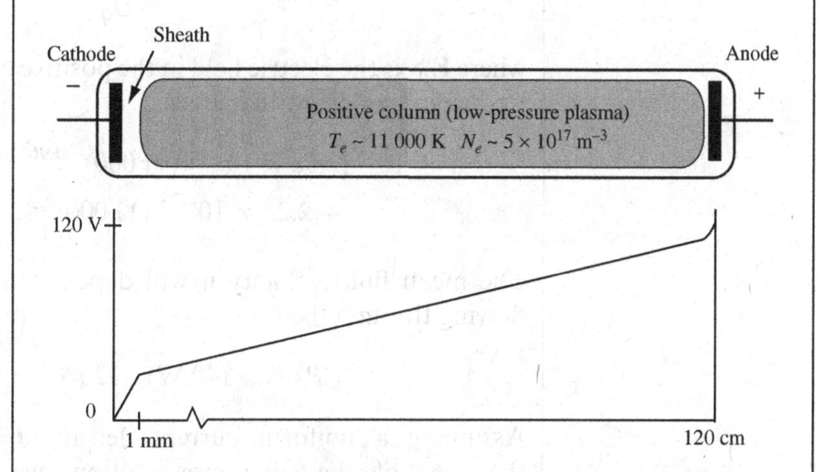

Figure 4.2 Simplified fluorescent lamp. A fluorescent lamp uses a low-pressure plasma to excite UV radiation that illuminates a phosphorous coating on the glass tube. The electric field in the main discharge is close to constant but has a sharp gradient near the cathode. Such lamps operate in the AC mode so anode and cathode switch during each cycle.

rays hit the phosphorous coating on the tube, white light is emitted. The majority of gas molecules in the tube remain as neutrals. Although the behavior of a glow discharge can be very complex, the main features in a typical lamp include a small plasma sheath around the cathode that is devoid of electrons, and a larger region of constant electric field known as the positive column that is responsible for most of the UV generation. The sheath is formed because the electrons are more mobile and are repelled from the cathode.[4] Almost all of the current flowing through the lamp is carried by electrons. Assuming the simplified model shown in Figure 4.2 and the following operational parameters, find the electric field in the positive column and the magnitude of the electron pressure term at the edge of the sheath region under steady-state conditions: power = 40 W; voltage = 120 V; length = 120 cm; diameter = 3.6 cm; $N_e = 5 \times 10^{17}\ \mathrm{m}^{-3}$; N_{gas} (argon) $= 10^{23}\ \mathrm{m}^{-3}$; $T_e = 11\,000$ K. The collision frequency between electrons and argon atoms can be estimated at $\nu = N_{Ar}\left(2.58 \times 10^{-12}\ T_e^{-0.96} + 2.25 \times 10^{-23}\ T_e^{2.29}\right)$ Hz [3].

Solution: Since the neutral density is much greater than the electron density we can use Equation (4.25). Within the positive column the electron temperature and density are constant so the pressure term (∇p) is zero, and for steady-state conditions there are no changes in time. Equation (4.25) thus simplifies to

$$\mathbf{E}_{pc} = \frac{m_e \nu \mathbf{u}}{q},$$

where $\mathbf{E}_{pc}$ is the electric field in the positive column. The collision frequency can be evaluated as

$$\nu = 10^{23}\,[2.58 \times 10^{-12}\,(11\,000)^{-0.96} + 2.25 \times 10^{-23}\,(11\,000)^{2.29}] = 4 \times 10^9\ \mathrm{s}^{-1}.$$

The mean fluid velocity $\mathbf{u}$ will depend on the total current I flowing through the tube:

$$I = P/V = (40\ \mathrm{W})/(120\ \mathrm{V}) = 333\ \mathrm{mA}.$$

Assuming a uniform current density $\mathbf{J} = I/A$, where A is the area of the tube cross-section, we can evaluate $\mathbf{u}$ as

4 A more accurate description of a fluorescent lamp includes a total of six separate regions: the cathode dark space (sheath), negative glow, Faraday dark space, positive column, anode glow, and anode dark space [1, 2].

$$u = \frac{J}{qN_e} = \frac{(0.333\ \text{A})}{(0.001\ \text{m}^2)(1.6\times10^{-19}\ \text{C})(5\times10^{17}\ \text{m}^{-3})}$$

$$= 4163\ \text{m}\,\text{s}^{-1}.$$

Plugging in to the expression for electric field above yields: $\mathbf{E}_{pc} = 93.84\ \text{V}\,\text{m}^{-1}$.

At the positive column-sheath boundary there will be a density gradient in the electron population since there are virtually no electrons in the sheath region. The density gradient will create a non-zero pressure term (∇p) which will need to be balanced by a stronger electric field in this region. We can obtain a crude estimate of $\mathbf{E}_s$, the electric field in the sheath, by dividing the remaining voltage drop (outside of the positive column) by the width of the sheath (1 mm):

$$\mathbf{E}_s = \frac{120\ \text{V} - (0.9384\ \text{V/cm})(119.9\ \text{cm})}{1\ \text{mm}} = 7486\ \text{V}\,\text{m}^{-1}.$$

Using Equation (4.25) again we have

$$\nabla p = qN_e\mathbf{E}_s - m_eN_e\nu\mathbf{u} = (1.6\times10^{-19})(5\times10^{17})(6000)$$

$$-(9.1\times10^{-31})(5\times10^{17})(4\times10^{9})(4163) = 591.3\ \text{N}\,\text{m}^{-3},$$

where variations in other parameters have been neglected for simplicity.

4.4 The second-order moment: energy transport equation

The second-order moment of the Boltzmann equation, i.e., the equation of energy conservation, is obtained by multiplying (3.9) by $\frac{1}{2}mv^2$ and integrating over velocity space. For ease of discussion we will not present the derivation here but will instead refer the interested reader to Appendix A. The energy-conservation equation can be written in several forms, one of which is

$$\frac{\partial}{\partial t}\left[N\frac{1}{2}mu^2\right] + \nabla\cdot\left[N\frac{1}{2}m\langle u^2\mathbf{u}\rangle\right] - Nq\langle\mathbf{E}\cdot\mathbf{u}\rangle$$

$$= \underbrace{\frac{m}{2}\int u^2\left(\frac{\partial f}{\partial t}\right)_{\text{coll}} d\mathbf{u}}_{\mathbf{S}_{\text{coll}}}, \tag{4.26}$$

which can be interpreted in a manner quite similar to (4.10) and (4.21), i.e., the first term is the rate of change of energy density, the second is the energy loss rate per volume element due to the energy flux (or heat transfer), and the third is the power fed into the system by the electric field (i.e., $\mathbf{E}\cdot\mathbf{J}$), noting that no work is done by the magnetic field. Note also that the collision term is non-zero since energy transfer between species occurs in collisions, although the total energy in the system is conserved. The energy-conservation equation can be also written as

$$\frac{\partial\left[\frac{1}{2}Nm\langle w^2\rangle\right]}{\partial t}+\nabla\cdot\left(\frac{1}{2}Nm\,\langle w^2\rangle\,\mathbf{u}\right)+(\Psi\cdot\nabla)\cdot\mathbf{u}+\nabla\cdot\mathbf{q}=\mathbf{S}_{\text{coll}}.$$

For an isotropic plasma the pressure tensor Ψ reduces to the scalar pressure p; considering that the average energy of the plasma is $\frac{1}{2}m\langle\mathbf{w}\cdot\mathbf{w}\rangle=\frac{3}{2}k_{\text{B}}T$ and using $p=Nk_{\text{B}}T$, we find $\frac{3}{2}p=\frac{1}{2}Nm\langle\mathbf{w}\cdot\mathbf{w}\rangle$, which is the energy density (in J m^{-3}). The energy conservation equation then reduces to

$$\frac{\partial\left(\frac{3}{2}p\right)}{\partial t}+\nabla\cdot\left(\frac{3}{2}p\,\mathbf{u}\right)-p\nabla\cdot\mathbf{u}+\nabla\cdot\mathbf{q}=\mathbf{S}_{\text{coll}}.\tag{4.27}$$

The quantity $\frac{3}{2}p\mathbf{u}$ represents the flow of energy density at the fluid velocity, or the macroscopic energy flux (in units of W m^{-2}). The third term, $p\nabla\cdot\mathbf{u}$ (in W m^{-3}), represents the heating or cooling of the fluid due to compression or expansion of its volume. The new quantity $\mathbf{q}$ is the heat-flow (or heat-flux) vector (in W m^{-2}), which represents microscopic energy flux and is related to the particle random velocity by $\mathbf{q}=(N/2)m\langle w^2\mathbf{w}\rangle$. For steady-state, low-pressure discharges, the macroscopic energy flux is balanced against the collisional processes, resulting in a simpler equation,

$$\nabla\cdot\left(\frac{3}{2}p\,\mathbf{u}\right)=\mathbf{S}_{\text{coll}}.\tag{4.28}$$

4.5 Systems of macroscopic equations: cold- and warm-plasma models

It is worthwhile to review the moments of the Boltzmann equation that we have derived so far. Each of the moments is a transport equation describing the dynamics of a quantity associated with a given power of v:

Continuity equation
mass or charge transport
$$\boxed{\frac{\partial N}{\partial t} + \nabla \cdot [N\mathbf{u}] = 0} \tag{4.29}$$

Momentum transport
$$\boxed{mN\left[\frac{\partial \mathbf{u}}{\partial t} + (\mathbf{u}\cdot\nabla)\mathbf{u}\right] = -\nabla\cdot\Psi + qN(\mathbf{E} + \mathbf{u}\times\mathbf{B}) + \mathbf{S}_{ij}} \tag{4.30}$$

Energy transport
$$\boxed{\frac{\partial}{\partial t}\left[N\frac{1}{2}mu^2\right] + \nabla\cdot\left[N\frac{1}{2}m\langle u^2\mathbf{u}\rangle\right] - Nq\langle\mathbf{E}\cdot\mathbf{u}\rangle = \underbrace{\frac{m}{2}\int u^2\left(\frac{\partial f}{\partial t}\right)_{\text{coll}} d\mathbf{u}}_{\mathbf{S}_{\text{coll}}}}. \tag{4.31}$$

We could in principle proceed by evaluating higher and higher-order moments of the Boltzmann equation. However, moments that involve the third or higher powers of the particle velocity v lack simple physical meaning and are generally useful in only specialized cases. Accordingly, we will not evaluate or discuss higher-order moments. The equations of conservation of particle number, momentum, and energy are useful in making general statements about plasmas, but they cannot be considered as a closed system of plasma equations. It is worth remembering that our motivation in finding the moments of the Boltzmann equation was to avoid having to solve this equation directly for the velocity space distribution function $f(\mathbf{r}, \mathbf{v}, t)$. So instead of attempting to solve for $f(\mathbf{r}, \mathbf{v}, t)$ explicitly, we set out to determine its moments. In calculating each moment of the Boltzmann equation, however, we always obtained an equation that contained the next moment. In the zeroth-order moment the change in particle density was expressed as a function of the mean fluid velocity. In the first-order moment, the change in mean fluid velocity was expressed as a function of the pressure tensor. The second-order moment is an expression for the change in the pressure tensor, but brings in a new heat-flow term. Every time we obtain a new equation a new unknown appears, so that the number of equations is never sufficient for the determination of all the macroscopic quantities. The number of unknowns always exceeds the number of equations. Because of this, it is necessary to truncate the system of equations at some point in the hierarchy of moments by making simplifying assumptions. Among the several different sets of macroscopic equations used to describe plasma dynamics, the two most commonly used are the so-called cold-plasma and warm-plasma models, briefly described below.

4.5.1 The cold-plasma model

The simplest set of macroscopic equations can be obtained by introducing the truncation at the momentum-transfer equation, (4.30). The physical assumption adopted is to neglect the thermal motions of the particles, which is achieved by setting the kinetic pressure tensor to zero, i.e., $\Psi = 0$. The only remaining macroscopic variables are then the number density N and fluid velocity $\mathbf{u}$, which are described by the two equations

$$\frac{\partial N}{\partial t} + \nabla \cdot [N\mathbf{u}] = 0 \tag{4.32a}$$

$$mN\left[\frac{\partial \mathbf{u}}{\partial t} + (\mathbf{u} \cdot \nabla)\mathbf{u}\right] = qN(\mathbf{E} + \mathbf{u} \times \mathbf{B}) + \mathbf{S}_{ij}. \tag{4.32b}$$

In general, a suitable method for evaluation of the collision term $\mathbf{S}_{ij}$ is needed in order for (4.32) to be useful in analysis of plasma dynamics. One common method is to view the collisions as an impediment to motion, causing a rate of decrease in momentum as determined by an "effective" collision frequency, in which case we use

$$\mathbf{S}_{ij} = -m\,N\,\nu_{\text{eff}}\,\mathbf{u}.$$

4.5.2 The warm-plasma model

An alternative set of macroscopic equations is obtained by introducing truncation at the energy-conservation equation. Thermal motions are accounted for but it is assumed that the kinetic pressure tensor is diagonal, with equal diagonal terms, so that $\nabla \cdot \Psi = \nabla p$. Physically, this means that viscous forces are neglected. We then have

$$\frac{\partial N}{\partial t} + \nabla \cdot [N\mathbf{u}] = 0 \tag{4.33a}$$

$$mN\left[\frac{\partial \mathbf{u}}{\partial t} + (\mathbf{u} \cdot \nabla)\mathbf{u}\right] = -\nabla p + qN(\mathbf{E} + \mathbf{u} \times \mathbf{B}) + \mathbf{S}_{ij}. \tag{4.33b}$$

The system (4.33) still does not form a closed set, since the scalar pressure is now a third variable. In principle, the energy-conservation equation (4.27) is needed to determine p, but it contains a fourth unknown, $\mathbf{q}$. However, it is often possible to truncate the system of equations at this point by adopting simplifying assumptions which either make the energy-transport equation unnecessary or reduce it to simpler forms expressed in terms of p.[5] The simplest method of truncation is to assume a thermodynamic equation of state in order to relate p to the number density N. The actual form of the equation of state varies from case to case. The two most common equations of state are the so-called *isothermal* and *adiabatic* ones. The isothermal equation of state is

$$p = Nk_{\mathrm{B}}T \quad \text{or} \quad \nabla p = k_{\mathrm{B}}T\nabla N, \tag{4.34}$$

and holds for relatively slow time variations, allowing temperatures to reach equilibrium. In this case, the plasma fluid can exchange energy with its surroundings, and the simpler version of the energy-conservation equation, (4.28), is also required. Alternatively, we can use the adiabatic equation of state given by

$$pN^{-\gamma} = C \quad \text{or} \quad \frac{\nabla p}{p} = \gamma\frac{\nabla N}{N}, \tag{4.35}$$

where C is a constant and γ is the ratio of specific heat at constant pressure to that at constant volume. Typically, $\gamma = 1 + 2/n_d$, where n_d is the number of degrees of freedom. The adiabatic relation holds for fast time variations, as in the case of plasma waves, when the plasma fluid does not exchange energy with its surroundings; thus a separate energy-conservation equation is not needed, since it leads to (4.34). The use of the adiabatic gas law to close the system of equations is equivalent to assuming that there is no energy interchange due to collisional interactions and that there is no heat flow. In cases where the explicit use of the energy conservation equation is required, the heat-flow (or heat-flux) vector $\mathbf{q}$ would be the outstanding unknown, and we would need to adopt a physical assumption in order to close the system of moment equations. For electrons, the approximation most commonly used is one which is derived from thermodynamics:

$$\mathbf{q} = -K\nabla T,$$

where K is the thermal conductivity.

4.6 Summary

In this chapter we set out to avoid solving the Boltzmann equation for the velocity distribution function since this is often not straightforward and since the averages obtained by integrating over

[5] For simplifications particularly suited to ionospheric and magnetospheric plasmas, see Section 2.4.5 of [4].

the distribution function are more useful in practical applications. Instead, we set out to solve for the moments of the distribution, which are averages of different powers of particle velocity. The moments of the distribution function represent average quantities such as particle density (zeroth-order moment) and mean velocity (first-order moment) and kinetic energy (third-order moment). These average bulk quantities are the primary variables in modeling a plasma as a fluid, which we will discuss in the next chapters. Taking any moment of the Boltzmann equation yields an expression that contains the next-highest moment. In principle, the velocity distribution function is known once we know all of its moments. However, finding a large number of successively higher moments is not practical.

The power of the moment approach lies in truncating a finite set of moment equations using an approximation for the highest moment appearing in the system. Such a truncation yields a closed set of equations that can be solved, and are known as a plasma model. We presented two of the most commonly used plasma models: the cold- and warm-plasma approximations. In the cold-plasma model, thermal motions are assumed to be negligible, which means that the pressure term in the moment equations is equal to zero. In the warm-plasma model, thermal motions are taken into account using either an isothermal or an adiabatic approximation, and the heat-flow term is set to be zero. In the warm-plasma model the pressure term is a scalar value like that used to describe non-ionized gases. Although not covered in this text, it is possible to develop more complicated plasma models where truncation is done at higher-level moments.

4.7 Problems

4-1. Using the ionospheric parameters given in Example 4-1, use a numerical technique to find how much time after sunrise it will take the electron density to change from its nighttime ambient value to its daytime value. Make a plot of the electron density versus time. You may ignore any plasma flows.

4-2. In a laboratory plasma experiment a plasma density of $10^{19}\ \mathrm{m}^{-3}$ is created by a rapid burst of ultraviolet radiation. The plasma density is observed to decay to half of its original value in 10 ms. Find the value of the recombination coefficient ν_α, assuming that attachment is negligible.

4-3. The strength of the Earth's magnetic field at the surface is approximately 30 μT. Show that this field has a negligible effect on the physics of the fluorescent lamp discussed in Example 4.2.

4-4. For the fluorescent lamp discussed in Example 4.2, calculate the Debye length for the electrons in the positive column region and compare it to the dimensions of the lamp.

4-5. An engineer proposes to double the length of the 120 cm fluorescent lamp in Example 4-2. If making the tube longer causes the electric field in the positive column to drop by a factor of two, calculate the power the lamp will draw if the electron and gas density, temperature, and voltage all remain the same.

4-6. Consider the one-dimensional plasma configuration shown below, which is clearly not in equilibrium. Calculate an electric field that could be used to maintain the density profile, assuming mobile electrons and stationary ions.

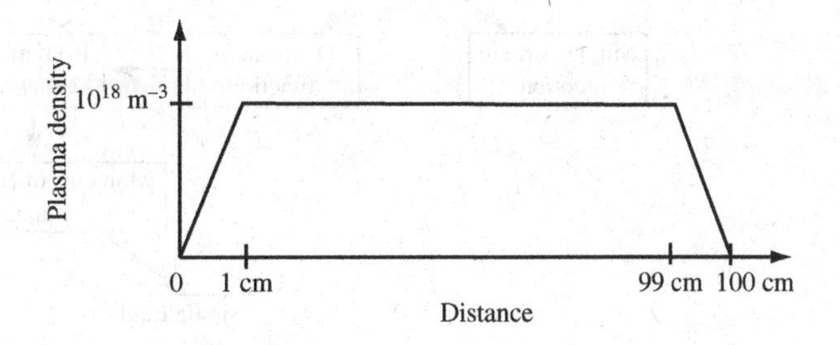

4-7. Consider a 200 V m^{-1} electric field applied to a partially ionized plasma with an electron density of 10^{15} m^{-3}. The effective collision frequency for the electrons is 3.5 GHz. Using the cold-plasma model and ignoring ion motion, (a) find the steady-state electron fluid velocity; (b) use the energy transport equation to find the energy dissipated, ignoring convective terms and interactions with other species.

References

[1] G. G. Lister, Low pressure gas discharge modelling. *J. Phys. D: Appl. Phys.*, **25** (1992), 1649–80.

[2] G. G. Lister, J. E. Lawler, W. P. Lapatovich, and V. A. Godyak, The physics of discharge lamps. *Rev. Mod. Phys.*, **76** (2004), 541–98.

[3] P. Baille, J.-S. Chang, A. Claude, *et al.*, Effective collision frequency of electrons in noble gases. *J. Phys. B: At., Mol. Opt. Phys.*, **14** (1981), 1485–95.

[4] T. E. Cravens, *Physics of Solar System Plasmas* (Cambridge: Cambridge University Press, 1997).

CHAPTER

5 Multiple-fluid theory of plasmas

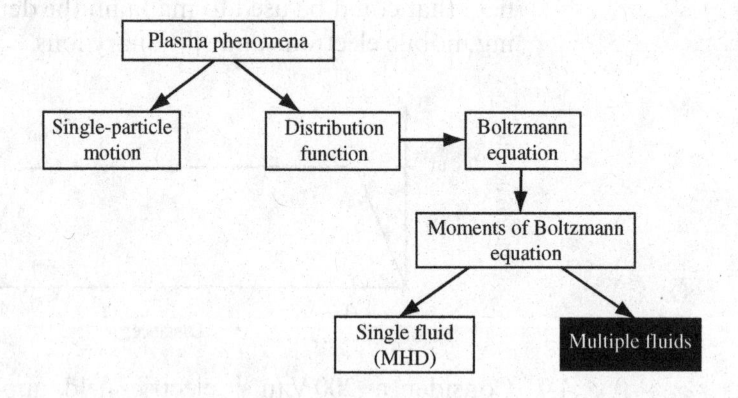

5.1 Introduction

In this chapter we begin the treatment of plasmas using fluid theory. Practically speaking, a fluid approach is nothing more than a description of the dynamics of bulk properties of a plasma, and in this context can be characterized as a rather crude description of a medium that is composed of very large numbers of individual particles. However, despite its inherent simplifications, the fluid approach is remarkably successful in describing plasma behavior and is perhaps the most widely used treatment of plasma phenomena. Most of the remaining topics in this text, with the exception of hot plasmas, are founded on a fluid description. Fluid theory follows directly from the moments of the Boltzmann equation that were derived in the previous chapter. We begin by presenting a description of a plasma as two interpenetrating fluids of electrons and ions. In Chapter 6 a single-fluid description will be introduced.

5.2 Complete set of two-fluid equations

Using a truncated set of moment equations, we can write the complete set of electrodynamic equations consisting of the continuity and momentum equations (4.29) and (4.30) and Maxwell's equations. For simplicity, we consider a fully ionized, isotropic (i.e., $\nabla \cdot \Psi = \nabla p$), and collisionless (i.e., $\mathbf{S}_{\text{ie}}$ or $\mathbf{S}_{\text{ei}} = 0$) plasma with only two species: electrons and ions. We have

$$m_e N_e \left[\frac{\partial \mathbf{u}_e}{\partial t} + (\mathbf{u}_e \cdot \nabla)\mathbf{u}_e \right] = -\nabla p_e + q_e N_e (\mathbf{E} + \mathbf{u}_e \times \mathbf{B}) \tag{5.1a}$$

$$m_i N_i \left[\frac{\partial \mathbf{u}_i}{\partial t} + (\mathbf{u}_i \cdot \nabla)\mathbf{u}_i \right] = -\nabla p_i + q_i N_i (\mathbf{E} + \mathbf{u}_i \times \mathbf{B}) \tag{5.1b}$$

$$\nabla \times \mathbf{E} = -\frac{\partial \mathbf{B}}{\partial t} \tag{5.1c}$$

$$\epsilon_0 \nabla \cdot \mathbf{E} = \underbrace{N_i q_i + N_e q_e}_{\rho} \tag{5.1d}$$

$$\frac{1}{\mu_0} \nabla \times \mathbf{B} = \underbrace{N_i q_i \mathbf{u}_i + N_e q_e \mathbf{u}_e}_{\mathbf{J}} + \epsilon_0 \frac{\partial \mathbf{E}}{\partial t} \tag{5.1e}$$

$$\nabla \cdot \mathbf{B} = 0 \tag{5.1f}$$

$$\frac{\partial N_e}{\partial t} + \nabla \cdot [N_e \mathbf{u}_e] = 0 \tag{5.1g}$$

$$\frac{\partial N_i}{\partial t} + \nabla \cdot [N_i \mathbf{u}_i] = 0 \tag{5.1h}$$

$$p_e = k_B T_e N_e \quad \text{or} \quad p_e = C_e N_e^{\gamma} \tag{5.1i}$$

$$p_i = k_B T_i N_i \quad \text{or} \quad p_i = C_i N_i^{\gamma}. \tag{5.1j}$$

Note that the velocities in question are those of fluid elements expressed as $\mathbf{u}$, as opposed to the velocities of individual particles expressed as $\mathbf{v}$ in previous chapters. The system in (5.1) constitutes four vector equations ((5.1)a, b, c, e) plus six scalar ones, amounting to a total of 18 scalar equations. On the other hand, Equations (5.1) contain four scalar unknowns (p_e, p_i, N_e, N_i), and four vector unknowns ($\mathbf{u}_i$, $\mathbf{u}_e$, $\mathbf{E}$, $\mathbf{B}$), amounting to a total of 16 unknowns. However, one of the two continuity equations (5.1g) and (5.1h) is redundant, since it can be derived by using the other and (5.1d), together with the divergence of (5.1e). Also, (5.1f) is redundant, since it can be derived from (5.1c). Thus, we actually have 16 scalar equations and 16 scalar unknowns, providing a self-consistent set of equations that describe the dynamics of the electron and ion fluids under the influence of electromagnetic fields in a fully ionized, collisionless, isotropic, two-species plasma. The two-fluid model described by these equations is applicable to the dynamics of the plasma on all time scales, ranging from the rapid motions of electrons (e.g., plasma oscillations at frequencies comparable to $\omega_{pe} = \sqrt{N_e q_e^2/(\epsilon_0 m_e)}$), and the much slower motions of ions (occurring at time scales comparable to $1/\omega_{pi}$, where $\omega_{pi} = \sqrt{N_i q_i^2/(\epsilon_0 m_i)}$). We will use Equations (5.1) in later chapters to derive the properties of electromagnetic waves which exist in such plasmas over a wide range of frequencies. Here, we will consider two interesting aspects of fluid motions, namely fluid drifts perpendicular to **B** and parallel pressure balance.

Example 5-1 Plasma discharge for IC manufacture

The manufacture of integrated circuits (IC) involves etching with plasmas, typically produced between two electrodes, as shown in Figure 5.1 [1]. The time-averaged electron and ion densities between the electrode plates are shown in the graph; we approximate their distribution using the following polynomial expressions:

$$N_i(x) = 10^{16}\left[\left(-6.803\times10^8\right)x^4 + \left(1.361\times10^7\right)x^3 + \left(-9.809\times10^4\right)x^2 + 300.6x + 0.712\right]\ \text{m}^{-3}$$

$$N_e(x) = 10^{16}\left[\left(-17.69\times10^8\right)x^4 + \left(3.538\times10^7\right)x^3 + \left(-25.43\times10^4\right)x^2 + 774.0x + 0.207\right]\ \text{m}^{-3},$$

where x is in units of m. Calculate and plot the time-averaged electric field and potential across the plasma, assuming both electrode plates are at zero potential.

Solution: To find the electric field we need to use Equation (5.1d), whose one-dimensional form is

$$\epsilon_0 \frac{dE_x}{dx} = N_i(x)q_i + N_e(x)q_e.$$

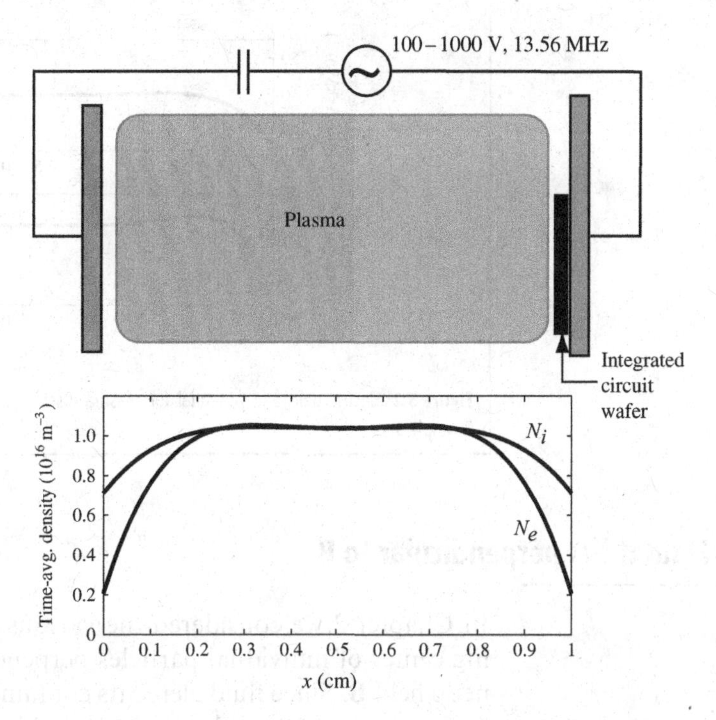

Figure 5.1 Basic setup for RF-induced plasma etching of integrated circuits. The bottom panel shows time-averaged electron and ion densities.

Integrating and plugging in the polynomial expressions yields

$$E_x(x) = \frac{|q_e|}{\epsilon_0} \int_0^x \left[N_i(x') - N_e(x')\right] dx'$$

$$= 1.808 \times 10^8 \Big[2.177 \times 10^8 x^5 - 5.443 \times 10^6 x^4 + 5.207 \times 10^4 x^3 - 236.7 x^2 + 0.503 x\Big] + C_0 \, \mathrm{V\,m^{-1}}.$$

We find the constant C_0 by noting that the electric field in the center of the discharge where the electron and ion densities are equal must be zero $E_x(0.5\,\mathrm{cm}) = 0$. This yields $C_0 = -6.958 \times 10^4$. The potential can be found in a straightforward manner using $\frac{d\Phi}{dx} = -E_x$ and performing another integration. Plots of the electric field and potential are shown in Figure 5.2. The regions near the electrodes where the ion and electron densities diverge are known as the *sheath* regions and result from higher electron mobility. See Chapters 1 and 4 for an introduction to the concept of the plasma sheath, and Chapter 13 for a quantitative discussion of the subject.

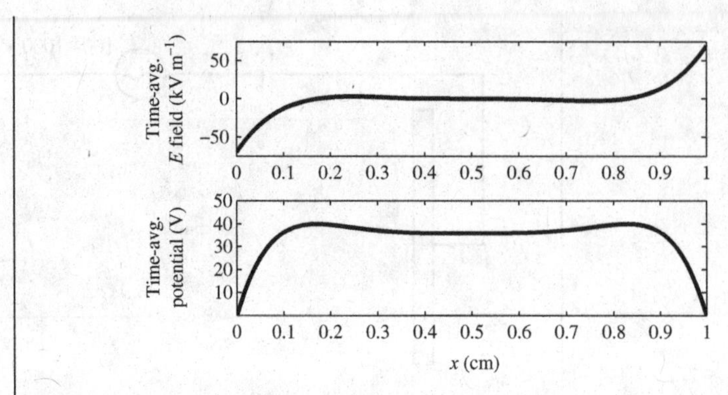

Figure 5.2 Electric field and potential for plasma discharge shown in Figure 5.1 and discussed in Example 5-1.

5.3 Fluid drifts perpendicular to B

In Chapter 2 we considered the various drift motions of the guiding center of individual particles perpendicular to an applied magnetic field **B**. Since fluid elements contain many individual particles, we would expect that they would exhibit the drifts perpendicular to **B** whenever such drifts are experienced by the individual guiding centers. This expectation is true for the $\mathbf{E} \times \mathbf{B}$ and curvature drifts, but not for the gradient drift; furthermore, an additional type of drift called *diamagnetic drift* occurs for fluid elements as a result of the ∇p term. Examining the drift motions using fluid theory is very instructive for deeper understanding of the fluid approach. Consider the momentum equation valid for each of the species:

$$mN\left[\frac{\partial \mathbf{u}}{\partial t} + (\mathbf{u}\cdot\nabla)\mathbf{u}\right] = -\nabla p + qN(\mathbf{E} + \mathbf{u}\times\mathbf{B}). \tag{5.2}$$

Considering drifts which are slow compared to the time scale corresponding to the gyrofrequency ω_c,[1] and neglecting second-order terms, (5.2) reduces to

$$0 \simeq qN[\mathbf{E} + \mathbf{u}_\perp \times \mathbf{B}] - \nabla p, \tag{5.3}$$

where we have noted that $\mathbf{u}\times\mathbf{B} = (\mathbf{u}_\perp + \mathbf{u}_\parallel)\times\mathbf{B} = \mathbf{u}_\perp\times\mathbf{B}$. We now take the cross-product of (5.3) with **B** to find

$$0 = qN[\mathbf{E}\times\mathbf{B} + (\mathbf{u}_\perp\times\mathbf{B})\times\mathbf{B}] - \nabla p\times\mathbf{B}$$

[1] If we consider variations of fluid velocity **u** at a frequency ω, the first term on the left-hand side of (5.2) is of order $mN|\partial\mathbf{u}/\partial t| = mN(j\omega\mathrm{u})$ while the last term on the right is of order $|qN\mathbf{u}\times\mathbf{B}| = qN\mathrm{u}B$, so that the ratio of the two terms is (ω/ω_c).

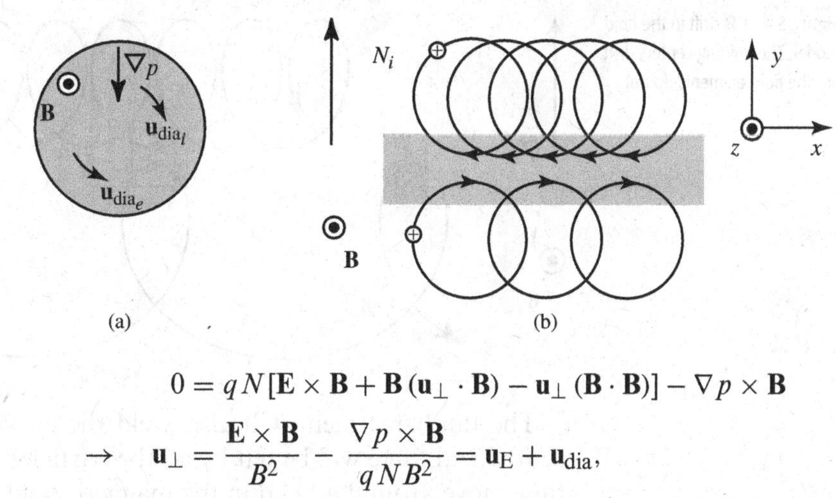

Figure 5.3 Diamagnetic drift. (a) Cylindrical plasma column. (b) Gyromotion of ions in the presence of a density gradient.

$$0 = qN[\mathbf{E} \times \mathbf{B} + \mathbf{B}(\mathbf{u}_\perp \cdot \mathbf{B}) - \mathbf{u}_\perp(\mathbf{B} \cdot \mathbf{B})] - \nabla p \times \mathbf{B}$$

$$\rightarrow \quad \mathbf{u}_\perp = \frac{\mathbf{E} \times \mathbf{B}}{B^2} - \frac{\nabla p \times \mathbf{B}}{qNB^2} = \mathbf{u}_\mathrm{E} + \mathbf{u}_\mathrm{dia},$$

where we have used the usual identity $(\mathbf{A} \times \mathbf{B}) \times \mathbf{C} \equiv \mathbf{B}(\mathbf{A} \cdot \mathbf{C}) - \mathbf{C}(\mathbf{A} \cdot \mathbf{B})$. Thus we see that the fluid elements experience $\mathbf{E} \times \mathbf{B}$ drift given by the same expression as was found for individual particles in Chapter 2. Additionally, we see that there is a new drift, the diamagnetic drift, given by

$$\text{Diamagnetic drift} \quad \boxed{\mathbf{u}_\mathrm{dia} = -\frac{\nabla p \times \mathbf{B}}{qNB^2}}. \tag{5.4}$$

The diamagnetic drift is in opposite directions for electrons and for ions, and in a cylindrical plasma (see Figure 5.3) gives rise to currents that tend to reduce the magnetic field inside the plasma, which is why this phenomenon is referred to as "diamagnetic."

The physical interpretation of diamagnetic drift is shown in Figure 5.3b. For this case we have assumed a constant-temperature plasma so that the pressure gradient ∇p manifests itself as a density gradient ∇N. There are a larger number of particles at higher values of y, as indicated by the larger number of gyrating particles per unit area in Figure 5.3b. We note that, through any fixed volume element such as the shaded region shown, there are more ions moving to the left than to the right, amounting to a drift of the fluid element, even though the guiding centers of individual particles are stationary.

The diamagnetic drifts of ions and electrons in opposite directions give rise to a diamagnetic current given by

$$\mathbf{J}_\mathrm{dia} = N_i q_i \mathbf{u}_{\mathrm{dia}_i} - N_e q_e \mathbf{u}_{\mathrm{dia}_e} = \frac{\mathbf{B} \times \nabla(p_i + p_e)}{B^2}. \tag{5.5}$$

For an isothermal ($p = k_\mathrm{B} T N$) plasma with $N_i = N_e = N$, the diamagnetic current is given by

$$\mathbf{J}_\mathrm{dia} = (k_\mathrm{B} T_i + k_\mathrm{B} T_e) \frac{\mathbf{B} \times \nabla N}{B^2}. \tag{5.6}$$

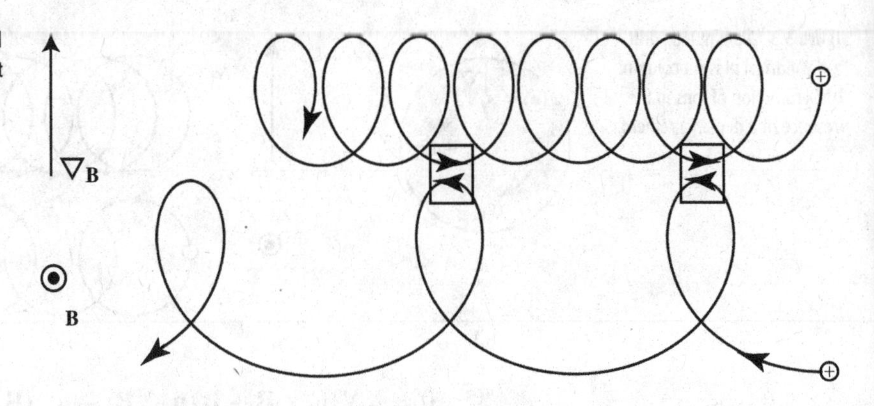

Figure 5.4 ∇**B** drift in the fluid model. The guiding centers drift but the fluid elements do not.

The fluid treatment will also yield the curvature drift since the centrifugal force will be felt by all the particles in a fluid element as they move around a bend in the magnetic field. The curvature drift can be obtained by adding a term

$$\langle F_{\mathrm{cf}} \rangle = \left\langle N m v_{\parallel}^2 \right\rangle / R_c = N k_{\mathrm{B}} T_{\parallel} / R_c,$$

which is the fluid analog to Equation (2.24). However, the gradient drift (∇**B**) does not exist in the fluid picture. In a reversed symmetry to diamagnetic drift, for gradient drift there is no fluid drift even though there is a drift of individual guiding centers. The lack of a fluid drift in the presence of a magnetic field gradient can most easily be seen by considering the motion of the two particles in Figure 5.4. Since there is no electric field the gyroradius changes only because of the gradient of **B**. Since the magnetic field alone cannot change the energy of the particle, the particle velocities in the fluid element marked by the boxes in the figure will cancel exactly. Since the fluid picture assumes an even distribution of particles in each local element defined by N, all fluid elements at all positions will have a net zero velocity if there is no density gradient.

The seeming contradiction between the fluid model and the single-particle approach begs the question of which model is more correct. It is important to realize that both the single-particle and the fluid pictures are simplified models of complex interactions. The single-particle approach tracks a single particle, in complete isolation, which is not realistic. The fluid approach, on the other hand, gives us net average values over a small but finite volume. Both approaches are approximations to full knowledge of all parameters and variables. In the physical sciences the highest measure of correctness is agreement with observations, and in this context the preferred model depends on the situation. There are situations where particle densities are very small (e.g., in the high-energy particle populations of the Earth's radiation belts) and the single-particle approach is appropriate since collective effects are negligible. However, in most practical situations plasma densities are high enough that the fluid approach exhibits better agreement with observations over any reasonable volume of interest. For such cases a comparison can be made to electrical current flowing in a simple wire or resistor. An individual electron in such a wire or resistor could have an extremely complicated trajectory, with multiple changes in direction and velocity, yet the collective behavior of the current is accurately described by Ohm's law ($V = IR$). Therefore, outside specialized situations the fluid approach is more reliable even though it yields results that can seem counterintuitive at first.

5.4 Parallel pressure balance

We now consider the component of the momentum equation (5.2) which is parallel to the magnetic field $\mathbf{B} = B\hat{\mathbf{z}}$. Neglecting the convective term, we have

$$mN\frac{\partial u}{\partial t} \simeq qNE_z - \frac{\partial p}{\partial z}. \tag{5.7}$$

This shows that the fluid element is accelerated along $\mathbf{B}$ under the combined influence of electrostatic force and the pressure-gradient force. When these two forces are balanced, the fluid experiences no acceleration, i.e., $\partial u_z/\partial t = 0$. Assuming a constant temperature T along the field line, and expressing the parallel electric field as the negative gradient of an electrostatic potential ($E_z = -\partial\Phi/\partial z$), we can write this pressure balance condition as

$$0 = -qN\frac{\partial \Phi}{\partial z} - k_B T\frac{\partial N}{\partial z}. \tag{5.8}$$

Integrating, we find

$$q\Phi + k_B T \ln N = \text{constant} \quad \rightarrow \quad N = N_0 \exp\left[\frac{-q\Phi}{k_B T}\right], \tag{5.9}$$

which is the so-called *Boltzmann factor* for electrons that we saw earlier in Equations (1.6) and (3.39), and which is the equilibrium solution of the Boltzmann equation in the presence of an external force. Physically this result reaffirms the strong tendency of the plasma to remain electrically neutral as a result of the very small charge-to-mass ratio of the electrons. Electrons have a tendency to move rapidly in response to an external force (e.g., an electrostatic potential gradient); however, once they begin to leave a region they leave behind large ion charges, producing electric fields which pull them back, with the final distribution of electrons being determined by the balance between electrostatic and pressure (or density)-gradient forces.

5.5 Summary

In this chapter we introduced the two-fluid model of a plasma, which consists of the truncated moment equations and Maxwell's equations. Application of the model to drift motions yielded similar results to the single-particle approach described in Chapter 2 in that $\mathbf{E} \times \mathbf{B}$ drift and curvature drift also exist in the fluid model. The fluid description also yields a new drift motion, called diamagnetic drift, which cannot be obtained in the single-particle approach since it is based on a density gradient. At the same time the fluid model does not yield a drift perpendicular to a static magnetic field gradient. The next two chapters explore a single-fluid approach to plasma, but we will return to the two-fluid model and use it to describe plasma conductivity, diffusion, and wave behavior in subsequent chapters.

5.6 Problems

5-1. An isothermal plasma is confined between the planes $x = \pm a$ in a magnetic field $\mathbf{B} = B_0\hat{\mathbf{z}}$. The density distribution is $N(x) = N_0(1 - x^2/a^2)$. (a) Derive an expression for the electron diamagnetic drift velocity $\mathbf{u}_{\text{dia}_e}$ as a function of x. (b) Draw a diagram showing the density profile and the direction of $\mathbf{u}_{\text{dia}_e}$ on both sides of the midplane if $\mathbf{B}$ points out of the paper. (c) Evaluate $\mathbf{u}_{\text{dia}_e}$ for $x = a/2$, $B_0 = 0.2$ T, $k_B T_e = 2$ eV, and $a = 4$ cm.

5-2. A cylindrically symmetric plasma column of radius $r = a$ extends along the z axis and is immersed in a uniform magnetic field $\mathbf{B} = \hat{\mathbf{z}} B_0$. The plasma is under the influence of an electrostatic potential $\Phi(r)$ so that

$$N_e = N_0 e^{-q_e \Phi/(k_B T)},$$

and the plasma density varies radially as

$$N_e(r) = N_0 e^{-r^2/a^2}.$$

Determine the $\mathbf{E} \times \mathbf{B}$ and diamagnetic drift velocities and compare their relative magnitudes.

5-3. Consider a partially ionized plasma with density $N = 10^{12}\ \mathrm{m^{-3}}$ through which a 100 V m^{-1} electric field is applied. Calculate the total current density if the electron–neutral collision frequency is 3 GHz and the ion–neutral collision frequency is 5 GHz.

5-4. Draw a diagram similar to Figure 5.4 but for the trajectory of two ions experiencing an $\mathbf{E} \times \mathbf{B}$ drift. Explain why the two trajectories will not cancel out for the $\mathbf{E} \times \mathbf{B}$ drift even though they do so for the gradient drift.

Reference

[1] M. A. Lieberman and A. J. Lichtenberg, *Principles of Plasma Discharges and Materials Processing* (New York: John Wiley & Sons, 1994), 1–22.

CHAPTER

6 Single-fluid theory of plasmas: magnetohydrodynamics

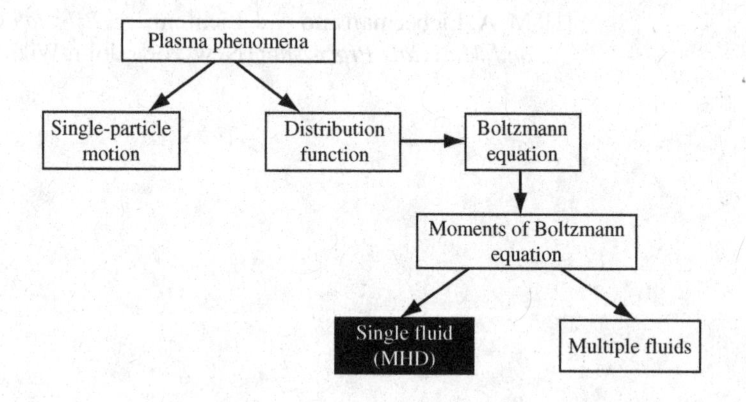

6.1 Introduction

In the previous chapter a multiple-fluid description of a plasma was introduced, in which electrons and various species of ions were governed by separate continuity and force equations. However, under certain conditions it is appropriate to consider the entire plasma population as a single fluid without differentiating between ions or even between ions and electrons. This approach, known as *magnetohydrodynamics* (abbreviated MHD), is in general a method for modeling highly conductive fluids (of which salt water and mercury are additional examples). As will be shown, the single-fluid approach to plasmas is appropriate for dealing with slowly varying conditions when the plasma is highly ionized (i.e., neutrals do not play a role) and electrons and ions are forced to act in unison, either because of frequent collisions or by the action of a strong external magnetic field. The MHD description of a plasma is very useful for describing a wide range of plasma behavior and illustrates important results not easily obtained from multiple-fluid models. Certain plasma phenomena, notably in fusion and space physics, are modeled exclusively using a single-fluid treatment, making MHD one of the most important subfields in plasma physics.

6.2 Single-fluid equations for a fully ionized plasma

We begin by showing how the multiple-fluid equations presented in Chapter 5 can be combined into a set of equations for a single meta-fluid. Assuming for simplicity a two-species plasma of electrons and a single type of ions, the corresponding electron and ion equations are

$$\frac{\partial N_e}{\partial t} + \nabla \cdot [N_e \mathbf{u}_e] = 0 \tag{6.1a}$$

$$m_e N_e \left[\frac{\partial \mathbf{u}_e}{\partial t} + (\mathbf{u}_e \cdot \nabla)\mathbf{u}_e \right] = -\nabla \cdot \Psi_e + q_e N_e (\mathbf{E} + \mathbf{u}_e \times \mathbf{B}) + \mathbf{S}_{\text{ei}} \tag{6.1b}$$

and

$$\frac{\partial N_i}{\partial t} + \nabla \cdot [N_i \mathbf{u}_i] = 0 \tag{6.2a}$$

$$m_i N_i \left[\frac{\partial \mathbf{u}_i}{\partial t} + (\mathbf{u}_i \cdot \nabla)\mathbf{u}_i \right] = -\nabla \cdot \Psi_i + q_i N_i (\mathbf{E} + \mathbf{u}_i \times \mathbf{B}) + \mathbf{S}_{\text{ie}}, \tag{6.2b}$$

where the subscripts e and i indicate, respectively, the quantities associated with electrons and with ions. The collision term $\mathbf{S}_{\text{ei}}$ ($\mathbf{S}_{\text{ie}}$) represents the total momentum transferred to the electrons (ions) per unit volume per unit time by collisional interactions with the ions (electrons). For a fully ionized two-species plasma with no neutral gas, the total momentum must be conserved in collisional interactions between the two species, i.e.,

$$\mathbf{S}_{\text{ei}} = -\mathbf{S}_{\text{ie}}. \tag{6.3}$$

In view of the substantial difference between the electron and ion masses, we expect that the characteristic time scales of the phenomena described by (6.1) and (6.2) will be substantially different. The characteristic frequencies of the plasma, such as plasma frequency or cyclotron frequency, are much larger for electrons than for ions. In general, the separate set of equations (6.1) and (6.2) must be solved simultaneously in order to properly account for the coupling between ion and electron motions. Such coupling often leads to non-linear effects, which we will discuss briefly in later chapters.

When we deal with plasma phenomena that are large-scale ($L \gg \lambda_D$) and have relatively low frequencies ($\omega \ll \omega_{pe}$ and $\omega \ll \omega_{ce}$), the plasma is on the average electrically neutral ($N_i \simeq N_e$) and the movement of electrons independently of ions is not important. In such cases, if is useful to treat the plasma as a single conducting fluid whose inertia is provided primarily by the mass of the ions. In this regime, the dynamics of the plasma is no different than any other conductive fluid, for example, liquid mercury. The macroscopic variables of the electrons and the ions are lumped together to obtain macroscopic parameters describing the ionized gas as a whole, instead of two separate constituents. The governing equations for the new set of macroscopic parameters can be obtained by combining the constituent equations (6.1) and (6.2). We first define appropriate macroscopic parameters for the plasma fluid:

Mass density
$$\rho_m \equiv N_e m_e + N_i m_i \tag{6.4a}$$

Electric current
$$\mathbf{J} \equiv N_e q_e \mathbf{u}_e + N_i q_i \mathbf{u}_i \tag{6.4b}$$

Mass velocity
$$\mathbf{u}_m \equiv \frac{N_e m_e \mathbf{u}_e + N_i m_i \mathbf{u}_i}{N_e m_e + N_i m_i} \tag{6.4c}$$

Total pressure tensor
$$\Psi \equiv \Psi_e + \Psi_i. \tag{6.4d}$$

Note that if the plasma were only partially ionized, (6.4a), (6.4c), and (6.4d) should also contain terms corresponding to neutral particles. Also, in such a case (6.3) would no longer be true. Retaining the assumption of full ionization, however, we now proceed to obtain the macroscopic equations for the single-fluid plasma. This exercise will be nothing more than making linear combinations of (6.1) and (6.2). To make the analysis tractable we will first neglect the $(\mathbf{u} \cdot \nabla)\mathbf{u}$ term in (6.1b) and (6.2b). This can be justified on the grounds that we are primarily dealing with small perturbations for which the fluid velocity ($\mathbf{u}$) multiplied by any gradients (the $\mathbf{u} \cdot \nabla$ operator) is assumed to be negligible. Along the same lines, we also distinguish between values that are perturbed and those that are assumed to remain relatively constant, denoting the latter with an additional "0" subscript. This simplification results in the following momentum equations for electrons and ions:

$$m_e N_{e0} \frac{\partial \mathbf{u}_e}{\partial t} \simeq -\nabla \cdot \Psi_e + q_e N_{e0}(\mathbf{E} + \mathbf{u}_e \times \mathbf{B}) + \mathbf{S}_{\mathrm{ei}} \qquad (6.5a)$$

$$m_i N_{i0} \frac{\partial \mathbf{u}_i}{\partial t} \simeq -\nabla \cdot \Psi_i + q_i N_{i0}(\mathbf{E} + \mathbf{u}_i \times \mathbf{B}) + \mathbf{S}_{\mathrm{ie}}. \qquad (6.5b)$$

6.2.1 Equations of mass and charge conservation

To obtain the equation of continuity of electric current (i.e., the equation of charge conservation), we multiply (6.1a) and (6.2a) respectively by q_e and q_i, and add. We find

$$\boxed{\frac{\partial \rho}{\partial t} + \nabla \cdot \mathbf{J} = 0}, \qquad (6.6)$$

where $\mathbf{J}$ is the electric current density, given by

$$\mathbf{J} = N_{e0} q_e \mathbf{u}_e + N_{i0} q_i \mathbf{u}_i, \qquad (6.7)$$

while ρ is the electric charge density,

$$\rho = N_e q_e + N_i q_i. \qquad (6.8)$$

By multiplying (6.1a) and (6.2a) respectively by m_e and m_i, we find the equation of mass conservation, namely

$$\boxed{\frac{\partial \rho_{\mathrm{m}}}{\partial t} + \rho_{\mathrm{m0}} \nabla \cdot \mathbf{u}_{\mathrm{m}} = 0}, \qquad (6.9)$$

where

$$\rho_{\mathrm{m0}} = N_{e0} m_e + N_{i0} m_i \qquad (6.10)$$

is the single-fluid mass density and $\mathbf{u}_{\mathrm{m}}$ is the linearized fluid mass velocity

$$\mathbf{u}_{\mathrm{m}} = \frac{N_e m_e \mathbf{u}_e + N_i m_i \mathbf{u}_i}{N_{e0} m_e + N_{i0} m_i}. \qquad (6.11)$$

6.2.2 Equation of motion

The equation of motion for the bulk plasma gas can be obtained by adding the individual momentum transport equations (6.1b) and (6.2b). We find

$$(N_{e0} m_e + N_{i0} m_i) \frac{\partial \mathbf{u}_{\mathrm{m}}}{\partial t}$$

$$= -\nabla \cdot (\Psi_e + \Psi_i) + (N_{e0} q_e + N_{i0} q_i)\, \mathbf{E} + \mathbf{J} \times \mathbf{B}_0. \qquad (6.12)$$

Note that we have taken advantage (6.3). The second term on the right is proportional to the ambient value of the electric charge density, $\rho_0 = N_{e0}q_e + N_{i0}q_i$, which is zero since the plasma is macroscopically neutral. We thus arrive at the equation of motion

$$\boxed{\underbrace{(N_{e0}m_e + N_{i0}m_i)}_{\rho_{\rm m0}} \frac{\partial \mathbf{u}_{\rm m}}{\partial t} = -\nabla \cdot (\Psi_e + \Psi_i) + \mathbf{J} \times \mathbf{B}_0}, \qquad (6.13)$$

which is identical to the general equation of motion for an arbitrary conducting fluid, such as mercury, if we substitute for the mass density $\rho_{\rm m} = N_{e0}m_e + N_{i0}m_i$ and the total pressure tensor $\Psi = \Psi_e + \Psi_i$. Note that for an isotropic plasma we have $\nabla \cdot \Psi = \nabla p$, so that the first term on the right in (4.30) is the negative gradient of the total pressure $p = p_e + p_i$.

6.2.3 Generalized Ohm's law

The final single-fluid equation describes the variation of current density $\mathbf{J}$, and is obtained by multiplying Equations (6.1b) and (6.2b) respectively by q_e/m_e and q_i/m_i, and adding them together. Since q_e and q_i are opposite in sign, this amounts to the difference between the two momentum equations:

$$\frac{\partial \mathbf{J}}{\partial t} = -\frac{q_e}{m_e}\nabla \cdot \Psi_e - \frac{q_i}{m_i}\nabla \cdot \Psi_i + \left(\frac{N_{e0}q_e^2}{m_e} + \frac{N_{i0}q_i^2}{m_i}\right)\mathbf{E}$$

$$+ \left(\frac{N_{e0}q_e^2}{m_e}\mathbf{u}_e + \frac{N_{i0}q_i^2}{m_i}\mathbf{u}_i\right) \times \mathbf{B}_0 + \frac{q_e}{m_e}\mathbf{S}_{\rm ei} + \frac{q_i}{m_i}\mathbf{S}_{\rm ie}. \qquad (6.14)$$

Noting that for an electrically neutral plasma we have $|q_e N_{e0}| \simeq |q_i N_{i0}|$, we can use (6.7) and (6.11) to rewrite (6.14):[1]

$$\frac{\partial \mathbf{J}}{\partial t} = -\frac{q_e}{m_e}\nabla \cdot \Psi_e - \frac{q_i}{m_i}\nabla \cdot \Psi_i + \left(\frac{N_{e0}q_e^2}{m_e} + \frac{N_{i0}q_i^2}{m_i}\right)(\mathbf{E} + \mathbf{u}_{\rm m} \times \mathbf{B}_0)$$

$$+ \left(\frac{q_e}{m_e} + \frac{q_i}{m_i}\right)(\mathbf{J} \times \mathbf{B}_0) + \left(\frac{q_e}{m_e} - \frac{q_i}{m_i}\right)\mathbf{S}_{\rm ei}. \qquad (6.15)$$

[1] Using $\rho_{\rm m0} = N_{e0}m_e + N_{i0}m_i$ and $\rho_0 = N_{e0}q_e + N_{i0}q_i$, we can write

$$N_{e0} = \frac{m_i\rho_0 - q_i\rho_{\rm m0}}{q_e m_i - q_i m_e} \quad \text{and} \quad N_{i0} = \frac{-m_e\rho_0 + q_e\rho_{\rm m0}}{q_e m_i - q_i m_e},$$

We now make further approximations, noting that the ion mass m_i is much larger than the electron mass m_e, so $(q_e/m_e) \gg (q_i/m_i)$, and that $\left(N_{e0}q_e^2/m_e\right) \gg \left(N_{i0}q_i^2/m_i\right)$. We can also assume that the plasma is near thermal equilibrium, so that the kinetic pressures of electrons and of ions are of similar magnitude (i.e., $\Psi_e \simeq \Psi_i$), so that $(q_e/m_e)\Psi_e \gg (q_i/m_i)\Psi_i$. With these assumptions, (6.15) reduces to

$$\frac{\partial \mathbf{J}}{\partial t} = -\frac{q_e}{m_e}\nabla \cdot \Psi_e + \frac{N_{e0}q_e^2}{m_e}(\mathbf{E} + \mathbf{u}_{\mathrm{m}} \times \mathbf{B}_0) + \frac{q_e}{m_e}(\mathbf{J} \times \mathbf{B}_0) + \frac{q_e}{m_e}\mathbf{S}_{\mathrm{ei}}. \tag{6.16}$$

We are now forced to deal with the collision term $\mathbf{S}_{\mathrm{ei}}$, which has so far not been related to macroscopic parameters. Solely on physical grounds, it is reasonable to expect that the total momentum transferred to the electrons per unit volume per unit time (as a result of collisions with ions) is proportional to the relative average velocity difference between the species $(\mathbf{u}_i - \mathbf{u}_e)$. Since the collisions in a fully ionized plasma are Coulomb collisions, we expect $\mathbf{S}_{\mathrm{ei}}$ to be proportional to the Coulomb force or $q_e q_i = q^2$, as well as being proportional to the density of electrons N_{e0} and the density of scattering centers $N_{i0} = N_{e0}$. Thus we can write

from which it follows that

$$\frac{N_{e0}q_e^2}{m_e} + \frac{N_{i0}q_i^2}{m_i} = \left(\frac{q_e}{m_e} + \frac{q_i}{m_i}\right)\rho_0 - \frac{q_e q_i}{m_e m_i}\rho_{\mathrm{m}0},$$

which for a macroscopically neutral plasma ($\rho_0 \simeq 0$) reduces to

$$\frac{N_{e0}q_e^2}{m_e} + \frac{N_{i0}q_i^2}{m_i} \simeq -\frac{q_e q_i}{m_e m_i}\rho_{\mathrm{m}0}.$$

Using a similar procedure, the simultaneous solution of (6.4b) and (6.4c) for $\mathbf{u}_e$ and $\mathbf{u}_i$ results in

$$\mathbf{u}_e = \frac{\rho_{\mathrm{m}0}}{m_e m_i N_{e0}\left(\dfrac{q_e}{m_e} - \dfrac{q_i}{m_i}\right)}\left[\frac{m_i}{\rho_0}\mathbf{J} - q_i\,\mathbf{u}_{\mathrm{m}}\right]$$

$$\mathbf{u}_i = \frac{\rho_{\mathrm{m}0}}{m_e m_i N_{e0}\left(\dfrac{q_e}{m_e} - \dfrac{q_i}{m_i}\right)}\left[-\frac{m_e}{\rho_0}\mathbf{J} + q_e\,\mathbf{u}_{\mathrm{m}}\right].$$

Using the above equations, we can write

$$\frac{N_{e0}q_e^2}{m_e}\mathbf{u}_e + \frac{N_{i0}q_i^2}{m_i}\mathbf{u}_i = \left(\frac{q_e}{m_e} + \frac{q_i}{m_i}\right)\mathbf{J} - \frac{q_e q_i}{m_e m_i}\rho_{\mathrm{m}0}\,\mathbf{u}_{\mathrm{m}}.$$

Using these relationships in (6.14) leads to (6.15).

$$\mathbf{S}_{\text{ei}} = \eta q^2 N_{e0}^2(\mathbf{u}_i - \mathbf{u}_e), \tag{6.17}$$

where η is a constant of proportionality known as the *specific resistivity* of the plasma. It will become clear below that η has dimensions of resistivity (i.e., Ω m or V m A^{-1}). It is often useful to express resistivity in terms of a collision frequency ν_{ei}:

$$\eta = \frac{m_e \nu_{\text{ei}}}{N_{e0} q^2},$$

where ν_{ei} is the so-called collision frequency for momentum transfer from the ions to the electrons. Note that ν_{ei} is not necessarily the actual rate at which collisions between particles occur; its function is in fact more as a "fudge factor," accounting for those aspects of the collision process that we do not explicitly account for, such as collision cross-sections, effectiveness of momentum transfer, etc. Noting that $q_i = -q_e$ and $N_{e0} = N_{i0}$, we can use (6.7) to rewrite (6.17) as

$$\mathbf{J} = N_{e0} q_e (\mathbf{u}_e - \mathbf{u}_i) \quad \rightarrow \quad \mathbf{S}_{\text{ei}} = -N_{e0} q_e \eta \mathbf{J}. \tag{6.18}$$

More generally, as we will show later, a magnetized plasma is anisotropic, so that the resistivity is a tensor $\overleftrightarrow{\eta}$ and we can write

$$\mathbf{S}_{\text{ei}} = -N_{e0} q_e \overleftrightarrow{\eta} \cdot \mathbf{J}. \tag{6.19}$$

Substituting in (6.16), we arrive at the so-called *generalized Ohm's law*:

$$\boxed{\frac{\partial \mathbf{J}}{\partial t} = -\frac{q_e}{m_e}\nabla \cdot \Psi_e + \frac{N_{e0} q_e^2}{m_e}(\mathbf{E} + \mathbf{u}_{\text{m}} \times \mathbf{B}_0) + \frac{q_e}{m_e}(\mathbf{J} \times \mathbf{B}_0) - \frac{N_{e0} q_e^2}{m_e}\overleftrightarrow{\eta} \cdot \mathbf{J}}. \tag{6.20}$$

For a steady current in a uniform plasma with no static magnetic field, we have $\partial \mathbf{J}/\partial t = 0$, $\nabla \cdot \Psi = 0$, and $\mathbf{B}_0 = 0$, and the resistivity tensor reduces to a scalar so that (6.19) becomes

$$\mathbf{E} = \eta \mathbf{J} \quad \rightarrow \quad \mathbf{J} = \frac{1}{\eta}\mathbf{E},$$

underscoring our use of "generalized Ohm's law" for (6.20) and "resistivity" for η. The electric field $\mathbf{E}$ can be found explicitly from (6.20):

$$\mathbf{E} = \underbrace{-\mathbf{u}_{\text{m}} \times \mathbf{B}_0}_{\text{motional } \mathbf{E}} - \underbrace{\frac{\mathbf{J} \times \mathbf{B}_0}{N_{e0} q_e}}_{\text{Hall effect}} + \underbrace{\frac{\nabla \cdot \Psi_e}{N_{e0} q_e}}_{\text{ambipolar}} + \underbrace{\overleftrightarrow{\eta} \cdot \mathbf{J}}_{\text{ohmic loss}} + \underbrace{\frac{m_e}{N_{e0} q_e}\frac{\partial \mathbf{J}}{\partial t}}_{\text{electron inertia}}. \tag{6.21}$$

On the right-hand side, the first term is the motional electric field term, associated with the frame of reference of the fluid. In a frame of reference moving with the fluid at the speed $\mathbf{u}_{\mathrm{m}}$, the electric field does not include this term and is given by $\mathbf{E}' = \mathbf{E} + \mathbf{u}_{\mathrm{m}} \times \mathbf{B}_0$. The second term is the so-called Hall effect term and is often neglected. The third term is the so-called ambipolar polarization term and describes the electric field that arises from gradients in plasma density along the magnetic field. The fourth term is the ohmic loss (or Joule heating) term, and the last term can be interpreted as the contribution of electron inertia to the current flow. Also note from (6.20) that for a collisionless plasma the resistivity is zero and the conductivity is infinite, and the plasma fluid behaves much like mercury or other liquid metal. Equations (6.6), (6.9), (6.13), and (6.20) are the fundamental equations for the treatment of a plasma as a single conducting fluid. They form the basis of magnetohydrodynamics and the related concepts of "frozen-in" magnetic field lines and magnetic pressure (see Sections 6.4 and 6.5).

6.3 Magnetohydrodynamics plasma model

Magnetohydrodynamics is a powerful tool for representating the dynamics of many different types of plasma phenomena, ranging from the generation and evolution of magnetic fields within stellar interiors to the magnetic confinement of thermonuclear plasmas. MHD is particularly useful in modeling large-scale, relatively low-frequency plasma phenomena for which the plasma can be treated as a single conducting fluid. At a more general level, MHD is a framework for the description of the dynamics of an electrically conducting fluid (e.g., mercury) in the presence of a magnetic field, either externally applied or produced by a current flowing in the fluid. As noted above, the fundamental equations of MHD are (6.6), (6.9), (6.13), and (6.20), coupled with Maxwell's equations.

A single-fluid description of a plasma is only valid if the plasma is collision-dominated, since only then are the particles restricted in their motions sticking together so that their motion may be represented by that of the local center of mass, superimposed on an isotropic distribution of velocities. If the plasma is not collision-dominated, particles nearby in space at any given time may have completely different velocity vectors and fly away in different directions to remote parts of the plasma, in which case the motion of the local center of mass does not correspond to any real mass motion. While this is generally the case, collisionless plasmas under the influence of a very strong magnetic field can also be described in the context of a single-fluid model. This is because if the magnetic field is sufficiently strong, and its variation in space and time is slow, we know from Chapter 2 that the particles are forced into nearly circular orbits around a drifting guiding center. The main motion of the guiding center (due to electric-field-driven drift) is common to all particles, and can be viewed as the local mass motion, while the gyrating motion of particles is similar to an isotropic velocity distribution. In other words, the resultant effect of the strong magnetic field is similar to that of collisions. Although we can thus accept that there might be "thermal equilibrium" in the transverse direction, the motions in the parallel direction remain far from equilibrium, resulting in an anisotropic pressure tensor. At first we will exclusively consider application of MHD to collision-dominated plasmas. Many of the concepts to be developed can be extended to collisionless plasmas with a strong magnetic field, as we will see in the final sections of the chapter.

In a collision-dominated plasma, the particles species are in a Maxwellian state. By adopting an MHD description, we are implicitly interested in slow variations, so that thermal equilibrium can be established over time scales of interest. Furthermore, the spatial scales of interest are assumed to be much larger than the Debye length, so that the plasma can be assumed to be macroscopically neutral.

6.4 Simplified MHD equations

The single-fluid equations derived above here are reproduced (Equations (6.9), (6.13), (6.20)):

$$\frac{\partial \rho_{\mathrm{m}}}{\partial t} + \rho_{\mathrm{m}0} \nabla \cdot \mathbf{u}_{\mathrm{m}} = 0$$

$$\rho_{\mathrm{m}} \frac{\partial \mathbf{u}_{\mathrm{m}}}{\partial t} = -\nabla \cdot \Psi + \mathbf{J} \times \mathbf{B}$$

$$\frac{\partial \mathbf{J}}{\partial t} = -\frac{q_e}{m_e} \nabla \cdot \Psi_e + \frac{N_{e0} q_e^2}{m_e} (\mathbf{E} + \mathbf{u}_{\mathrm{m}} \times \mathbf{B})$$

$$+ \frac{q_e}{m_e} (\mathbf{J} \times \mathbf{B}) - \frac{N_{e0} q_e^2}{m_e} \overleftrightarrow{\eta} \cdot \mathbf{J}.$$

At this point, we make several simplifying approximations. First, we consider only isotropic fluids, so that $\nabla \cdot \Psi = \nabla p$, where p is the scalar pressure, and the plasma resistivity tensor $\overleftrightarrow{\eta}$ reduces to the scalar resistivity η. Second, we neglect the Hall effect and the ambipolar polarization terms in the generalized Ohm's law, since these are only important under special circumstances. We also neglect the $\partial\mathbf{J}/\partial t$ term in (6.20), since for low frequencies the time variations are small (note, however, that we are retaining the time-varying terms in other MHD equations). The generalized Ohm's law then reduces to

$$0 = \frac{N_{e0}q_e^2}{m_e}(\mathbf{E} + \mathbf{u}_\mathrm{m} \times \mathbf{B}) - \frac{N_{e0}q_e^2}{m_e}\eta\mathbf{J} \quad \rightarrow \quad \mathbf{J} = \sigma(\mathbf{E} + \mathbf{u}_\mathrm{m} \times \mathbf{B}), \tag{6.22}$$

where $\sigma = 1/\eta$. Note that we can write (6.22) as $\mathbf{J} = \sigma\mathbf{E}'$, where $\mathbf{E}' = (\mathbf{E} + \mathbf{u}_\mathrm{m} \times \mathbf{B})$ is the electric field felt by an observer (or a particle) moving with the fluid element. The simplified MHD equations can thus be written as

Simplified MHD equations

$$\mathbf{J} = \sigma(\mathbf{E} + \mathbf{u}_\mathrm{m} \times \mathbf{B}) \tag{6.23a}$$

$$\frac{\partial \rho_\mathrm{m}}{\partial t} + \rho_{\mathrm{m}0}\nabla\cdot\mathbf{u}_\mathrm{m} = 0 \tag{6.23b}$$

$$\rho_\mathrm{m}\frac{\partial \mathbf{u}_\mathrm{m}}{\partial t} = -\nabla p + \mathbf{J}\times\mathbf{B} \tag{6.23c}$$

We consider only highly conducting fluids, since most MHD plasmas resemble fluids with very large (nearly infinite) conductivity. Equations (6.23) are complemented by Maxwell's equations, governing the electromagnetic fields. Since we are considering fluids of high conductivity, we neglect displacement currents (compared to conduction currents) and also assume that there is no accumulation of space charge inside the fluid (i.e., $\rho = 0$). Maxwell's equations then reduce to

Reduced Maxwell's equations

$$\nabla\times\mathbf{B} = \mu_0\mathbf{J} \tag{6.24a}$$

$$\nabla\times\mathbf{E} = -\frac{\partial\mathbf{B}}{\partial t} \tag{6.24b}$$

$$\nabla\cdot\mathbf{E} = 0 \tag{6.24c}$$

$$\nabla\cdot\mathbf{B} = 0. \tag{6.24d}$$

The description of the fluid behavior is still not complete until we specify the relationship between p and density N. This relationship is the equation of state; we adopt the adiabatic assumption given in (4.35), which can also be expressed as

$$\frac{d}{dt}\left(p\,\rho_\mathrm{m}^{-\gamma}\right) = 0. \tag{6.25}$$

Equations (6.22) through (6.25) form a closed system which can be solved for any of the fluid or electromagnetic variables. It is common to eliminate the electric field and reduce the number of variables to four, namely ρ_m, $\mathbf{u}_m$, p, and $\mathbf{B}$. Using (6.24a) together with (6.23c) we can write

$$\rho_m \frac{\partial \mathbf{u}_m}{\partial t} = -\nabla p + \frac{1}{\mu_0}(\nabla \times \mathbf{B}) \times \mathbf{B} = -\nabla\left(p + \frac{B^2}{2\mu_0}\right) + \frac{(\mathbf{B}\cdot\nabla)\mathbf{B}}{\mu_0}, \tag{6.26}$$

where we have used the identity

$$(\nabla \times \mathbf{B}) \times \mathbf{B} = -\mathbf{B} \times (\nabla \times \mathbf{B}) \equiv -(\mathbf{B}\cdot\mathbf{B})\nabla + (\mathbf{B}\cdot\nabla)\mathbf{B}.$$

Some further manipulation[2] is needed to prove the equality

$$\frac{1}{\mu_0}(\nabla \times \mathbf{B}) \times \mathbf{B} = -\nabla\left(\frac{B^2}{2\mu_0}\right) + \frac{(\mathbf{B}\cdot\nabla)\mathbf{B}}{\mu_0}$$

$$\text{or } (\nabla \times \mathbf{B}) \times \mathbf{B} = -\tfrac{1}{2}\nabla B^2 + (\mathbf{B}\cdot\nabla)\mathbf{B}.$$

[2] We first note the component form of the second term on the right,

$$(\mathbf{B}\cdot\nabla)\mathbf{B} = \left[B_x\frac{\partial}{\partial x} + B_y\frac{\partial}{\partial y} + B_z\frac{\partial}{\partial z}\right]\mathbf{B} = \hat{\mathbf{x}}\left[B_x\frac{\partial B_x}{\partial x} + B_y\frac{\partial B_x}{\partial y} + B_z\frac{\partial B_x}{\partial z}\right] + \hat{\mathbf{y}}[\cdots] + \hat{\mathbf{z}}[\cdots],$$

while the component form of the ∇B^2 term on the right is

$$\nabla B^2 = \nabla\left(B_x^2 + B_y^2 + B_z^2\right) = \hat{\mathbf{x}}\left[\frac{\partial B_x^2}{\partial x} + \frac{\partial B_y^2}{\partial x} + \frac{\partial B_z^2}{\partial x}\right] + \hat{\mathbf{y}}[\cdots] + \hat{\mathbf{z}}[\cdots].$$

We now examine the component form of the left-hand side:

$$(\nabla \times \mathbf{B}) \times \mathbf{B} = \begin{vmatrix} \hat{\mathbf{x}} & \hat{\mathbf{y}} & \hat{\mathbf{z}} \\ \frac{\partial B_z}{\partial y} - \frac{\partial B_y}{\partial z} & \frac{\partial B_x}{\partial z} - \frac{\partial B_z}{\partial x} & \frac{\partial B_y}{\partial x} - \frac{\partial B_x}{\partial y} \\ B_x & B_y & B_z \end{vmatrix}$$

$$= \hat{\mathbf{x}}\left[\underbrace{B_x\frac{\partial B_x}{\partial x} + B_y\frac{\partial B_x}{\partial y} + B_z\frac{\partial B_x}{\partial z}}_{[(\mathbf{B}\cdot\nabla)\mathbf{B}]_x} - B_x\frac{\partial B_x}{\partial x} - B_y\frac{\partial B_y}{\partial x} - B_z\frac{\partial B_z}{\partial x}\right] + \cdots$$

$$= \hat{\mathbf{x}}\left\{[(\mathbf{B}\cdot\nabla)\mathbf{B}]_x - \frac{1}{2}\frac{\partial B_x^2}{\partial x} - \frac{1}{2}\frac{\partial B_y^2}{\partial x} - \frac{1}{2}\frac{\partial B_z^2}{\partial x}\right\} + \hat{\mathbf{y}}\{\cdots\} + \hat{\mathbf{z}}\{\cdots\}$$

$$(\nabla \times \mathbf{B}) \times \mathbf{B} = \hat{\mathbf{x}}\left\{-\frac{1}{2}\left[\nabla B^2\right]_x + [(\mathbf{B}\cdot\nabla)\mathbf{B}]_x\right\} + \hat{\mathbf{y}}\{\cdots\} + \hat{\mathbf{z}}\{\cdots\} \qquad \text{Q.E.D.}$$

Substituting (6.22) in (6.24b) and using (6.24a) and (6.24d), we find

$$\frac{\partial \mathbf{B}}{\partial t} = \nabla \times \left(\mathbf{u}_{\mathrm{m}} \times \mathbf{B} - \frac{\mathbf{J}}{\sigma} \right) = \underbrace{\nabla \times (\mathbf{u}_{\mathrm{m}} \times \mathbf{B})}_{\text{flow}} + \underbrace{\frac{\nabla^2 \mathbf{B}}{\mu_0 \sigma}}_{\text{diffusion}}, \tag{6.27}$$

where we have used

$$\nabla \times (\nabla \times \mathbf{B}) \equiv \nabla(\nabla \cdot \mathbf{B}) - \nabla^2 \mathbf{B},$$

with $\nabla \cdot \mathbf{B} = 0$ from (6.24d). Equations (6.23b), (6.24d), (6.26), and (6.27) are sufficient to solve for the four variables ρ_{m}, $\mathbf{u}_{\mathrm{m}}$, p, and $\mathbf{B}$.

Equation (6.27) expresses the sensitivity of the magnetic field to plasma fluid motion (flow term) and plasma conductivity (diffusion term). For a perfectly conducting fluid ($\sigma = \infty$) (6.27) can be written as

$$\frac{\partial \mathbf{B}}{\partial t} = \nabla \times (\mathbf{u}_{\mathrm{m}} \times \mathbf{B}). \tag{6.28}$$

Note that for $\sigma = \infty$ (6.22) reduces to

$$\mathbf{E}' = \mathbf{E} + \mathbf{u}_{\mathrm{m}} \times \mathbf{B} = 0, \tag{6.29}$$

as expected since the electric field within a perfect conductor must be zero. The validity of (6.28) is based on the fact that we can neglect the second term in the parentheses in (6.27). This term is negligible if

$$\left| \frac{\mathbf{J}}{\sigma} \right| \ll |\mathbf{u}_{\mathrm{m}} \times \mathbf{B}|.$$

With displacement currents neglected, (6.24a) is valid and we can write

$$|\mathbf{B}| \simeq \mu_0 L |\mathbf{J}|,$$

where L is a characteristic dimension of the system. Thus, the inequality above is equivalent to

$$R_{\mathrm{M}} = \sigma |\mathbf{u}_{\mathrm{m}}| \mu_0 L \gg 1,$$

where R_{M} is a quantity often referred to as the *magnetic Reynolds number* by analogy with the hydrodynamic Reynolds number, which also scales with L and $\mathbf{u}_{\mathrm{m}}$. The Reynolds number is very useful in determining whether a system is diffusion- or flow-dominated. When $R_{\mathrm{M}} \gg 1$, flow dominates and the magnetic field simply moves with the flow, as shown in the next section.

6.4.1 Frozen-in magnetic flux lines

An interesting application of MHD equations in general and of (6.28) in particular is the demonstration that the lines of magnetic flux are frozen into a perfectly conducting fluid and are thus transported with it. Note that Equation (6.28) describes the coupling between the magnetic field and the fluid motion for a perfectly conducting fluid. Consider an arbitrary surface S enclosed by a contour C, as shown in Figure 6.1. We can write the integral of (6.28) over this arbitrary surface as

$$\frac{\partial}{\partial t}\int_S \mathbf{B}\cdot d\mathbf{s} = \int_S [\nabla\times(\mathbf{u}_{\mathrm{m}}\times\mathbf{B})]\cdot d\mathbf{s}$$

$$\frac{\partial}{\partial t}\int_S \mathbf{B}\cdot d\mathbf{s} = \oint_C (\mathbf{u}_{\mathrm{m}}\times\mathbf{B})\cdot d\mathbf{l}$$

$$\frac{\partial}{\partial t}\int_S \mathbf{B}\cdot d\mathbf{s} - \oint_C (\mathbf{u}_{\mathrm{m}}\times\mathbf{B})\cdot d\mathbf{l} = 0, \qquad (6.30)$$

where we have used Stokes's theorem. Equation (6.30) can be written as

$$\frac{\partial \psi_{\mathrm{m}}}{\partial t} + \oint_C \mathbf{B}\cdot(\mathbf{u}_{\mathrm{m}}\times d\mathbf{l}) = 0 \qquad (6.31a)$$

$$\frac{d}{dt}\left(\int_S \mathbf{B}\cdot d\mathbf{s}\right) = 0, \qquad (6.31b)$$

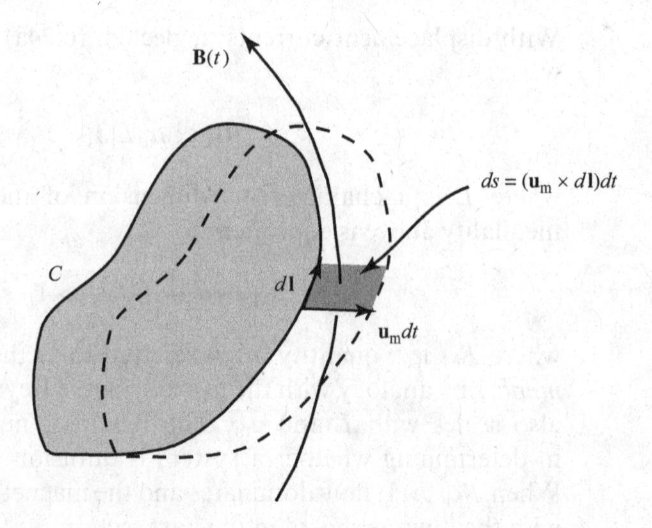

Figure 6.1 Illustration of the "frozen-in" concept. A highly conducting fluid in motion acts to transport magnetic field lines.

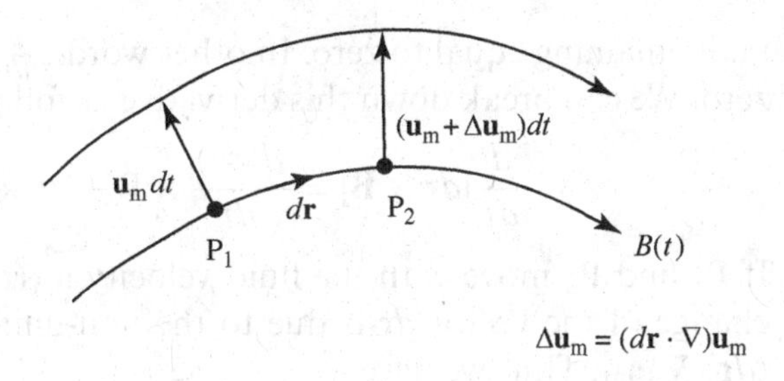

Figure 6.2 In a MHD plasma with infinite conductivity, two points on the same field line remain on the same field line even though they move with the fluid.

where ψ_m is the magnetic flux.[3] The first term in (6.31a) represents the rate of change of flux through the fixed surface S, while the second is the additional increment of flux swept out per unit time by the movement of the periphery with the local fluid velocity $\mathbf{u}_m$. The left-hand side of (6.31a) is thus the total rate of change of flux through a "material" surface fixed to and moving with the fluid, as restated in (6.31b). Hence (6.31) is a statement of the constancy of magnetic flux through any material surface in a perfectly conducting fluid. Another way of stating this is to say that the field line is attached to or "frozen in" the fluid and moves with it. We can also arrive at the same result by considering two points P_1 and P_2 on a field line and showing that they remain on the same field line as they move with the fluid. Using the identity

$$\nabla \times (\mathbf{u}_m \times \mathbf{B}) \equiv \mathbf{u}_m(\nabla \cdot \mathbf{B}) + (\mathbf{B} \cdot \nabla)\mathbf{u}_m - \mathbf{B}(\nabla \cdot \mathbf{u}_m) - (\mathbf{u}_m \cdot \nabla)\mathbf{B},$$

taking advantage of our earlier definition of the convective derivative (3.31), and noting (6.24c), we can write (6.28) as

$$\underbrace{\frac{d\mathbf{B}}{dt} - (\mathbf{u}_m \cdot \nabla)\mathbf{B}}_{\partial\mathbf{B}/\partial t} = 0 + (\mathbf{B} \cdot \nabla)\mathbf{u}_m - \mathbf{B}\,(\nabla \cdot \mathbf{u}_m) - (\mathbf{u}_m \cdot \nabla)\mathbf{B}$$

$$\frac{d\mathbf{B}}{dt} = (\mathbf{B} \cdot \nabla)\mathbf{u}_m - \mathbf{B}\,(\nabla \cdot \mathbf{u}_m). \qquad (6.32)$$

Let P_1 and P_2 be any two neighboring points on the same magnetic field line, with the distance between them represented by the vector $d\mathbf{r}$, as shown in Figure 6.2. If the points are on the same magnetic field line, then by definition $d\mathbf{r} \times \mathbf{B} = 0$ since $d\mathbf{r}$ and $\mathbf{B}$ are parallel vectors. So P_1 and P_2 remaining on the same magnetic field line even as they move with the fluid is the equivalent of $d\mathbf{r} \times \mathbf{B}$ not changing

[3] We have used the identity $\mathbf{B} \cdot (\mathbf{u}_m \times d\mathbf{l}) = -(\mathbf{u}_m \times \mathbf{B}) \cdot d\mathbf{l}$.

and remaining equal to zero. In other words, $\frac{d}{dt}[d\mathbf{r} \times \mathbf{B}]$ must equal zero. We can break down this derivative as follows:

$$\frac{d}{dt}[d\mathbf{r} \times \mathbf{B}] = \frac{d(d\mathbf{r})}{dt} \times \mathbf{B} + d\mathbf{r} \times \frac{d\mathbf{B}}{dt}. \tag{6.33}$$

If P_1 and P_2 move with the fluid velocity $\mathbf{u}_m(\mathbf{r}, t)$, then the rate of change of the vector $d\mathbf{r}$ is due to the non-uniformity of $\mathbf{u}_m$ and is $(d\mathbf{r} \cdot \nabla)\mathbf{u}_m$. Thus we have

$$\frac{d(d\mathbf{r})}{dt} = (d\mathbf{r} \cdot \nabla)\mathbf{u}_m. \tag{6.34}$$

Using (6.34) and (6.32), Equation (6.33) becomes

$$\frac{d}{dt}[d\mathbf{r} \times \mathbf{B}] = (d\mathbf{r} \cdot \nabla)\mathbf{u}_m \times \mathbf{B} + d\mathbf{r} \times [(\mathbf{B} \cdot \nabla)\mathbf{u}_m - \mathbf{B}(\nabla \cdot \mathbf{u}_m)] \tag{6.35a}$$

$$= (d\mathbf{r} \cdot \nabla)\mathbf{u}_m \times \mathbf{B} + d\mathbf{r} \times (\mathbf{B} \cdot \nabla)\mathbf{u}_m - d\mathbf{r} \times \mathbf{B}(\nabla \cdot \mathbf{u}_m) \tag{6.35b}$$

$$= (\mathbf{B} \cdot \nabla)\mathbf{u}_m \times d\mathbf{r} + d\mathbf{r} \times (\mathbf{B} \cdot \nabla)\mathbf{u}_m \tag{6.35c}$$

$$= 0, \tag{6.35d}$$

where because the last term in (6.35b) is zero $d\mathbf{r} \times \mathbf{B} = 0$. Likewise, the fact that $d\mathbf{r}$ and $\mathbf{B}$ are parallel allows them to be interchanged in the first term in (6.35b); the two terms in (6.35c) then cancel since they are equal and opposite. Thus, if P_1 and P_2 were initially on the same line of force, they will remain on it.

The concept of magnetic lines of force is an abstraction, and in general no identity can be attached to these lines. However, we can very conveniently regard the lines as attached to and convected with the fluid. The "frozen-in" concept implies that all particles and all magnetic flux contained in a certain flux tube at a certain instant stay inside the flux tube at all instants, independent of any motion of the flux tube or any change in the form of its bounding surface. Additional insight into the concept of frozen-in field lines can be found by seeing it as a manifestation of Lenz's law for electromagnetic induction. Lenz's law applies to the classic experiment of electrodynamics in which a conductor is moved across a static magnetic field. The law states that a current is induced in the moving conductor so as to oppose the change in magnetic flux through the conductor. In the limit where the conductor is a continuous plasma with infinite conductivity, the induced currents prohibit any change in magnetic flux, thus transporting the same constant magnetic field lines with the plasma.

Example 6-1 The solar wind
A prominent example of the frozen-in field concept occurs in the interaction of the solar wind with the interplanetary magnetic field (IMF). Both the IMF and the solar wind originate from the Sun. The solar wind is a highly conducting plasma with density $\sim 10^7\,\mathrm{m^{-3}}$ and fluid speed of $u_{sw} = 350\,\mathrm{km\,s^{-1}}$ in an outward radial direction. The IMF is generally in the plane of the Earth's orbit and, because of the Sun's rotation and the frozen-in field concept as driven by the solar wind, its field lines take on a spiral shape as shown in Figure 6.3. Given that the angular velocity of the Sun's rotation is $\omega = 2.7 \times 10^{-6}$ radians $\mathrm{s^{-1}}$, calculate the spiral angle Ψ shown in Figure 6.3.

Solution: The spiral angle is given by

$$\tan(\Psi) = \frac{r\omega}{u_{sw}},$$

where r is the radial distance from the Sun, equal to 149.6×10^9 m.

$$\tan(\Psi) = \frac{(149.6 \times 10^9\,\mathrm{m})(2.7 \times 10^{-6}\,\mathrm{rad\,s^{-1}})}{3.5 \times 10^5\,\mathrm{m\,s^{-1}}}$$

$$\Psi = 49^\circ.$$

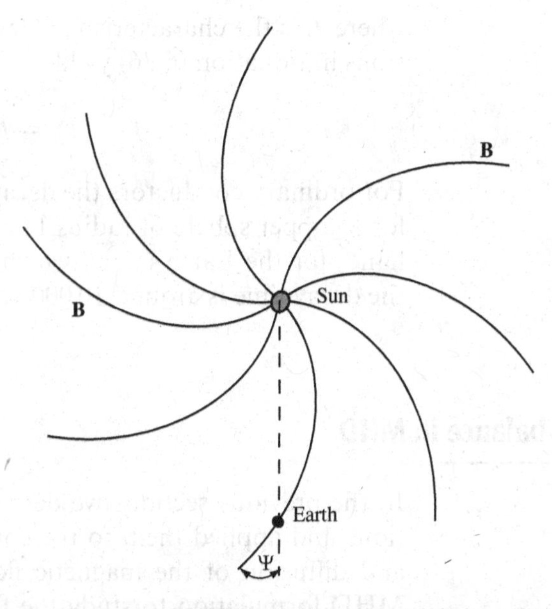

Figure 6.3 Interplanetary magnetic field lines frozen in the solar wind plasma result in a spiral shape.

During periods of solar disturbance, the solar wind speed is known to increase to 600–700 km s^{-1}, which changes the spiral angle sometimes called the "garden hose" angle).

6.4.2 Diffusion of magnetic field lines

When $R_M = \sigma |\mathbf{u}_m| \mu_0 L \ll 1$, the diffusion term in Equation (6.27) dominates and the expression reduces to

$$\frac{\partial \mathbf{B}}{\partial t} = \frac{\nabla^2 \mathbf{B}}{\mu_0 \sigma}. \tag{6.36}$$

Equation (6.36) is an expression for a magnetic field in a stationary but imperfect ($\sigma < \infty$) conductor, in which the magnetic field is shown to decay with time. A characteristic decay time τ_D for the magnetic field can be obtained by approximating the spatial and time derivatives:

$$|\partial \mathbf{B}/\partial t| \approx \frac{B}{\tau_D}$$

$$\left|\nabla^2 \mathbf{B}\right| \approx \frac{B}{L^2},$$

where L is the characteristic length. Substituting the above expressions in Equation (6.36) yields

$$\tau_D = L^2 \mu_0 \sigma.$$

For ordinary conductors the decay time is very small. For example, for a copper sphere of radius 1 m, τ_D is less than 10 s. On the other hand, for the Earth's core, which is considered to be molten iron, the decay time is around 10 000 years.

6.5 Force balance in MHD

In the previous sections we derived the fundamental MHD equations and applied them to the concepts of frozen-in magnetic flux and diffusion of the magnetic field. In this section we apply the MHD formulation to study the forces exerted by a magnetic field in a fluid, and the related concept of magnetic pressure. Before we proceed, we list the MHD equations with the inclusion of collision and gravitational terms of importance for space plasmas:

$$\frac{\partial \rho_{\rm m}}{\partial t} + \rho_{\rm m} \nabla \cdot \mathbf{u}_{\rm m} = 0 \tag{6.37}$$

$$\rho_{\rm m} \frac{\partial \mathbf{u}_{\rm m}}{\partial t} = -\nabla \cdot \Psi + \mathbf{J} \times \mathbf{B} + \rho_{\rm m}\mathbf{g} - \rho_{\rm m} \nu_{\rm eff} \mathbf{u}_{\rm m} \tag{6.38}$$

$$\frac{\partial \mathbf{B}}{\partial t} = \underbrace{\nabla \times (\mathbf{u}_{\rm m} \times \mathbf{B})}_{\text{flow}} + \underbrace{\frac{\nabla^2 \mathbf{B}}{\mu_0 \sigma}}_{\text{diffusion}}, \tag{6.39}$$

where $\mathbf{g}$ is the gravitational acceleration vector and $\nu_{\rm eff}$ is an effective collision frequency representing collisions of charged species (electrons and ions, but mostly electrons) with neutrals, which are assumed to be stationary. In our discussions below, we will neglect the gravitational and collision terms and assume an isotropic fluid, so that (6.38) becomes

$$\rho_{\rm m} \frac{\partial \mathbf{u}_{\rm m}}{\partial t} = -\nabla p + \mathbf{J} \times \mathbf{B}, \tag{6.40}$$

where p is the total pressure. Note that (6.39) was derived from the simplified version of Ohm's law, namely (Equation (6.22))

$$\mathbf{J} = \sigma(\mathbf{E} + \mathbf{u}_{\rm m} \times \mathbf{B}),$$

and Maxwell's equation (6.24a). The approximations we made in arriving at (6.22) were not fully justified, and in fact it is often difficult to quantitatively justify neglecting the $\mathbf{J} \times \mathbf{B}$ and $\partial \mathbf{J}/\partial t$ terms and the isotropic pressure tensor (see [1]). However, consider the comparison between the $-\nabla p$ and $\mathbf{u}_{\rm m} \times \mathbf{B}$ terms in (6.20):

$$\frac{\left|\dfrac{q_e \nabla p}{m_e}\right|}{\left|\dfrac{N_e q_e^2}{m_e}(\mathbf{u}_{\rm m} \times \mathbf{B})\right|} \simeq \frac{\dfrac{q_e}{m_e}\dfrac{N_e}{3} m_e \left\langle v_{\rm th}^2 \right\rangle}{\dfrac{N_e q_e^2}{m_e}|\mathbf{u}_{\rm m}|B} \simeq \frac{v_{\rm th}^2}{3L|\mathbf{u}_{\rm m}|\omega_c} \simeq \frac{\dfrac{v_{\rm th}}{|\mathbf{u}_{\rm m}|}}{\dfrac{L}{r_c}}, \tag{6.41}$$

where we have assumed $\nabla p \simeq pL^{-1}$, with L being the scale height of the system, and used $p = N_e k_{\rm B} T = N_e \frac{1}{3} m_e \left\langle v_{\rm th}^2 \right\rangle$, as well as $r_c = m v_{\rm th}/(q_e B)$. $v_{\rm th}$ is the magnitude of the random thermal motion, previously denoted as $\mathbf{w}$, i.e., $v_{\rm th} = |\mathbf{w}|$. Thus, the pressure term is negligible in Ohm's law when $L \gg r_c$ and $v_{\rm th} \ll |\mathbf{u}_{\rm m}|$ or, more precisely, when $(L/r_c) \gg (v_{\rm th}/|\mathbf{u}_{\rm m}|)$. This condition is well justified in most cases for which an MHD treatment is used, when the scale sizes are very large and the particle thermal motions can be neglected in comparison to the drift motions. For example, the solar wind speed $\mathbf{u}_m$ is of order 300 to 1400 km s^{-1}, while the electron thermal velocity v_{th} is comparable, and obviously $L \gg r_c$. Note that neglecting the $-\nabla p$ term in (6.20) does not mean that it is also dropped in (6.40). In fact, as we will see in the following sections, the $-\nabla p$ term in (6.40) is quite comparable to the $\mathbf{J} \times \mathbf{B}$ term, which it balances under conditions of equilibrium.

6.5.1 Magnetic forces

We now consider the forces that a magnetic field exerts on a fluid in which it is present. It is obvious that magnetic fields can carry forces in a medium; for example, consider the force-density term that appears in (6.40):

$$\mathbf{F}_m = \mathbf{J} \times \mathbf{B}.$$

Since current density and magnetic field are related through (6.24a), knowledge of $\mathbf{J}$ is not necessary for determining the magnetic force $\mathbf{F}_m$ at any given point, as long as we know the field at adjacent points, so that we can determine its curl. Using (6.24a) and (6.26), we can write

$$\mathbf{F}_m = \frac{1}{\mu_0}(\nabla \times \mathbf{B}) \times \mathbf{B} = \frac{1}{\mu_0}(\mathbf{B} \cdot \nabla)\mathbf{B} - \nabla \frac{B^2}{2\mu_0}, \tag{6.42}$$

where we have used the same vector identity as we used in arriving at (6.26). The magnetic force $\mathbf{F}_m$ is the fundamental force which sets a conducting fluid in motion (by accelerating it, as described by (6.40)) when it is in the presence of a magnetic field. This force, acting via (6.40), and Equation (6.39), which describes how the motion of the fluid in turn causes the field to evolve in time, describe the manner in which most large-scale plasma dynamics occur. To better understand the physical nature of the magnetic force $\mathbf{F}_m$, it is useful to expand it into its components parallel and perpendicular to the magnetic field. Of particular interest is the term $\mu_0^{-1}(\mathbf{B} \cdot \nabla)\mathbf{B}$, associated with the rate of change observed in the magnetic field as we move along it (note that $\mathbf{B} \cdot \nabla$ simply selects the projection of the ∇ operator along $\mathbf{B}$). Denoting by s the distance along the field line, we can write

$$(\mathbf{B} \cdot \nabla) = B\frac{\partial}{\partial s},$$

where B is the local intensity of $\mathbf{B}$, i.e., $B = |\mathbf{B}|$. Note that the vector $\mathbf{B}$ may change both in intensity and in direction. We can write $\mathbf{B}$ as

$\mathbf{B} = B\hat{\mathbf{b}}$, where $\hat{\mathbf{b}}$ is the unit vector locally in the direction of the magnetic field. We then have

$$\frac{1}{\mu_0}(\mathbf{B}\cdot\nabla)\mathbf{B} = \frac{1}{\mu_0}B\frac{\partial}{\partial s}(B\hat{\mathbf{b}}) = \frac{1}{\mu_0}\left[B\frac{\partial B}{\partial s}\hat{\mathbf{b}} + B^2\frac{\partial \hat{\mathbf{b}}}{\partial s}\right]$$

$$= \hat{\mathbf{b}}\frac{\partial}{\partial s}\left(\frac{B^2}{2\mu_0}\right) + \frac{B^2}{\mu_0}\frac{\partial \hat{\mathbf{b}}}{\partial s} \tag{6.43a}$$

$$= \hat{\mathbf{b}}\frac{\partial}{\partial s}\left(\frac{B^2}{2\mu_0}\right) + \frac{B^2}{\mu_0}\frac{\hat{\mathbf{n}}}{R_c}. \tag{6.43b}$$

The second term on the right in (6.43) originates from the change in the direction of the magnetic field as we move along it and must thus be related to the local radius of curvature R_c of the field lines. As shown in Figure 6.4, this term is a vector in the direction locally normal to the field line and has a magnitude inversely proportional to the radius of curvature, as given in (6.43b). The first term in (6.43b) is equal in magnitude and opposite in sign to the component along the field of the last term in (6.42). To better see this, it is convenient to separate the ∇ operator into its components parallel and perpendicular to the field line, i.e.,

$$\nabla = \nabla_\parallel + \nabla_\perp = \hat{\mathbf{b}}\frac{\partial}{\partial s} + \nabla_\perp,$$

where we have observed that $\nabla_\parallel = \hat{\mathbf{b}}(\partial/\partial s)$. We can then rewrite (6.42) as

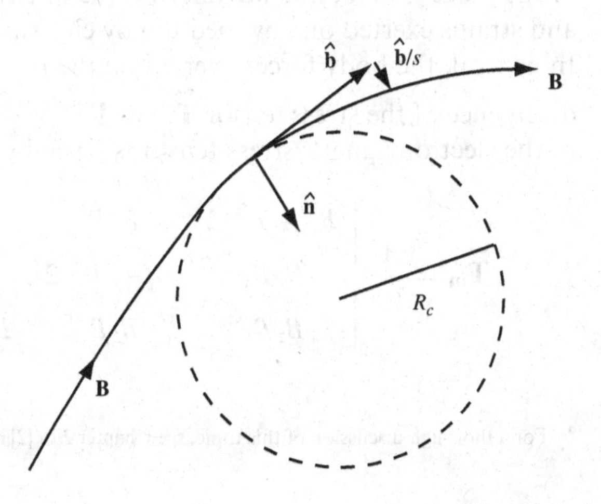

Figure 6.4 Forces in magnetic fields.

$$\mathbf{F}_{\mathrm{m}} = \frac{1}{\mu_0}(\mathbf{B}\cdot\nabla)\mathbf{B} - \nabla\frac{B^2}{2\mu_0}$$

$$= \hat{\mathbf{b}}\frac{\partial}{\partial s}\left(\frac{B^2}{2\mu_0}\right) + \frac{B^2}{\mu_0}\frac{\hat{\mathbf{n}}}{R_c} - \nabla_{\perp}\left(\frac{B^2}{2\mu_0}\right) - \hat{\mathbf{b}}\frac{\partial}{\partial s}\left(\frac{B^2}{2\mu_0}\right)$$

$$\mathbf{F}_{\mathrm{m}} = \frac{B^2}{\mu_0}\frac{\hat{\mathbf{n}}}{R_c} - \nabla_{\perp}\left(\frac{B^2}{2\mu_0}\right). \tag{6.44}$$

Equation (6.44) indicates that the magnetic field exerts two types of force on the fluid. First we observe that there is no component of the force along $\mathbf{B}$, as expected from the fact that the force is given by $\mathbf{J} \times \mathbf{B}$. In the plane normal to the field, there is a magnetic force: a perpendicular magnetic pressure which acts to drive the fluid away from regions of high magnetic field (just as the kinetic pressure term $-\nabla p$ in (6.40) drives the fluid away from regions of high pressure). The additional transverse force $R_c^{-1}(B^2/\mu_0)\hat{\mathbf{n}}$ points locally toward the center of curvature of the field line (Figure 6.4). This force pushes the fluid in the direction of the radius of curvature, acting as a tension. Since the magnetic flux is frozen in (see Section 6.4.1), the resulting fluid motion would pull the magnetic field line with it, thus acting to straighten the line. This tension force works much as if the field lines were elastic cords pulling against the fluid, having a tendency to reduce their curvature. This rubber band or elastic cord-like behavior of the magnetic field lines is precisely what leads to hydromagnetic wave motion in a magnetized fluid, which we will study in later chapters.

The magnetic body forces described above are not specific to plasmas or fluids. Rather, the case considered here is a special case of the more general and advanced topic of electromagnetic stresses and strains exerted on any medium by electric and magnetic fields.[4] In general, the body forces exerted on the medium are given by the divergence of the stress tensor $\overleftrightarrow{\mathbf{T}}$, i.e., $\mathbf{F} = \nabla \cdot \overleftrightarrow{\mathbf{T}}$. The magnetic part of the electromagnetic stress tensor is given by

$$\overleftrightarrow{\mathbf{T}}_{\mathrm{m}} = \frac{1}{\mu_0}\begin{bmatrix} B_x^2 - B^2/2 & B_x B_y & B_x B_z \\ B_y B_x & B_y^2 - B^2/2 & B_y B_z \\ B_z B_x & B_z B_y & B_z^2 - B^2/2 \end{bmatrix}, \tag{6.45}$$

[4] For a thorough discussion of this topic, see Chapter 2 of [2] or Chapters 6 and 10 of [3].

sometimes written in terms of its entries as $[\overleftrightarrow{\mathbf{T}}_{\mathrm{m}}]_{ij} = \mu_0^{-1}[B_i B_j - \delta_{ij} B^2/2]$, where $\delta_{ij} = 0$ for $i \neq j$ and $\delta_{ij} = 1$ for $i = j$. It is useful to rewrite (6.40) at this point, in order to compare the kinetic pressure gradient term and the magnetic force density term. We have

$$\begin{aligned}
\rho_{\mathrm{m}} \frac{\partial \mathbf{u}_{\mathrm{m}}}{\partial t} &= -\nabla p + \mathbf{J} \times \mathbf{B} \\
&= -\nabla \cdot \Psi + \mathbf{F}_{\mathrm{m}} \\
&= -\nabla \cdot \Psi + \nabla \cdot \overleftrightarrow{\mathbf{T}}_{\mathrm{m}} \\
&= -\nabla \cdot \Psi - \nabla \cdot \Psi_{\mathrm{m}}
\end{aligned} \tag{6.46}$$

where we have introduced a *magnetic pressure tensor* Ψ_{m}, which works in (6.40) in exactly the same manner as the kinetic pressure tensor, and which is the negative of the magnetic part of the electromagnetic stress tensor, i.e., $\Psi_{\mathrm{m}} = -\overleftrightarrow{\mathbf{T}}_{\mathrm{m}}$. For the special case of a unidirectional static magnetic field $\mathbf{B} = \hat{\mathbf{z}}B$, the magnetic pressure tensor is diagonal and given by

$$\Psi_{\mathrm{m}} = \frac{1}{\mu_0} \begin{bmatrix} B^2/2 & 0 & 0 \\ 0 & B^2/2 & 0 \\ 0 & 0 & -B^2/2 \end{bmatrix}. \tag{6.47}$$

In fact, (6.47) constitutes a general expression for the magnetic pressure tensor in any local magnetic coordinate system, where the first two coordinates are perpendicular to $\mathbf{B}$ and the third coordinate is along $\mathbf{B}$. We can interpret (6.47) as indicating that the stress caused by the magnetic field is in the form of an isotropic (same in all three directions) magnetic pressure $B^2/(2\mu_0)$ and a tension of B^2/μ_0 along the magnetic field lines. Note that tension along the magnetic field lines is consistent with the force acting along $\hat{\mathbf{n}}$ in Figure 6.4, tending to straighten the field lines. In general, the magnetic force can be written as

$$\mathbf{F}_{\mathrm{m}} = -\nabla\left(\frac{B^2}{2\mu_0}\right) + \frac{1}{\mu_0}\nabla \cdot (\mathbf{BB}), \tag{6.48}$$

where $\mathbf{BB}$ is a tensor, defined in the usual manner (see footnote 2 in Chapter 4).

6.6 Magnetohydrostatics

An important class of problems involves cases when the plasma fluid is stationary, or nearly stationary, so that $\mathbf{u}_m \simeq 0$, and static, so that all time derivatives are zero. In this case Equation (6.40) can be written as

$$0 = -\nabla p + \mathbf{J} \times \mathbf{B}. \tag{6.49}$$

Our interest in this section is to comment on plasma confinement, which is a central issue in controlled fusion experiments where highly energetic charged particles are forced to collide and trigger nuclear fusion reactions. In such experiments a hydrostatic equilibrium is determined by the balance between kinetic and magnetic pressure. In static equilibrium we thus have

$$\mathbf{J} \times \mathbf{B} = \nabla p \tag{6.50}$$

or, eliminating $\mathbf{J}$ using (6.24a),

$$\frac{1}{\mu_0}(\nabla \times \mathbf{B}) \times \mathbf{B} = \nabla p, \tag{6.51}$$

which can further be written as

$$\nabla\left(p + \frac{B^2}{2\mu_0}\right) = \frac{1}{\mu_0}(\mathbf{B} \cdot \nabla)\mathbf{B}, \tag{6.52}$$

where we see once again that the magnetic pressure $B^2/(2\mu_0)$ plays the same role as kinetic pressure p, while the term on the right-hand side describes the forces caused by curvature of the field lines. In general, we can show from (6.50) that

$$\mathbf{J} \cdot \nabla p = 0 \quad \text{and} \quad \mathbf{B} \cdot \nabla p = 0, \tag{6.53}$$

which states that pressure gradients cannot exist in the direction of magnetic fields or currents. In other words, the surfaces with $p =$ constant are both magnetic surfaces (i.e., they are made up of magnetic field lines) and current surfaces (i.e., they are made of current flow lines). Note that it follows from (6.24a) that

$$\nabla \cdot \mathbf{J} = 0. \tag{6.54}$$

Since both $\mathbf{B}$ and $\mathbf{J}$ are divergence-free, the magnetic field as well as the current lines either extend to infinity or must close on themselves. The magnetic and current surfaces with $p =$ constant take the form either of tubes extending to infinity or of toroids. Equations (6.50), (6.54), and (6.24d) are the key magnetohydrostatic equations, constituting the solution of the more general magnetohydrodynamic (MHD) equations for the special case of $\mathbf{u}_\mathrm{m} = 0$ and isotropic pressure. In the general case, the fluid plasma current required for equilibrium can be found by multiplying (6.50) by $\mathbf{B}$:

$$\nabla p = \mathbf{J} \times \mathbf{B}$$

$$\mathbf{B} \times \nabla p = \mathbf{B} \times (\mathbf{J} \times \mathbf{B}) \equiv (\mathbf{B} \cdot \mathbf{B})\mathbf{J} - (\mathbf{J} \cdot \mathbf{B})\mathbf{B}$$

$$\mathbf{B} \times \nabla p = B^2(\mathbf{J}_\perp + J_\| \hat{\mathbf{b}}) - (J_\| \hat{\mathbf{b}} \cdot B\hat{\mathbf{b}}) B\hat{\mathbf{b}}$$

$$\mathbf{B} \times \nabla p = B^2 \mathbf{J}_\perp + J_\| B^2 \hat{\mathbf{b}} - J_\| B^2 \hat{\mathbf{b}}$$

$$\mathbf{J}_\perp = \frac{\mathbf{B} \times \nabla p}{B^2},$$

which is the "diamagnetic current" that we saw earlier, in Equation (5.5). The parallel component of the current can be determined using (6.54). We have

$$\nabla \cdot \mathbf{J} = 0 \quad \rightarrow \quad \nabla \cdot (\mathbf{J}_\perp + J_\| \hat{\mathbf{b}}) = 0$$

$$\nabla \cdot \mathbf{J}_\perp + \nabla \cdot \left[\frac{J_\| \mathbf{B}}{B} \right] = 0$$

$$\nabla \cdot \mathbf{J}_\perp + \frac{J_\|}{B} \underbrace{\nabla \cdot \mathbf{B}}_{=0} + \mathbf{B} \cdot \nabla \left[\frac{J_\|}{B} \right] = 0$$

$$\nabla \cdot \mathbf{J}_\perp + \mathbf{B} \cdot \nabla \left[\frac{J_\|}{B} \right] = 0.$$

In most cases, the perpendicular current is divergence-free (no temporal variations of net charge density in the transverse plane) so that $J_\|$ can be zero. We now discuss some special equilibrium configurations.

6.6.1 The θ-pinch

Consider a system of magnetic field lines which are straight and parallel, for example, generated by an external solenoid of infinite length. We have $\mathbf{B} = B(x, y)\hat{\mathbf{z}}$, i.e., the intensity of the magnetic field does not vary in the direction of the field lines. This configuration, shown in Figure 6.5, is called the θ-pinch configuration.

Noting that the right-hand side of (6.52) vanishes, we have

$$\nabla \left(p + \frac{B^2}{2\mu_0} \right) = 0. \qquad (6.55)$$

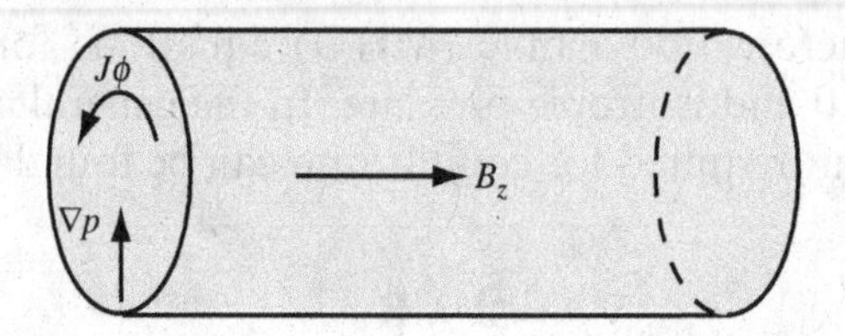

Figure 6.5 The θ-pinch configuration.

We also note that (6.24d) is automatically satisfied, i.e., $\nabla \cdot \mathbf{B} = 0$. According to (6.55), the magnetic field is reduced in the region occupied by the plasma, a particularly good example of the general *diamagnetism* of a plasma.[5] Assuming a bounded plasma column, if the distribution of the fluid pressure $p(x, y)$ is given, the magnetic pressure can be determined from (6.55). If the intensity of the field outside the fluid is B_0, then the maximum pressure which can be confined is $p_{\max} = B_0^2/(2\mu_0)$. For the special case where a finite region may be filled with plasma of uniform pressure $p_{\max}$ (with no fluid anywhere else), a sharp boundary exists between a region of no fluid with magnetic field strength B_0 and a region in which the magnetic field is zero.

The beta parameter

We see from the above that hydrodynamic equilibrium is determined by the balance between kinetic and magnetic pressure. The ratio of the plasma pressure to the magnetic field pressure (the latter normally measured either outside the plasma or at its boundary) is usually denoted by β and referred to as the beta parameter:

$$\beta = \frac{2\mu_0 p}{B^2}. \tag{6.56}$$

The quantity β is a measure of the degree to which a magnetic field holds a plasma in equilibrium. In a low-β plasma, the force balance is mainly between different magnetic forces. At $\beta \simeq 1$, magnetic and pressure forces are nearly in balance, whereas for $\beta \gg 1$ the magnetic field plays a minor role in the plasma dynamics. Magnetically confined laboratory plasmas tend to have beta values in the range of a few percent, while astrophysical plasmas can have β values approaching unity or even $\beta \gg 1$. Examples of both low- and

[5] The fundamental reason for the diamagnetism of a plasma is that the individual particle motions give rise to currents which induce a magnetic flux in a direction opposite to the applied magnetic field. The electric current depends on the number density of the charged particles and on their velocities. The kinetic pressure also depends on these same parameters. Thus, as the kinetic pressure increases; the electric current increases; so does the induced magnetic flux, and thus the diamagnetic effect.

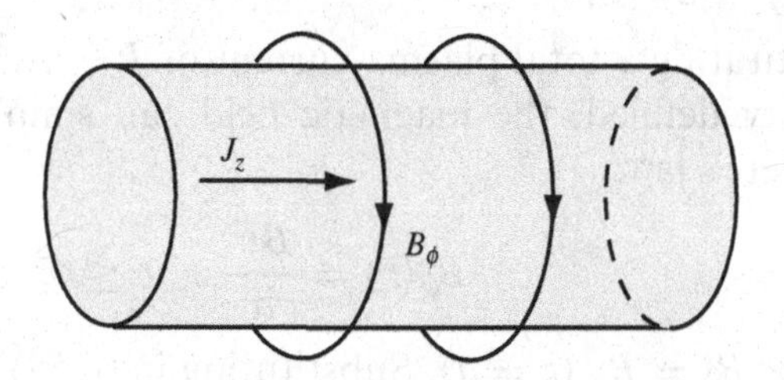

Figure 6.6 The cylindrical pinch configuration.

high-beta plasmas are encountered in the Earth's magnetosphere and ionosphere.

6.6.2 The cylindrical pinch

Another interesting configuration of hydrostatic equilibrium with a structure of infinite extent is the so-called cylindrical pinch geometry. For this configuration, the magnetic field is azimuthal (i.e., B_ϕ only), while the plasma current is axial (i.e., J_z only) and is the source of the magnetic field; see Figure 6.6. Since the magnetic field is curved in this case, the pressure balance equation must include the term on the right-hand side of (6.52), which describes the tension of the field lines which confine the plasma. Using (6.44), the radial component of the pressure balance equation (6.52) becomes

$$\frac{\partial}{\partial r}\left(p + \frac{B^2}{2\mu_0}\right) + \frac{B^2}{\mu_0 r} = 0, \tag{6.57}$$

since $(\mathbf{B}\cdot\nabla)\mathbf{B}$ simplifies to $\left(B_\phi \frac{1}{r}\frac{\partial}{\partial \phi}\right) B_\phi \hat{\phi} = -\frac{1}{r} B_\phi^2 \hat{\mathbf{r}}$.[6] Integrating (6.57) from 0 to r we find

$$p(r) = p_0 - \frac{B_\phi^2(r)}{2\mu_0} - \frac{1}{\mu_0}\int_0^r \frac{B_\phi^2(\zeta)}{\zeta} d\zeta. \tag{6.58}$$

where $p_0 = p\,(r = 0)$ is the maximum pressure, located at the center of the column. An infinite number of possible equilibria can satisfy (6.58). As an example, consider the case of a plasma carrying a uniform current, i.e.,

$$J_z(r) = \begin{cases} J_0 & r \le a \\ 0 & r > a \end{cases}, \tag{6.59}$$

[6] The geometry of $\hat{\phi}$ and $\partial\hat{\phi}/\partial\phi$ is the same as that of $\hat{b}$ and $\partial\hat{b}/\partial s$ as shown in Figure 6.4.

constituting a total plasma current of $I = \pi a^2 J_0$. With the current density defined, the magnetic field can simply be determined by Ampère's law:

$$B_\phi(r) = \frac{B_0 r}{a}, \quad r \leq a, \tag{6.60}$$

where $B_0 = B_\phi \; (r = a)$. Substituting in (6.58) and carrying out the integral, we find that it is precisely equal to the second term in (6.58), so the pressure within the plasma is given by

$$p(r) = p_0 - \frac{B_0^2 r^2}{\mu_0 a^2} \quad \rightarrow \quad p_0 = \frac{B_0^2}{\mu_0} = \frac{\mu_0 I^2}{4\pi^2 a^2}, \tag{6.61}$$

since the pressure $p(r)$ must vanish at the edge of the column, i.e., at $r = a$. The condition (6.61) is known as the pinch condition, describing a magnetically self-confined, current-carrying plasma. Note that in the case of the cylindrical pinch, the plasma current provides the entire magnetic field, unlike the θ-pinch where the magnetic field is provided externally. Establishing a cylindrical pinch can be accomplished by applying a very large voltage difference using electrodes, and driving the current in the plasma. Both the pressure gradient and the magnetic field are proportional to r. Although the static configuration of the cylindrical pinch appears ideal for plasma confinement, it can be shown both theoretically and experimentally that this configuration is highly unstable to small perturbations. In general, any departure from the delicate equilibrium of magnetohydrostatic confinement schemes leads to growth of the original perturbations and disintegration of the plasma geometry. Although discussion of stability is beyond the scope of this text, it is important to note that the challenge in confining high-energy plasmas has been one of the main obstacles to achieving controlled thermonuclear fusion.[7]

Example 6-2 Tokamak
The Joint European Torus (JET) is a tokamak-style plasma-confinement machine that has an axial magnetic field applied externally by toroidal coils as well as an azimuthal field created by the internal plasma current. The net result is a twisted field geometry, shown in Figure 6.7, that increases plasma stability. The tokamak can be modeled as an infinitely long cylinder. The externally applied field in the JET achieves a maximum value

[7] A discussion of confinement instability at an introductory level can be found in Chapter 13 of [4].

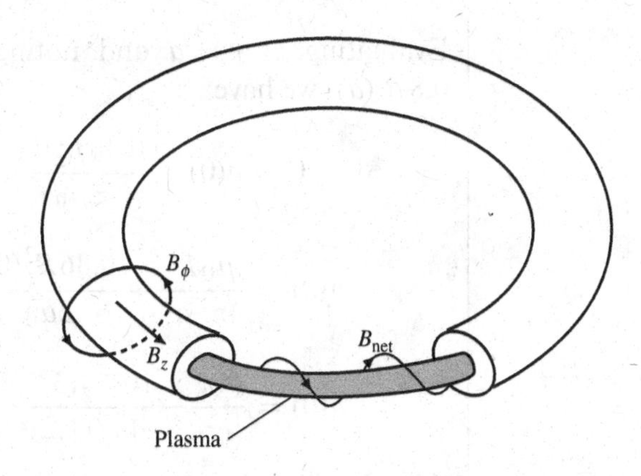

Figure 6.7 Tokamak plasma-confinement system, consisting of an externally applied field (B_z) and a field resulting from the plasma current (B_ϕ). The resulting field (B_{net}) has a helical shape.

of 3.45 T at the plasma boundary at $r = a = 1.2$ m and is estimated to be 20% lower at the center of the cylinder.[8] If the total axial plasma current is $I = 3.2$ MA and assumed to be uniformly distributed, find the maximum plasma pressure at the center ($r = 0$).

Solution: We start with Equation (6.52), which for the case of a magnetic field with both B_z and B_ϕ components reduces to

$$\frac{\partial}{\partial r}\left(p + \frac{B_\phi^2 + B_z^2}{2\mu_0}\right) + \frac{B_\phi^2}{\mu_0 r} = 0.$$

Integrating yields

$$p(r) = p(0) - \frac{B_\phi^2}{2\mu_0} - \frac{B_z^2}{2\mu_0} + \frac{B_z^2(0)}{2\mu_0} - \int_0^r \frac{B_\phi^2(\rho)}{\rho} d\rho.$$

For a uniform current $B_\phi(r) = \dfrac{B_0 r}{a}$, where $B_0 = \dfrac{\mu_0 I}{2\pi a}$,

$$p(r) = p(0) - \frac{1}{2\mu_0}\left(\frac{\mu_0 I r}{2\pi a^2}\right)^2 - \frac{B_z^2}{2\mu_0} + \frac{B_z^2(0)}{2\mu_0} - \frac{1}{2\mu_0}\left(\frac{\mu_0 I r}{2\pi a^2}\right)^2$$

$$p(r) = p(0) - \frac{B_z^2}{2\mu_0} + \frac{B_z^2(0)}{2\mu_0} - \frac{1}{\mu_0}\left(\frac{\mu_0 I r}{2\pi a^2}\right)^2.$$

[8] Note that this is a diamagnetic effect in which the plasma acts to reduce the externally applied field, as was seen with the θ-pinch.

Evaluating at $r = a$ and noting that $p(a) = 0$ and $B_z(0) = 0.8\,B_z(a)$, we have

$$0 = p(0) - \frac{0.36\,B_z^2(0)}{2\mu_0} - \frac{\mu_0 I^2}{4\pi^2 a^2}$$

$$p(0) = \frac{\mu_0 I^2}{4\pi^2 a^2} + \frac{0.36\,B_z^2(0)}{2\mu_0}$$

$$p(0) = \frac{(4\pi \times 10^{-7})(3.2 \times 10^6)^2}{4\pi^2 (1.2)^2} + \frac{0.36(3.45)^2}{2(4\pi \times 10^{-7})}$$

$$p(0) = 1.824 \times 10^6 \text{ Pa.}$$

6.7 Collisionless plasmas with strong magnetic field

It was mentioned above that the MHD formulation is strictly applicable only in collision-dominated plasmas, since the particles have to stay together in order for us to reasonably assign a velocity to their "center of mass." It turns out, however, that the MHD formulation can also be applied successfully to the dynamics of collisionless plasmas under the influence of strong magnetic fields, in which the gyrofrequency is much larger than the collision frequency. In this case, the effect of the magnetic field replaces to a certain extent the role of collisions in maintaining a local equilibrium, by not allowing particles to "stray" significantly in phase space in the transverse direction. An important distinction between collision-driven and magnetic field-driven local equilibria is that the magnetic field does not exert any force in the direction parallel to the field lines. The magnetic field thus makes the pressure exerted by the plasma anisotropic, i.e., different in directions parallel or perpendicular to the magnetic field. In the previous sections we treated the effect of the magnetic field as magnetic forces acting on a plasma with isotropic pressure. In the case of the strong magnetic field it is more appropriate to include the effect of the magnetic field directly in the pressure term. The notion of combining kinetic and magnetic pressure effects was alluded to in the discussion of Equation (6.46). Here we present a more thorough treatment and include certain approximations that must be made in order to have a closed set of MHD equations.

We start by observing that in the regime of high gyrofrequency (i.e., high magnetic field), particle kinetic energies are divided equally between the two perpendicular directions, but not necessarily along **B**. Thus we expect the kinetic pressure tensor to be diagonal as expressed in a local coordinate system, with the third coordinate along **B**:

$$\Psi = \begin{bmatrix} p_\perp & 0 & 0 \\ 0 & p_\perp & 0 \\ 0 & 0 & p_\parallel \end{bmatrix} = p_\perp \overset{\leftrightarrow}{\mathbf{I}} + (p_\parallel - p_\perp)\frac{(\mathbf{BB})}{B^2}, \tag{6.62}$$

where $\overset{\leftrightarrow}{\mathbf{I}}$ is the identity matrix and **BB** represents a tensor product. The momentum equation (4.30) can be reduced by using the tensor identity given in footnote 2 of Chapter 4. We have

$$\begin{aligned}
\rho_m \frac{d\mathbf{u}_m}{dt} &= -\nabla \cdot \Psi + \mathbf{J} \times \mathbf{B} \\
&= -\nabla \cdot \left(p_\perp \overset{\leftrightarrow}{\mathbf{I}}\right) - \nabla \cdot \left[(p_\parallel - p_\perp)\frac{(\mathbf{BB})}{B^2}\right] + \mathbf{J} \times \mathbf{B} \\
&= -\nabla p_\perp - \nabla \cdot \Bigg\{(\mathbf{B} \cdot \nabla)\left[(p_\parallel - p_\perp)\frac{\mathbf{B}}{B^2}\right] \\
&\qquad \underbrace{+ \left[(p_\parallel - p_\perp)\frac{\mathbf{B}}{B^2}\right]\underbrace{(\nabla \cdot \mathbf{B})}_{=0}}_{=0}\Bigg\} + \mathbf{J} \times \mathbf{B} \\
&= -\nabla p_\perp - \nabla \cdot \left\{(\mathbf{B} \cdot \nabla)\left[(p_\parallel - p_\perp)\frac{\mathbf{B}}{B^2}\right]\right. \\
&\qquad + \underbrace{\frac{1}{\mu_0}(\nabla \times \mathbf{B}) \times \mathbf{B}}_{\text{using (6.24a), i.e., } \nabla\times\mathbf{B} = \mu_0\mathbf{J}} \\
&= -\nabla p_\perp - \nabla \cdot \left\{(\mathbf{B} \cdot \nabla)\left[(p_\parallel - p_\perp)\frac{\mathbf{B}}{B^2}\right]\right. \\
&\qquad + \frac{1}{\mu_0}\underbrace{[(\mathbf{B} \cdot \nabla)\mathbf{B} - (\mathbf{B} \cdot \mathbf{B})\nabla]}_{\text{using } (\nabla\times\mathbf{C})\times\mathbf{D} = -\mathbf{D}\times(\nabla\times\mathbf{C}) \equiv (\mathbf{D}\cdot\nabla)\mathbf{C} - (\mathbf{C}\cdot\mathbf{D})\nabla}
\end{aligned}$$

$$= -\nabla p_\perp - \nabla \cdot \left\{ (\mathbf{B} \cdot \nabla) \left[(p_\parallel - p_\perp) \frac{\mathbf{B}}{B^2} \right] \right.$$

$$+ \frac{1}{\mu_0} \left[(\mathbf{B} \cdot \nabla)\mathbf{B} - \nabla \left(\frac{1}{2} B^2 \right) \right]$$

$$\rho_\mathrm{m} \frac{d\mathbf{u}_\mathrm{m}}{dt} = -\nabla \left(p_\perp + \frac{B^2}{2\mu_0} \right) + (\mathbf{B} \cdot \nabla) \left[\frac{\mathbf{B}}{\mu_0} - \frac{(p_\parallel - p_\perp)\mathbf{B}}{B^2} \right]. \tag{6.63}$$

Note that (6.63) can be rewritten in tensor form by simply combining (6.62) and (6.47). Equations (6.37), (6.63), and Maxwell's equations (6.24) are an almost-complete system of equations for the unknowns $\mathbf{E}$, $\mathbf{B}$, ρ_m, $p_\parallel$, $p_\perp$, and $\mathbf{u}_\mathrm{m}$, if we can express ρ_m in terms of $p_\parallel$ and $p_\perp$. The two equations of state relating ρ_m to the parallel and perpendicular scalar pressures are known as the *double adiabatic approximation* or the Chew–Goldberger–Low (CGL) equations [5]. To zeroth order, the velocity distribution is isotropic in the perpendicular plane and in the longitudinal direction, so that we can use the two- and one-dimensional versions of the adiabatic equation of state introduced in Section 4.6, provided there is no energy transport from one region of the plasma to another. With this basic assumption, which we make in truncating the series of moments of the Boltzmann equation at the heat-flow equation, we have

$$\frac{p_\parallel p_\perp^2}{\rho_\mathrm{m}^5} = \text{constant} \tag{6.64}$$

which takes the place of the adiabatic energy conservation equation for isotropic plasmas. The second equation of state can be obtained by recognizing the connection between the magnetic field strength and the perpendicular component of particle energy. Since the magnetic moment is conserved, we can write

$$\frac{\langle m w_\perp^2 \rangle}{B} = \frac{p_\perp}{\rho_\mathrm{m} B} = \text{constant}. \tag{6.65}$$

6.7.1 Mirror equilibrium

The possibility of an anisotropic plasma pressure allows for stable confinement of plasmas in the so-called mirror configuration.

Consider the component of the static equilibrium equation (6.50) along the field vector:

$$\hat{\mathbf{b}} \cdot (\mathbf{J} \times \mathbf{B}) = \hat{\mathbf{b}} \cdot \nabla p$$

$$0 = \hat{\mathbf{b}} \cdot \nabla p. \tag{6.66}$$

Taking $\mathbf{B} = \hat{\mathbf{z}} B$, we have

$$\frac{\partial p}{\partial z} = 0,$$

indicating that the plasma pressure must be constant along the field lines under equilibrium conditions. According to this, there cannot be any gradient of plasma pressure (and thus density) along the field lines, so magnetic field configurations with field lines that extend outside the region of the plasma cannot confine a plasma with isotropic pressure. On the other hand, based on our analyses of particle motions in earlier sections, we know well that the particles can be trapped quite effectively in a mirror geometry. This dilemma is resolved by noting that the plasma pressure in such cases must be anisotropic, with the pressure tensor given by (6.62):

$$\Psi = \begin{bmatrix} p_\perp & 0 & 0 \\ 0 & p_\perp & 0 \\ 0 & 0 & p_\parallel \end{bmatrix} = p_\perp \overleftrightarrow{\mathbf{I}} + (p_\parallel - p_\perp) \frac{(\mathbf{BB})}{B^2}.$$

The more general form of the equilibrium equation (6.66) is

$$0 = \hat{\mathbf{b}} \cdot \nabla \Psi, \tag{6.67}$$

which on manipulation (and using $\nabla \cdot \mathbf{B} = 0$) reduces to

$$0 = \hat{\mathbf{b}} \cdot \nabla p_\parallel + \left[\frac{p_\perp - p_\parallel}{B} \right] \hat{\mathbf{b}} \cdot \nabla B$$

$$0 = \frac{\partial p_\parallel}{\partial z} + \left[\frac{p_\perp - p_\parallel}{B} \right] \frac{\partial B}{\partial z}. \tag{6.68}$$

Many different solutions of (6.68) are possible, as long as $p_\perp > p_\parallel$. Typically, both $p_\perp$ and $p_\parallel$ decrease with distance z away from the central confinement region (at which B is a minimum).

It is worth mentioning that, in the mirror geometry above, the magnetic field lines cannot be strictly parallel to the $\hat{\mathbf{z}}$ axis and have a simultaneous parallel gradient $\left(\frac{\partial B}{\partial z} \neq 0\right)$. As can be seen in Figure 6.8, the parallel gradient causes a finite curvature of the field lines, requiring $\mathbf{B}$ to have components in the $\hat{\mathbf{x}}$ and $\hat{\mathbf{y}}$ directions, as was shown in Figure 2.8 in Chapter 2. The fundamental force

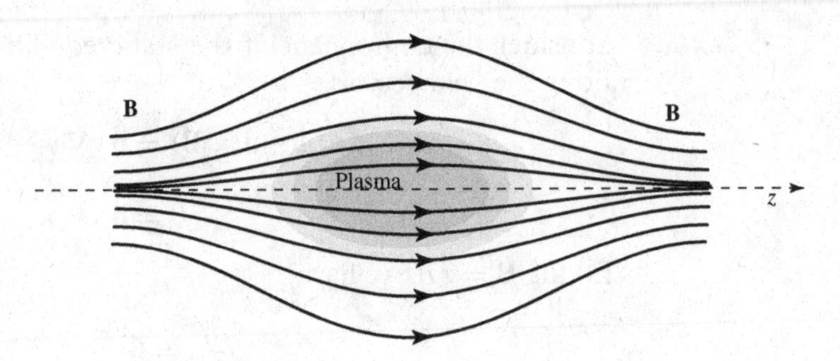

Figure 6.8 Equilibrium in magnetic "mirror" configuration.

at work in the mirror confinement is still the Lorentz force. The strength of the MHD approach and anisotropic pressure formulation is that the complex geometry and multiple particle trajectories can be ignored. The field lines can be treated as approximately parallel, and the problem reduces to a balance of macroscopic quantities as shown in Equation (6.68).

6.8 Summary

Magnetohydrodynamics (MHD) is an important subfield of plasma physics involving the treatment of a plasma as a single conducting fluid. The MHD plasma model is appropriate for low-frequency phenomena when the plasma is highly conductive and dense. The principal applications of MHD are in astrophysics and laboratory plasma confinement. A key requirement for applicability of the single-fluid approach is that the various plasma species are forced to act in unison under the influence of either frequent collisions or a strong magnetic field. The fundamental equations of MHD are derived by combining the individual fluid equations for each plasma species into a single set. Along with Maxwell's equations and an appropriate equation of state, the MHD equations form a closed set of expressions for the macroscopic parameters of total charge and mass densities, the mean fluid velocity, and the magnetic field. The MHD model is effective in describing complex phenomena that are not easily accessible using other plasma models. A remarkable property of an MHD plasma is the concept of frozen-in magnetic field lines, in which magnetic field lines are forced to remain unchanged in the highly conducting plasma and are therefore transported as part of a mobile fluid. Likewise, the MHD formalism illustrates the effects of magnetic forces exerted on a plasma. The action of magnetic forces leads to the concept of a magnetic pressure. Balance between magnetic and kinetic pressures allows for equilibrium configurations of plasma confinement. MHD plasmas also host a variety of wave phenomena that will be explored in later chapters.

6.9 Problems

6-1. Calculate the spiral angle (Ψ in Figure 6.3) that the interplanetary magnetic field makes at the Earth for a solar wind speed $u_{\rm sw} = 750\ {\rm km\,s^{-1}}$ occurring during a solar disturbance.

6-2. An axial magnetic field $\mathbf{B} = B_0\hat{\mathbf{z}}$ is applied externally to a cylindrical plasma column of radius a in magnetohydrostatic equilibrium and with a pressure profile given by

$$p(r) = p_0 \left[\cos\left(\frac{\pi r}{2a}\right)\right]^2.$$

(a) What is the maximum value of p_0 in terms of B_0? (b) Find an expression for $\mathbf{B}(r)$, the magnetic field magnitude as a function of the radius. (c) Calculate $\mathbf{J}(r)$, the current density in the plasma column.

6-3. Consider the cylindrical tokamak equilibrium described in Example 6-2. Show that for low plasma pressures $p(0) < B_\phi^2/2\mu_0$ the tokamak becomes paramagnetic, meaning that B_z will be stronger at the center than at the edge of the plasma.

6-4. The Earth's magnetic field in space exhibits a sharp boundary known as the magnetopause, resulting from the solar wind exerting pressure on the Earth's magnetic field, distorting its shape. Without the effect of the solar wind, the Earth's unperturbed field is close to that of a dipole, and in the equatorial plane can be approximated by

$$B_E(r) = B_0\left(\frac{R_E}{r}\right)^3 \hat{\mathbf{z}}$$

as a function of radial distance r from the Earth's center, where R_E is the radius of the Earth. (a) Find an expression for the magnetopause distance if the solar wind exerts a pressure of $\rho_{\rm sw}u_{\rm sw}^2$, where $\rho_{\rm sw}$ and $u_{\rm sw}$ are the solar wind mass density and speed, respectively. (b) At the magnetopause the solar wind pressure equals the magnetic pressure of the perturbed field. Repeat part (a) but with the inclusion of the additional magnetic pressure exerted by the solar wind as a result of the frozen-in field concept. The magnetic pressure of the solar wind is expressed as $B_s^2/(2\mu_0)$.

6-5. Consider a toroidal implementation of a θ-pinch with the mean radius of the torus r_m being much larger than the radius of the cylindrical section a. The magnetic field applied by external coils along the axis of the torus is given by the expression

$$\mathbf{B} = \hat{\phi}\,\frac{B_0 a}{r}.$$

Examine the type of charge separation that occurs because of the inhomogeneity of the B_ϕ field in the r direction. What is the direction of the $\mathbf{E} \times \mathbf{B}$ drift, where $\mathbf{E}$ is the electric field induced by the charge separation? Comment on whether the resultant effects of the gradient, curvature, and $\mathbf{E} \times \mathbf{B}$ drifts work to prevent confinement of a plasma in the magnetic field of the toroid.

6-6. A cylindrical confined plasma column of radius a contains a coaxial magnetic field $\mathbf{B} = \hat{\mathbf{z}} B_0$. The plasma has a pressure profile of

$$p = p_0 \cos\left(\frac{\pi r}{2a}\right).$$

(a) Calculate the maximum value of p_0. (b) Find an expression for the diamagnetic current.

6-7. The minimum density N and confinement time τ_c of a plasma necessary for achieving controlled thermonuclear fusion is given by the Lawson criterion, $N\tau_c \leq 10^{20}\ \mathrm{m}^{-3}\ \mathrm{s}$. For a plasma confined in a θ-pinch configuration, with a maximum applied magnetic field stength of 10 T, what is the minimum confinement time that will satisfy the Lawson criterion?

6-8. The Yamato 1 is an MHD propulsion ship that uses an applied magnetic field and direct electric current passed through seawater to propel the vessel. The applied magnetic field is 4 T. The spacing between the electrodes that drive the DC current is 175 mm and the total power dissipated is 3600 kW, yielding a force of 16 kN. Calculate the voltage that is applied across the electrodes.

References

[1] N. A. Krall and A. W. Trivelpiece, *Principles of Plasma Physics* (San Francisco: San Francisco Press, 1986), Section 3.6.

[2] J. A. Stratton, *Electromagnetic Theory* (New York: McGraw-Hill, 1941).

[3] W. K. H. Panofsky and M. Phillips, *Classical Electricity and Magnetism*, 2nd edn (Reading: Addison-Wesley, 1962).
[4] J. A. Bittencourt, *Fundamentals of Plasma Physics*, 3rd edn (New York: Springer-Verlag, 2004), 325–50.
[5] G. F. Chew, M. L. Goldberger, and F. E. Low, The Boltzmann equation and the one-fluid hydromagnetic equations in the absence of particle collisions. *Proc. R. Soc. London A*, **236** (1956), 112–18.

CHAPTER

7 Collisions and plasma conductivity

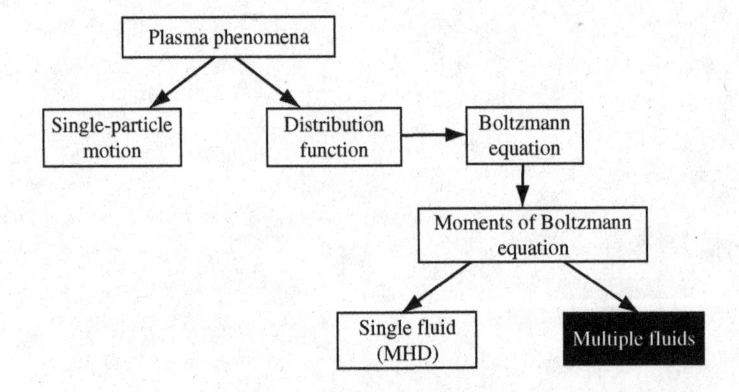

7.1 Introduction

Having outlined the different approaches to plasma phenomena, in this chapter we present a closer examination of the effects of particle collisions, which were mentioned only in passing in previous chapters. Although the physics of particle collisions occurs on the scale of individual particles, the collective effects of such collisions are manifested in the macroscopic parameters of plasmas, e.g., plasma conductivity. In general, plasmas can be classified into two types, collisional and collisionless. In collisionless plasmas, the collisional encounters between the constituents of the plasma gas are so infrequent that their effects on plasma dynamics can be neglected. Most of the plasma in the Earth's near-space environment, except at altitudes below ~500 km above the surface, can be treated as collisionless. In collisional plasmas, which are common in many technological applications, collisions occur frequently enough that the resultant momentum exchange between the particles is significant and may even dominate the plasma behavior.

Collisions facilitate various transport processes in a plasma. Consider a plasma with a spatial inhomogeneity, for example with a plasma density N that varies in space. As a result of their thermal motions, electrons collide with other particles. Although each electron individually moves at random, the collisions generate a tendency for the population to drift away from the high-density regions. Similar drift would occur if the spatial inhomogeneity involved a variation of plasma temperature with space. This type of drift of particles in inhomogeneous plasmas, facilitated by collisions, is called *diffusion*. If the spatial inhomogeneity provides a pressure or density gradient in a prescribed direction, the presence of such a drift makes the velocity distribution function anisotropic. However, such a non-equilibrium state of affairs can only be temporary; given enough time, an equilibrium should be reached, the average drift velocity goes to zero, and the distribution function becomes isotropic.

In the presence of an external force, such as an electric field, such drift of electrons (and ions) can occur even if the plasma is homogeneous. This type of drift is exactly analogous to conduction in an ordinary conductor. Individually the electrons move at random because of their finite temperature but, on the average, they drift in the direction opposite to the electric field. The parameter which describes the drift of particles in a plasma under the influence of an external electric field E is *mobility*, given by $\mu = |\mathbf{u}|/E$. Since the electrons have mass and carry charge, their drift corresponds to transport of mass and the conduction of electricity, i.e., electric current.

In an inhomogeneous plasma under the influence of an external field, particle transport is due to both diffusion and mobility. These two transport processes can be separated only if the average velocity due to the external electric field is small compared to the thermal velocity. It should be noted that since the electrons have kinetic energy, their drift also results in the transport of energy and conduction of heat; however, as before, we assume no significant energy or heat transport within the plasma, amounting to truncation of the moments of the Boltzmann equation, as discussed in previous chapters.

7.2 Collisions

Collisional plasmas can be further divided into two classes, partially ionized plasmas and fully ionized plasmas. In partially or weakly ionized plasmas, there exists a large background of relatively large neutral constituents, while fully ionized plasmas consist of only electrons and ions. In weakly ionized plasmas, the dominant collisional process is between electrons and neutrals, involving essentially head-on encounters, while in fully ionized plasmas the collisions are encounters between charged particles, governed by the Coulomb interaction force. In this section we briefly consider collisions with neutrals and those between charged particles.

7.2.1 Weakly ionized plasmas

Neutral particles in a weakly ionized plasma impede the motion of charged particles by their simple presence as heavy, compact obstacles. When an electron (or an ion) collides with a neutral atom, it may lose all of its momentum or only part of it, depending on the angle at which it rebounds. The probability of momentum loss can be expressed in terms of the equivalent molecular cross-section σ_n of the neutral atom. The electron–neutral collision frequency ν_en (the number of collisions per second) is then proportional to the number density N_n of neutral particles, the average (over velocity space) velocity $\langle v \rangle$, and σ_n. In other words,

$$\nu_\text{en} = N_n \sigma_\text{n} \langle v \rangle. \tag{7.1}$$

The molecular cross-section can be approximated as $\sigma_\text{n} = \pi a^2$, with a being the radius of the nuclei, so that we can often take $\sigma_\text{n} \simeq 10^{-19}$ m^2. An average over velocity space is needed since particles with different velocities may have different individual collision frequencies. In the most general cases, the cross-section σ_n may itself be a function of the velocity of the incident particle. Although the particle collision frequency ν is the only parameter that we will use in our analysis of conductivity and diffusion processes, it is useful to note some related quantities in passing. The so-called mean free path λ_en, given by

$$\lambda_\text{en} = \frac{1}{N_n \sigma_\text{n} \langle v \rangle},$$

is the average distance a particle travels before having a collision. The mean time τ between collisions is given by $\tau = N_n \sigma_\text{n} \langle v \rangle$.

7.2.2 Fully ionized plasmas: Coulomb collisions

In fully ionized plasmas, all collisions are between charged particles which interact by means of their electric Coulomb fields. Because of the long range of Coulomb forces, charged particles approaching one another are deflected at interparticle distances much larger than the atomic radius. In effect, the Coulomb interaction enhances the cross-section of the colliding particles; however, it also means that most particles will suffer only small-angle deflections. In a sufficiently dense plasma, the Coulomb potential is screened and the electric field is approximately confined to a Debye sphere, and we might think that the effective cross-section is approximately equal to $\pi\lambda_{\mathrm{D}}^2$. However, this is not accurate since particles with sufficient energy can penetrate the Debye sphere. Since the potential increases steeply with distance into the Debye sphere, most of the deflection occurs inside the sphere, with small-angle deflections being much more likely than large-angle ones. The Coulomb collision frequency in a fully ionized plasma can be written in a form identical to (7.1). The electron–ion collision frequency is given by

$$\nu_{\mathrm{ei}} = N_e \sigma_{\mathrm{c}} \langle v_e \rangle, \tag{7.2}$$

where we note that the total ion density $N_i = N_e$ because of charge neutrality. In order to estimate the numerical value of ν_{ei}, we must determine the Coulomb scattering cross-section σ_{c} on the basis of the microphysics of the collision. In the following, we present a simplified order-of-magnitude derivation of σ_{c} for the case of an electron colliding with a positively charged ion. Consider the situation shown in Figure 7.1, with an electron approaching a single heavy ion. To a first approximation, we can consider the ion to be at rest, with the electron trajectory deflected as a result of its attraction toward the ion. In a fully ionized plasma, the plasma temperature is high enough that the ion cannot trap the electron, which will simply move around the ion and escape, usually with a small deflection angle χ. The electron orbit is in fact a hyperbola, which can be

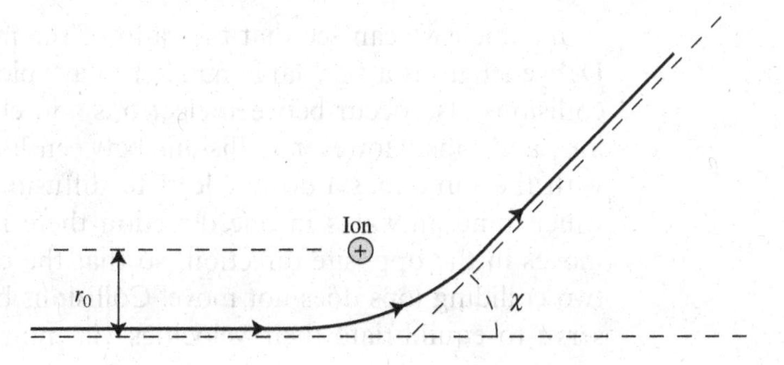

Figure 7.1 Orbit of an electron undergoing a Coulomb collision with an ion.

approximated as straight lines at large distances from the ion. The distance of closest approach in the absence of Coulomb attraction, denoted by r_0, is known as the impact parameter. The Coulomb force between the electron and the singly charged ion ($q_i = -q_e$) is

$$F_c = -\frac{q_e^2}{4\pi\epsilon_0 r_0^2}. \tag{7.3}$$

This force is felt by the electron only during an approximate average time $\tau_c \simeq r_0/v_e$, when it passes near the ion. The Coulomb collision cross-section σ_c is given by $\sigma_c = \pi r_0^2$. The change in momentum experienced by the electron during this time is given by

$$\Delta(m_e v_e) \simeq |F_c \tau_c| = \frac{q_e^2}{4\pi\epsilon_0 v_e r_0}. \tag{7.4}$$

From a detailed examination of the probability of large- versus small-angle collisions, the change in momentum $\Delta(m_e v_e)$ can be expressed as a fraction of the particle momentum $m_e v_e$, and it can be shown (see [1]) that

$$\nu_{ei} \simeq \frac{\sqrt{2}\omega_{pe}^4}{64\pi N_e}\left(\frac{k_B T_e}{m_e}\right)^{-\frac{3}{2}} \ln\Lambda, \tag{7.5}$$

where T_e appears when we replace v_e with $\langle v_e \rangle$ and use $\frac{1}{2}m_e\langle v_e^2\rangle = \frac{3}{2}k_B T_e$, and where $\Lambda = N_e\lambda_D^3$ is the so-called plasma parameter. The quantity $\ln\Lambda$ is sometimes referred to as the Coulomb logarithm; in most plasmas, since Λ is a very large number the value of $\ln\Lambda$ is in the range of 10 to 30. Using the definitions of various quantities, equation (7.5) can be simplified as

$$\nu_{ei} \simeq \frac{\omega_{pe}}{64\pi}\frac{\ln\Lambda}{\Lambda}. \tag{7.6}$$

The corresponding mean free path can be written as

$$\lambda_{ei} = \frac{\langle v_e \rangle}{\nu_{ei}} \simeq 64\pi\lambda_D \frac{\Lambda}{\ln\Lambda}, \tag{7.7}$$

from which we can see that the ratio of the mean free path to the Debye length is a very large number in a typical plasma. Coulomb collisions also occur between electrons and electrons and between ions and ions. However, collisions between like particles (particles with the same mass) do not lead to diffusion, since for each ion which random-walks in one direction there is another ion which moves in the opposite direction, so that the center of mass of the two colliding ions does not move. Collisions between like particles serve to equilibriate their velocities via momentum exchange, so that particles of the same species assume a well-defined average temperature. The temperatures of particles of differing masses can remain different for much longer times.

7.2.3 Specific resistivity

In Chapter 6, Equation (6.17) we expressed the electron–ion collision term in the generalized Ohm's law as

$$\mathbf{S}_{\text{ei}} = \eta q_e^2 N_e^2 (\mathbf{u}_i - \mathbf{u}_e)$$

and noted that the constant of proportionality η was known as the specific resistivity of the plasma. Alternatively, the loss of momentum due to collisions can be expressed in terms of the collision frequency (i.e., the number of collisions per second) ν_{ei} as

$$\mathbf{S}_{\text{ei}} = m_e N_e \nu_{\text{ei}} (\mathbf{u}_i - \mathbf{u}_e), \tag{7.8}$$

which implicitly assumes that the electrons lose all of their relative momentum in collisions with the much heavier ions. Comparison of (7.8) and (6.17) indicates that

$$\text{Specific resistivity} \quad \boxed{\eta = \frac{m_e \nu}{N_e q_e^2}}. \tag{7.9}$$

The relation (7.9) is valid for weakly ionized or fully ionized plasmas, as long as we use the appropriate collision frequency. In other words, for weakly ionized plasmas we can replace ν with the electron–neutral collision frequency ν_{en}, while for fully ionized plasmas ν is equal to the electron–ion collision frequency ν_{ei}. For a fully ionized plasma we can use (7.6) to write an expression for the resistivity in terms only of the parameter Λ and the plasma frequency ω_{pe}:

$$\eta \simeq \frac{1}{64\pi\epsilon_0\,\omega_{pe}} \frac{\ln\Lambda}{\Lambda}. \tag{7.10}$$

This resistivity of an isotropic, fully ionized plasma is referred to as the *Spitzer resistivity*. It is interesting to note that since ω_{pe} is directly proportional to $N_e^{\frac{1}{2}}$ while Λ is proportional to $N_e^{-\frac{1}{2}}$,[1] and

[1] From Chapter 1 we have

$$\omega_{pe} = \sqrt{\frac{N_e q_e^2}{m_e \epsilon_0}} \quad \text{and} \quad \lambda_{\text{D}} = \sqrt{\frac{\epsilon_0 k_{\text{B}} T_e}{N_e q_e^2}}.$$

With $\Lambda = N_e \lambda_{\text{D}}^3$, Λ is proportional to $N_e^{-\frac{1}{2}}$.

since $\ln \Lambda$ is roughly a constant, the Spitzer resistivity is independent of plasma density. This can be understood as follows: if we increase the plasma density by adding more charge carriers we do get an increase in current, but we also increase the collision frequency and the frictional drag, which decreases the drift velocity of the charge carriers.

In a weakly ionized plasma, the collision term in the electron momentum equation (4.30) can in general be written as

$$m_e N_e \left[\frac{\partial \mathbf{u}_e}{\partial t} + (\mathbf{u}_e \cdot \nabla)\mathbf{u} \right] = -\nabla \cdot \Psi_e + q_e N_e (\mathbf{E} + \mathbf{u}_e \times \mathbf{B}) + \underbrace{-\sum m_e N_e \nu_{ej} (\mathbf{u}_e - \mathbf{u}_j)}_{\mathbf{S}_{ej}} . \quad (7.11)$$

The dominant collision process in weakly ionized plasmas is between electrons and neutrals. To first order the rate of momentum transfer to electrons as a result of collisions with neutrals can be described as above, with the summation carried over all the neutral species j. The key assumptions under which we write (7.11) are that the number density of the neutral particles is much larger than the electron density N_e and that the mass of a neutral particle is essentially infinite (compared to m_e), so that the entire momentum of an electron is lost in its collision with the neutral particle. If ν_{ej} is the number per second (i.e., frequency) of such collisions, the average (over the ensemble) rate of loss of momentum of a fluid element containing N_e electrons is then $\nu_{ej} m_e N_e \mathbf{u}_e$. If the plasma is assumed to be isotropic, with the neutrals typically assumed to be stationary ($\mathbf{u}_j = 0$), the momentum transport equation reduces to

$$\underbrace{m_e N_e \left[\frac{\partial \mathbf{u}_e}{\partial t} + (\mathbf{u} \cdot \nabla)\mathbf{u} \right]}_{d\mathbf{u}/dt} = q_e N_e (\mathbf{E} + \mathbf{u}_e \times \mathbf{B}) - \nabla p - m_e N_e \nu \, \mathbf{u}_e, \quad (7.12)$$

where $\nu = \nu_{\text{eff}}$ is an *effective collision frequency* that is the sum of the different ν_{ej}, and is no longer necessarily the actual number of collisions per second with any neutral constituent. We typically drop the subscript "eff" for brevity, and will simply refer to this quantity as ν in the rest of this chapter and later on when we discuss the role of collisional processes in electromagnetic wave propagation in plasmas. Typically, the electron–neutral collision frequency in weakly ionized plasmas is a function of the electron temperature and the neutral density; application-specific examples of such functions are presented in Example 4-2 of Chapter 4 and Example 7-1 in this chapter.

7.3 Plasma conductivity

With the collision frequency determined as given in (7.1) or (7.6), the transport of each particle species is governed by the momentum transport equation for that species, which for electrons is given by (7.12). In this section we focus our attention on the phenomenon of conduction in homogeneous plasmas (i.e., $\nabla \cdot \Psi = 0$). The pressure-gradient term will be considered in the next chapter, when we study diffusion processes. To determine expressions for the conductivity of the plasma medium, we alternatively consider DC ($\partial/\partial t = 0$) and AC cases (quantities varying harmonically as $e^{j\omega t}$), and relate the current density $\mathbf{J}$ and the electric field $\mathbf{E}$, with $\mathbf{J} = q_e N_e \mathbf{u}_e$. For convenience and brevity, we neglect the contribution of ion motion to the electrical conductivity, essentially assuming the much heavier ions to be immobile, i.e., $\mathbf{u}_i = 0$. Later we will see that the inclusion of ion motions is actually quite straightforward. We start by linearizing the momentum equation (7.12), by assuming that the electric field $\mathbf{E}$ and fluid velocity $\mathbf{u}_e$ are small-signal quantities, while the magnetic field and electron density exhibit small-signal variations around large steady values. In other words,

$$\mathbf{B} = \mathbf{B}_0 + \mathbf{B}_1, \qquad N_e = N_{e0} + N_{e1},$$

where subscript "1" indicates small variations while subscript "0" indicates background (ambient) values. Note that we have $\mathbf{E}_0 = 0$ and $\mathbf{u}_{e0} = 0$, so that $\mathbf{E} = \mathbf{E}_1$ and $\mathbf{u}_e = \mathbf{u}_{e1}$. The linearized version of (7.12) is obtained by neglecting all terms involving products of any two small-signal quantities. We find

$$m_e N_{e0} \frac{\partial \mathbf{u}_e}{\partial t} \simeq q_e N_{e0}(\mathbf{E} + \mathbf{u}_e \times \mathbf{B}_0) - m_e N_{e0} \nu\, \mathbf{u}_e. \tag{7.13}$$

In the following we will often denote N_{e0} simply by N_e.

7.3.1 DC conductivity

We consider DC conductivity of the plasma under the influence of an applied electric field which is constant in time and spatially uniform. In such a case we have $\partial/\partial t = 0$, so that (7.13) reduces to

$$0 \simeq q_e N_e(\mathbf{E} + \mathbf{u}_e \times \mathbf{B}_0) - m_e N_e \nu\, \mathbf{u}_e. \tag{7.14}$$

In the case of an isotropic plasma with no magnetic field ($\mathbf{B}_0 = 0$), we see that the electric field force is balanced by the collision term:

$$q_e N_e \mathbf{E} \simeq m_e N_e \nu \, \mathbf{u}_e \quad \rightarrow \quad \mathbf{u}_e = \frac{q_e \mathbf{E}}{m_e \, \nu}. \tag{7.15}$$

The electric current density associated with this motion of the electron fluid is

$$\mathbf{J} = q_e N_e \mathbf{u}_e \quad \rightarrow \quad \mathbf{J} = \underbrace{\frac{N_e q_e^2}{m_e \nu}}_{\sigma_{\mathrm{dc}}} \mathbf{E}, \tag{7.16}$$

so that the DC conductivity of an unmagnetized (isotropic) plasma is

$$\text{DC conductivity} \quad \boxed{\sigma_{\mathrm{dc}} = \frac{N_e q_e^2}{m_e \nu}}. \tag{7.17}$$

The electron mobility μ_e is defined as the ratio of the fluid velocity to the applied electric field:

$$\mu_e \equiv \frac{|\mathbf{u}_e|}{|\mathbf{E}|} = \frac{q_e}{m_e \nu}. \tag{7.18}$$

As we saw in previous chapters, the presence of a static magnetic field $\mathbf{B}_0$ makes the plasma anisotropic, with electric current flow in directions other than that of the applied electric field, thus requiring that the conductivity be expressed as a tensor. From (7.14) we have

$$q_e N_e (\mathbf{E} + \mathbf{u}_e \times \mathbf{B}_0) \simeq m_e N_e \nu \, \mathbf{u}_e \quad \rightarrow$$

$$\mathbf{J} \equiv q_e N_e \mathbf{u}_e = \underbrace{\frac{N_e q_e^2}{m_e \nu}}_{\sigma_{\mathrm{dc}}} (\mathbf{E} + \mathbf{u}_e \times \mathbf{B}_0), \tag{7.19}$$

which is a simplified form of the generalized Ohm's law (6.20) that we also encountered as Equation (6.22). We now seek to write $\mathbf{J}$ in terms of $\mathbf{E}$ using a tensor conductivity $\overleftrightarrow{\sigma}$:

$$\mathbf{J} = \overleftrightarrow{\sigma} \cdot \mathbf{E}. \tag{7.20}$$

To determine $\overleftrightarrow{\sigma}$, we align our coordinate system so that the z axis is along the magnetic field, i.e., $\mathbf{B}_0 = \hat{\mathbf{z}} B_0$. From (7.19) we have

$$\mathbf{J} = \sigma_{\mathrm{dc}} \mathbf{E} + \frac{\sigma_{\mathrm{dc}} B_0}{q_e N_e} (\mathbf{J} \times \hat{\mathbf{z}}) = \sigma_{\mathrm{dc}} \mathbf{E} + \frac{\sigma_{\mathrm{dc}} B_0}{q_e N_e} (\hat{\mathbf{x}} J_y - \hat{\mathbf{y}} J_x), \tag{7.21}$$

which can be written in component form as

$$J_x = \sigma_{\text{dc}} E_x + \frac{\omega_{ce}}{\nu} J_y \tag{7.22a}$$

$$J_y = \sigma_{\text{dc}} E_y - \frac{\omega_{ce}}{\nu} J_x \tag{7.22b}$$

$$J_z = \sigma_{\text{dc}} E_z, \tag{7.22c}$$

where $\omega_{ce} = q_e B_0 / m_e$. We can combine (7.22a) and (7.22b) to rewrite (7.22) as

$$J_x = \frac{\nu^2}{\omega_{ce}^2 + \nu^2} \sigma_{\text{dc}} E_x + \frac{\nu \omega_{ce}}{\omega_{ce}^2 + \nu^2} \sigma_{\text{dc}} E_y \tag{7.23a}$$

$$J_y = \frac{-\nu \omega_{ce}}{\omega_{ce}^2 + \nu^2} \sigma_{\text{dc}} E_x + \frac{\nu^2}{\omega_{ce}^2 + \nu^2} \sigma_{\text{dc}} E_y \tag{7.23b}$$

$$J_z = \sigma_{\text{dc}} E_z, \tag{7.23c}$$

which can be expressed in tensor form as

$$\begin{bmatrix} J_x \\ J_y \\ J_z \end{bmatrix} = \underbrace{\begin{bmatrix} \dfrac{\sigma_{\text{dc}} \nu^2}{\omega_{ce}^2 + \nu^2} & \dfrac{\sigma_{\text{dc}} \nu \omega_{ce}}{\omega_{ce}^2 + \nu^2} & 0 \\ \dfrac{-\sigma_{\text{dc}} \nu \omega_{ce}}{\omega_{ce}^2 + \nu^2} & \dfrac{\sigma_{\text{dc}} \nu^2}{\omega_{ce}^2 + \nu^2} & 0 \\ 0 & 0 & \sigma_{\text{dc}} \end{bmatrix}}_{\overleftrightarrow{\sigma}} \begin{bmatrix} E_x \\ E_y \\ E_z \end{bmatrix} = \underbrace{\begin{bmatrix} \sigma_\perp & \sigma_{\text{H}} & 0 \\ -\sigma_{\text{H}} & \sigma_\perp & 0 \\ 0 & 0 & \sigma_\parallel \end{bmatrix}}_{\overleftrightarrow{\sigma}} \begin{bmatrix} E_x \\ E_y \\ E_z \end{bmatrix}, \tag{7.24}$$

where $\sigma_\parallel \equiv \sigma_{\text{dc}}$. The various entries in $\overleftrightarrow{\sigma}$ represent the conductivity of the magnetized plasma for current flow in various directions, as shown in Figure 7.2. The *parallel conductivity* (also referred to as the *longitudinal conductivity*) is equal to the DC conductivity and represents current flow in the direction parallel to $\mathbf{B}_0$. The conductivity labeled $\sigma_\perp$, the so-called *Pedersen conductivity*, represents current flow along the component of the electric field which is normal to the magnetic field, while the *Hall conductivity* σ_{H} represents current flow in the direction perpendicular to both $\mathbf{E}$ and $\mathbf{B}_0$. Note that for $\mathbf{B}_0 = 0$ we have $\omega_{ce} = 0$, so that $\sigma_\perp = \sigma_{\text{dc}}$, and $\sigma_{\text{H}} = 0$, so that the tensor conductivity $\overleftrightarrow{\sigma}$ reduces to the scalar σ_{dc}.

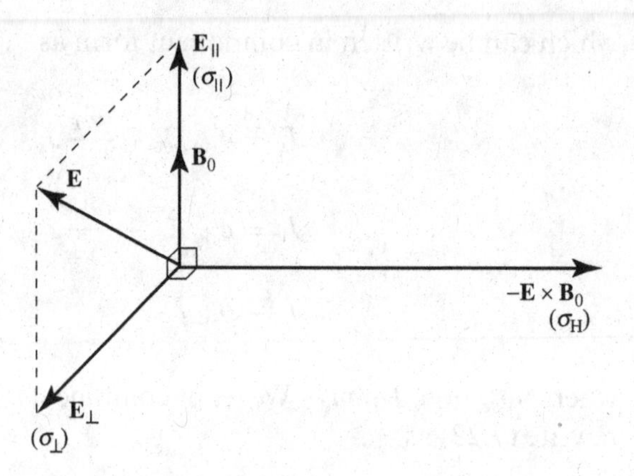

Figure 7.2 Current flows in an anisotropic magnetoplasma. The electric field is decomposed into its components parallel and perpendicular to $\mathbf{B}_0$. Current-flow directions are represented by the components of $\overleftrightarrow{\sigma}$.

Example 7-1 Ionospheric heating

The HAARP[2] ionospheric heating facility in Gakona, Alaska, can change the conductivity of the overhead plasma in the lower ionospheric D-region (altitude 75–95 km) by heating the electrons with high-power (3.6 MW) radio waves in the 2–9 MHz band. The Earth's magnetic field at an altitude of 85 km has an intensity of 54 μT, the daytime electron density is 10^9 m^{-3}, and the electron–neutral collision frequency is a function of temperature, given by

$$\nu_{\text{en}}(T_e) = 5278\sqrt{T_e} + 2752T_e - 0.31T_e^2 \quad \text{(units of Hz)}.$$

If HAARP can change the electron temperature from 200 K to 600 K, calculate the change in conductivity for all parts of the DC conductivity tensor.

Solution: The DC conductivity tensor is given by Equation (7.24):

$$\overleftrightarrow{\sigma} = \begin{bmatrix} \dfrac{\sigma_{\text{dc}}\nu^2}{\omega_{ce}^2+\nu^2} & \dfrac{\sigma_{\text{dc}}\nu\omega_{ce}}{\omega_{ce}^2+\nu^2} & 0 \\ \dfrac{-\sigma_{\text{dc}}\nu\omega_{ce}}{\omega_{ce}^2+\nu^2} & \dfrac{\sigma_{\text{dc}}\nu^2}{\omega_{ce}^2+\nu^2} & 0 \\ 0 & 0 & \sigma_{\text{dc}} \end{bmatrix} = \begin{bmatrix} \sigma_\perp & \sigma_H & 0 \\ -\sigma_H & \sigma_\perp & 0 \\ 0 & 0 & \sigma_\parallel \end{bmatrix}.$$

[2] HAARP stands for High Frequency Active Auroral Research Program.

The change in conductivity is given by $\Delta\overleftrightarrow{\sigma} = \overleftrightarrow{\sigma}(200\text{ K}) - \overleftrightarrow{\sigma}(600\text{ K})$. The collision frequency ($\nu = \nu_{en}$) and σ_{dc} both change with temperature. Evaluating all parameters,

$$\omega_{ce} = q_e B/m_e = \frac{(1.6 \times 10^{-19})(54 \times 10^{-6})}{9.1 \times 10^{-31}} = 9.57 \times 10^6 \text{ rad s}^{-1}$$

$$\nu_{en}(200\text{ K}) = 6.13 \times 10^5 \text{ Hz}$$

$$\nu_{en}(600\text{ K}) = 1.67 \times 10^6 \text{ Hz}$$

$$\sigma_{dc}(200\text{ K}) = \frac{N_e q_e^2}{m_e \nu} = \frac{(10^9)(1.6 \times 10^{-19})^2}{(9.1 \times 10^{-31})(6.13 \times 10^5)}$$

$$= 4.59 \times 10^{-5} \text{ S m}^{-1}$$

$$\sigma_{dc}(600\text{ K}) = \frac{N_e q_e^2}{m_e \nu} = \frac{(10^9)(1.6 \times 10^{-19})^2}{(9.1 \times 10^{-31})(1.67 \times 10^6)}$$

$$= 1.68 \times 10^{-5} \text{ S m}^{-1},$$

which allows evaluation of $\Delta\overleftrightarrow{\sigma}$:

$$\Delta\sigma_{\perp} = (1.88 \times 10^{-7}) - (4.96 \times 10^{-7}) = -3.08 \times 10^{-7} \text{ S m}^{-1}$$

$$\Delta\sigma_{H} = (2.93 \times 10^{-6}) - (2.85 \times 10^{-6}) = 8.08 \times 10^{-8} \text{ S m}^{-1}$$

$$\Delta\sigma_{\parallel} = (4.59 \times 10^{-5}) - (1.68 \times 10^{-5}) = 2.91 \times 10^{-5} \text{ S m}^{-1}.$$

7.3.2 AC conductivity

We now consider the conductivity of the magnetized plasma for an applied electric field which is time-harmonic in nature, i.e., varying as $e^{j\omega t}$. The linearized momentum equation (7.13) can then be written

$$j\omega m_e \mathbf{u}_e = q_e(\mathbf{E} + \mathbf{u}_e \times \mathbf{B}_0) - m_e \nu \mathbf{u}_e, \tag{7.25}$$

which can be rewritten

$$0 \simeq q_e(\mathbf{E} + \mathbf{u}_e \times \mathbf{B}) - m_e(\nu + j\omega)\mathbf{u}_e. \tag{7.26}$$

This is identical to (7.14) if we make the substitution $\nu \rightarrow (\nu + j\omega)$. Thus, the entries of the AC conductivity tensor can be obtained directly by making this same substitution:

$$\sigma_\perp = \frac{(\nu + j\omega)^2 \, \sigma_{\rm dc}}{(\nu + j\omega)^2 + \omega_{ce}^2} \tag{7.27a}$$

$$\sigma_{\rm H} = \frac{(\nu + j\omega)\omega_{ce} \, \sigma_{\rm dc}}{(\nu + j\omega)^2 + \omega_{ce}^2} \tag{7.27b}$$

$$\sigma_{\rm dc} = \sigma_\parallel = \frac{N_e q_e^2(\nu - j\omega)}{m_e^2(\nu^2 + \omega^2)}. \tag{7.27c}$$

7.3.3 Conductivity with ion motion

Although we have so far neglected ion motions for the sake of simplicity, the inclusion of ion effects is quite straightforward. Basically, ion motions contribute to the current density in an additive manner, i.e.,

$$\mathbf{J} = N_e q_e \mathbf{u}_e + \sum_k N_k q_k \mathbf{u}_k,$$

where the summation is over the different types of ions, each having their own charge, density, and fluid velocity. Thus, for a plasma composed of electrons and several species of ions, the components of the conductivity tensor become

$$\sigma_\perp = \epsilon_0 \left[\frac{\omega_{pe}^2(\nu_e + j\omega)}{(\nu_e + j\omega)^2 + \omega_{ce}^2} + \sum_k \frac{\omega_{pk}^2(\nu_k + j\omega)}{(\nu_k + j\omega)^2 + \omega_{ck}^2} \right] \tag{7.28a}$$

$$\sigma_{\rm H} = \epsilon_0 \left[\frac{\omega_{pe}^2 \omega_{ce}}{(\nu_e + j\omega)^2 + \omega_{ce}^2} + \sum_k \frac{\omega_{pk}^2 \omega_{ck}}{(\nu_k + j\omega)^2 + \omega_{ck}^2} \right] \tag{7.28b}$$

$$\sigma_\parallel = \epsilon_0 \left[\frac{\omega_{pe}^2}{(\nu_e + j\omega)} + \sum_k \frac{\omega_{pk}^2}{(\nu_k + j\omega)} \right], \tag{7.28c}$$

where ω_{pk} and ω_{ck} are, respectively, the ion-plasma and ion-cyclotron frequencies of the various ion species. Note that the DC conductivity with ion motion included can be obtained by simply setting $\omega = 0$ in Equation (7.28).

7.4 Summary

In this chapter we explored particle collisions in partially and fully ionized plasmas. In weakly ionized plasmas the dominant collision process is between electrons and neutral molecules, while in fully ionized plasmas collisions are governed by Coulomb forces. In both cases the collisions can be represented by an effective collision frequency which is a function of average particle velocity and an equivalent cross-section. Having an expression for plasma collisions allows us to calculate plasma conductivity by finding the relation between the current density (**J**) and the electric field (**E**), which can be done for DC ($\partial/\partial t = 0$) and AC (quantities varying harmonically as $e^{j\omega t}$) cases. The presence of a static magnetic field introduces anisotropy and requires that the conductivity be expressed as a tensor.

7.5 Problems

7-1. Consider a toroidal plasma-confinement system in which a current is driven in a fully ionized plasma by an electric field applied along **B**, which is azimuthal – along the toroid. What is the intensity of the electric field that must be applied in order to drive a total current of 150 kA if the plasma has a temperature of $k_B = 400$ eV and a cross-sectional area of 85 cm^2?

7-2. Estimate the mean free path of electrons in an atmospheric-pressure glow discharge (weakly ionized plasma).

7-3. Calculate the specific resistivity of a fully ionized plasma of density $N = 10^{12}$ m^{-3} and temperature 1 keV.

7-4. Consider a fully ionized electron–proton plasma in a fusion reactor with density $N = 10^{21}$ m^{-3} and temperature of 10 keV. (a) Find the electron–ion collision frequency. (b) Find the impact parameter r_0 and compare it to the radius of a proton, which is approximately 10^{-15} m.

7-5. Evaluate the change in the AC conductivity tensor in the ionosphere created by the HAARP HF heater described in Example 7-1, if the frequency in question is 2 kHz.

7-6. Measurements using a rocket indicate that horizontal and vertical electric fields (i.e., the $\hat{\mathbf{x}}$ and $\hat{\mathbf{z}}$ components) in the ionosphere at 130 km altitude are ~0.5 mV m^{-1} and ~10 mV m^{-1}, respectively, while the Pedersen and Hall conductivities at this altitude are measured as $\sigma_\perp \simeq 3 \times 10^4$ S m^{-1} and $\sigma_H \simeq 4 \times 10^4$, respectively. The Earth's magnetic field at that altitude is $\mathbf{B}_0 = \hat{\mathbf{z}}3 \times 10^{-5}$ T. (a) Calculate the collision frequency ν, electron-cyclotron frequency ω_c and DC conductivity σ_{dc}. (b) Estimate the magnitude of the three components of the ionospheric current density $\mathbf{J}$.

Reference

[1] J. A. Bittencourt, *Fundamentals of Plasma Physics*, 3rd edn (New York: Springer-Verlag, 2004), 560–88.

CHAPTER

8 Plasma diffusion

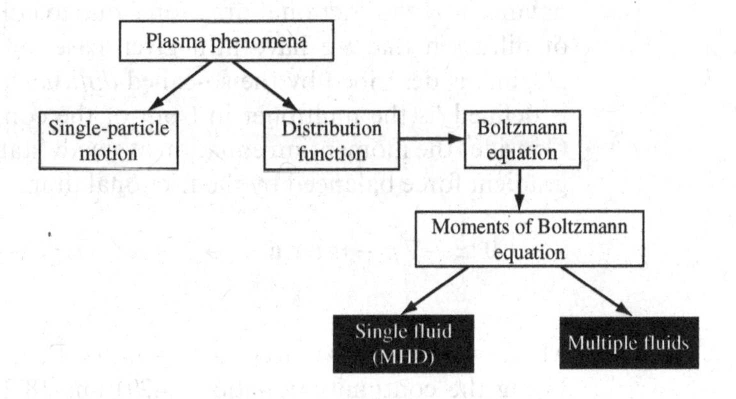

8.1 Introduction

Diffusion is the term used to describe plasma motion due to a non-uniform distribution of charged particles or a non-uniform distribution of temperature in a plasma. When a charged particle collides with another particle (either charged or neutral), its velocity vector undergoes an abrupt change, causing the particle to move from one collisionless orbit to another. After many such collisions, the particle would have wandered a significant distance away from its original location. In a uniform plasma, this process would not have caused net migration of particles since, on average, another particle would have moved into the original location of the first particle. In an inhomogeneous plasma, however, the result of the collisional motion is a net migration of particles from high-density regions to low-density regions, i.e., diffusion. Diffusion is a random-walk process. The particles take one step at a time, each step being (on average) Δx, and the time between steps being Δt. Consider very many similar particles all starting initially at position $x = 0$. As time passes, the average position of the particles is not expected to change, so that the quantity $\langle x \rangle$ remains zero. However, on probabilistic grounds alone it can be shown that $\langle x^2 \rangle$, the variance of the particle position, increases linearly with time:

$$\frac{d\left[\langle x^2 \rangle\right]}{dt} = \frac{(\Delta x)^2}{\Delta t} \quad \rightarrow \quad \langle x^2 \rangle = \left[\frac{(\Delta x)^2}{\Delta t}\right] t, \qquad (8.1)$$

indicating that the root-mean-square spread in the positions of the particles increases as $\sqrt{t}$.[1] We will show below that (8.1) is the natural solution of the diffusion equation for plasmas. The basic cause of diffusion is the $F = -\nabla p$ force, which in diffusive motion is balanced by the frictional drag force due to collisions. The amount of diffusion that we have in a given case, or the diffusivity of a plasma, is described by the so-called *diffusion coefficient*, D, which is defined as the multiplier in front of the concentration gradient. Consider the momentum equation at steady state, with the pressure-gradient force balanced by the frictional drag:

$$0 \simeq -\nabla p - mN\nu\,\mathbf{u} \quad \rightarrow \quad \Gamma \equiv N\mathbf{u} = -\underbrace{\frac{k_B T}{m\nu}}_{D} \nabla N, \qquad (8.2)$$

where we have introduced the quantity Γ, the flux of particles.[2] Using the continuity equation (4.29) and (8.2) we obtain the so-called *diffusion equation*, which governs the time evolution of particle density:

$$\frac{\partial N}{\partial t} + \nabla \cdot (N\mathbf{u}) = 0 \quad \rightarrow \quad \frac{\partial N}{\partial t} + \nabla \cdot \Gamma = 0. \qquad (8.3)$$

[1] For this purpose, we can consider a total of n random steps, k of which are to the right (i.e., involve increases in x) and $n - k$ to the left. While the probability of each unique sequence of n steps is 2^{-n}, the total number of sequences which involve k steps to the right is $n!/[k!(n-k)!]$, meaning that the probability of having k steps (out of n) to the right is $P_n(k) = 2^{-n}\, n!/[k!(n-k)!]$. After taking the n steps in a time $t = n\Delta t$, the net distance traveled by the particle is $x = k\Delta x - (n-k)\Delta x = (2k - n)\Delta x$. The ensemble average of the quantity x is then given as

$$\langle x^2 \rangle = (\Delta x)^2 \sum_{k=0}^{n} (2k-n)^2 \underbrace{\frac{2^{-n}\, n!}{k!(n-k)!}}_{P_n(k)} = n\,(\Delta x)^2 = \left[\frac{(\Delta x)^2}{\Delta t}\right] t,$$

where some degree of manipulation can be used to show that the summation is simply equal to n.

[2] The equation $\Gamma = -D\nabla N$ is known as *Fick's first law* of organized motions, relating the flux of a species of particles to the gradient of density.

Substituting (8.2) in (8.3) we obtain the diffusion equation

$$\text{Diffusion equation} \quad \boxed{\frac{\partial N}{\partial t} = D\nabla^2 N}, \tag{8.4}$$

the one-dimensional version of which is

$$\frac{\partial N}{\partial t} = D\frac{\partial^2 N}{\partial x^2}. \tag{8.5}$$

For the case when all of the particles start at $x=0$ at $t=0$, the exact solution of this equation is

$$N(x, t) = \left[\frac{N^2}{4\pi Dt}\right]^{1/2} e^{-x^2/(4Dt)}. \tag{8.6}$$

The ensemble average of the square of the spread in particle positions is then

$$\langle x^2 \rangle = \frac{1}{N}\int x^2 N(x, t)\, dx = 2\,Dt. \tag{8.7}$$

Comparing (8.7) and (8.1) we see that

$$D = \frac{(\Delta x)^2}{2\Delta t}. \tag{8.8}$$

On a heuristic basis, we expect the diffusion coefficient to be related to the time interval between collisions, $\tau = \nu^{-1}$, and to Δx, the ensemble average distance traveled between collisions, often called the *mean free path*, λ. Thus, it makes sense that D is proportional to $(\Delta x)^2/\Delta t$, or $\nu\lambda^2$. Since the mean free path is obtained by dividing the average velocity by the mean time between collisions, and since the average velocity squared is related to temperature, we see that D is also proportional to $k_B T \Delta t/m$. In magnetized plasmas, charged particles move freely along the magnetic field, unimpeded except for collisions. If there are density gradients along **B**, diffusion will occur in the same manner as in an unmagnetized plasma, since **B** does not affect the motion along **B**. Accordingly, we would expect the parallel diffusion coefficient $D_\parallel$ to be proportional to $k_B T/(m\nu)$.

Note that for diffusion along the magnetic field, as well as for diffusion in an unmagnetized plasma, collisions *impede* transport. On the other hand, for diffusion across the magnetic field, collisions *facilitate* transport. If there were no collisions, particles would not migrate across **B** and would simply continue to gyrate indefinitely about the same field line. There may of course be drifts across **B** due to gradients, curvature, or electric fields perpendicular to **B**; however, in practice these are often arranged to form closed drift orbits within a bounded plasma. For example, in a cylindrical plasma column the electric field and the gradients are in the radial direction, so that the drifts are in the azimuthal (ϕ) direction and do not carry the particles away from the plasma. When there are collisions, on the other hand, particles migrate across **B** by a random-walk process. Assuming that particles complete at least one gyration before colliding, we will show later that the step size in their random walk is no longer λ but rather the gyroradius r_c, and that the perpendicular diffusion coefficient $D_\perp$ is proportional to νr_c^2. The role of the collision frequency ν is reversed between $D_\parallel$ and $D_\perp$, as expected on the basis that collisions are the only means for perpendicular diffusion. In the following, we separately consider diffusion in weakly and fully ionized plasmas.

8.2 Diffusion in weakly ionized plasmas

We first consider diffusion in the absence of a magnetic field, realizing that the same analysis is valid for diffusion along **B** in a magnetized plasma. We then discuss diffusion across the **B** field.

8.2.1 Ambipolar diffusion in an unmagnetized plasma

The fluid equation of motion including collisions, valid for each species, is

$$mN\frac{d\mathbf{u}}{dt} = mN\left[\frac{\partial \mathbf{u}}{\partial t} + (\mathbf{u}\cdot\nabla)\mathbf{u}\right] = q\mathbf{E} - \nabla p - mN\nu\mathbf{u}. \tag{8.9}$$

We assume that ν is a known constant. First consider steady-state conditions such that $\partial\mathbf{u}/\partial t = 0$. If $\mathbf{u}$ is sufficiently small, or ν is sufficiently large, we may assume that a fluid element will not move into regions of different **E** or ∇p within a time $\tau = \nu^{-1}$, so that the convective term in (8.9) can also be neglected. We can then write

$$\mathbf{u} = \frac{1}{mN\nu}(q\mathbf{E} - k_B T\nabla N) = \underbrace{\frac{q}{m\nu}}_{\mu}\mathbf{E} - \underbrace{\frac{k_B T}{m\nu}}_{D}\frac{\nabla N}{N}, \tag{8.10}$$

where we have also assumed an isothermal plasma (i.e., $p = Nk_B T$). Note that we have identified the coefficients in front of the two motion components respectively as the mobility μ and the diffusion coefficient D. The relationship between mobility and diffusion coefficient is known as the Einstein relation:

$$\mu = \frac{qD}{k_B T}. \tag{8.11}$$

In diffusion processes we are typically concerned with the flux Γ of particles given by

$$\Gamma = N\mathbf{u} = \mu N\mathbf{E} - D\nabla N. \tag{8.12}$$

It can be seen from (8.10) that the diffusion coefficient is inversely proportional to the particle mass m. One might then think that the electrons would diffuse much more quickly than the ions. However, in a plasma, which must remain charge-neutral to a very high degree of accuracy, net motion of electrons and ions at separate rates cannot occur. If the plasma is to remain neutral, the fluxes of electrons and ions will somehow adjust themselves so that the two species diffuse at the same rate. The means of this adjustment is the electric field which arises as soon as a slight charge imbalance occurs. In unmagnetized plasmas, or for diffusion along the field in a magnetized plasma, the electrons undergo faster diffusion and tend to leave the ions behind, creating an electric field directed outward from the plasma, of such a magnitude that the preferential loss of electrons is impeded, although the loss of ions will necessarily be enhanced by the same electric field. For diffusion across the $\mathbf{B}$ field, we expect the ions to migrate more rapidly in view of their much larger gyroradii. The resultant electric field should thus be inwardly directed, i.e., toward the region of highest plasma density, to impede the preferential loss of ions. The process which involves the simultaneous migration of electrons and ions in a plasma is referred to as *ambipolar diffusion*. The electric field required for ambipolar diffusion can be found by setting $\Gamma_e = \Gamma_i$, which in view of charge neutrality in turn means $\mathbf{u}_e = \mathbf{u}_i$. We then have, from (8.12),

$$\mu_i N\mathbf{E} - D_i\nabla N = -\mu_e N\mathbf{E} - D_e\nabla N \quad \rightarrow \quad \mathbf{E} = \frac{D_i - D_e}{\mu_i + \mu_e}\frac{\nabla N}{N}. \tag{8.13}$$

The flux of particles is then given by (note that $\Gamma = \Gamma_i = \Gamma_e$)

$$\Gamma = \mu_i \frac{D_i - D_e}{\mu_i + \mu_e}\nabla N - D_i\nabla N = -\underbrace{\frac{\mu_i D_e + \mu_e D_i}{\mu_i + \mu_e}}_{D_\mathrm{a}}\nabla N, \tag{8.14}$$

identifying D_a as the ambipolar diffusion coefficient.

Ambipolar (unmagnetized) diffusion coefficient

$$\boxed{D_\mathrm{a} = \frac{\mu_i D_e + \mu_e D_i}{\mu_i + \mu_e}}. \tag{8.15}$$

In studying the diffusion of a plasma, for example to determine how a plasma created in a container decays by diffusion to the walls, the governing equations are the continuity equation and the equation of motion. Our neglect of the fluid velocity time derivative in the equation of motion, but not in the continuity equation, is the equivalent of assuming that diffusion is a slow process, which is usually the case. Specifically, if the characteristic time scale of diffusion $\tau_D \gg \tau = \nu^{-1}$ then the $\partial\mathbf{u}/\partial t$ term can be neglected.

Noting that $\mu_e \gg \mu_i$ (assuming that electron and ion temperatures are equal, and noting that the thermal velocities are proportional to $m^{-\frac{1}{2}}$, we can see that ν is proportional to $m^{-\frac{1}{2}}$, so $\mu = q/(m\nu)$ is also proportional to $m^{-\frac{1}{2}}$), we can write D_a as

$$D_\mathrm{a} \simeq D_i + \frac{\mu_i}{\mu_e} D_e = D_i + \frac{T_e}{T_i} D_i \simeq 2D_i. \tag{8.16}$$

Thus the effect of the ambipolar electric field is to enhance the diffusion of ions by a factor of two.

8.2.2 Free diffusion across a magnetic field

We now consider the case of a weakly ionized plasma in a magnetic field. Since the diffusion in the parallel direction is the same as for an unmagnetized plasma, we only consider the perpendicular component of motion. We have

$$mN\frac{d\mathbf{u}_\perp}{dt} = q_e N(\mathbf{E} + \mathbf{u}_\perp \times \mathbf{B}) - k_\mathrm{B} T \nabla N - mN\nu\mathbf{u}_\perp. \tag{8.17}$$

Once again considering steady-state conditions ($\partial/\partial t = 0$), and neglecting the convective term, and after separating $\mathbf{u}_\perp$ into its components and some algebra (similar to that in Section 5.3), it can be shown that

$$\mathbf{u}_\perp = \mu_\perp \mathbf{E} - D_\perp \frac{\nabla N}{N} + \omega_c^2 \frac{\mathbf{u}_\mathrm{E} + \mathbf{u}_\mathrm{D}}{\omega_c^2 \nu^2}, \tag{8.18}$$

where $\mathbf{u}_\mathrm{E} \equiv (\mathbf{E} \times \mathbf{B})/B^2$ and $\mathbf{u}_\mathrm{D} \equiv -(\nabla p \times \mathbf{B})/(qNB^2)$ are respectively the $\mathbf{E} \times \mathbf{B}$ and diamagnetic drift velocities. The perpendicular mobility and diffusion coefficient are given by

$$\mu_\perp = \frac{(q/m\nu)}{1 + \omega_c^2 \tau^2} \quad \text{and} \quad D_\perp = \frac{k_\mathrm{B} T/(m\nu)}{1 + \omega_c^2 \tau^2}. \tag{8.19}$$

Note that (8.18) applies separately for each species. Note that the perpendicular motion of particles is composed of two parts: the usual drifts, which are impeded by collisions, and the mobility and diffusion drifts along the gradients in electric potential and particle density. These drifts have the same form as in the case for no magnetic field, but the coefficients μ and D are reduced by the factor $1+\omega_c^2\tau^2$. When $\omega_c^2\tau^2 \ll 1$ the magnetic field has little effect on diffusion. On the other hand, when $\omega_c^2\tau^2 \gg 1$ the magnetic field significantly retards the rate of diffusion across **B**. In the limit $\omega_c^2\tau^2 \gg 1$ we have

$$D_\perp \simeq \frac{k_B T \nu}{m\omega_c^2}, \tag{8.20}$$

confirming our heuristic expectation that the role of collision frequency in the diffusion is reversed with respect to the case in an unmagnetized plasma. In other words, $D_\perp$ is proportional to ν, since collisions are needed for migration. To first order, we can write (8.20) as

$$D_\perp \simeq \frac{k_B T \nu}{m\omega_c^2} \simeq v_{th}^2 \frac{r_c^2}{v_{th}^2} \nu \simeq r_c^2 \nu,$$

indicating that diffusion across the magnetic field is a random walk with a step length r_c, rather than the mean free path λ.

Note that (8.17) can also be rewritten in tensor form for both parallel and perpendicular components, as

$$\Gamma = -\nabla \cdot \left[\overleftrightarrow{D} N \right], \tag{8.21}$$

where

$$\overleftrightarrow{D} = \begin{bmatrix} D_\perp & D_H & 0 \\ -D_H & D_\perp & 0 \\ 0 & 0 & D_\parallel \end{bmatrix} \tag{8.22}$$

is the tensor diffusion coefficient, analogous to the conductivity tensor discussed in Chapter 7 (see Equation (7.24)). The additional components are

$$D_\parallel = \frac{k_B T}{m\nu}, \quad D_H = \frac{\omega_c k_B T/(m\nu^2)}{1+\omega_c^2\tau^2}.$$

Just as in the case of the conductivity tensor, the D_H components of the diffusion tensor represent flow in directions perpendicular to the magnetic field but also perpendicular to the direction of density gradients.

Ambipolar diffusion across B

Since the diffusion and mobility coefficients are anisotropic, the problem of ambipolar diffusion in a magnetic field is not straightforward. Consider, for example, particle fluxes across **B**. Since $D_\perp$ is larger for ions, a transverse electric field would be set up so as to aid electron diffusion and retard ion diffusion. However, this electric field can be short-circuited by an imbalance of fluxes parallel to **B**: the negative charge that stays behind because ions diffuse faster in the transverse direction can be taken away by rapid diffusion of electrons along **B**. In other words, although the total diffusion must be ambipolar, the transverse portion does not need to be ambipolar. Whether the diffusion across **B** or along **B** is dominant depends on the particular configuration of the plasma and experimental conditions. In plasmas trapped in a mirror geometry, the loss of electrons along **B** is typically much faster than the loss of ions across the field lines. In closed plasma configurations (e.g., a toroid) there is no possibility of escape along the field lines and the cross-field loss of ions is the dominant factor.

8.3 Diffusion in fully ionized plasmas

The case of diffusion in fully ionized plasmas can best be studied by considering the plasma as a single conducting fluid. At steady state, the magnetic and pressure-gradient forces are balanced so that, from (6.23c), we have

$$0 \simeq -\nabla p + \mathbf{J} \times \mathbf{B}. \tag{8.23}$$

Also valid is (6.22):

$$\mathbf{J} = \sigma(\mathbf{E} + \mathbf{u} \times \mathbf{B}).$$

Taking the cross-product of (6.22) with **B** gives

$$\mathbf{J} \times \mathbf{B} = \sigma(\mathbf{E} \times \mathbf{B} - B^2\mathbf{u}_\perp) \quad \rightarrow \quad \mathbf{u}_\perp = \underbrace{\frac{\mathbf{E} \times \mathbf{B}}{B^2}}_{\mathbf{u}_\mathrm{E}} - \frac{\nabla p}{\sigma B^2}, \tag{8.24}$$

indicating that the plasma fluid drifts at a velocity $(\mathbf{E} \times \mathbf{B})/B^2$ while diffusing under the influence of the pressure gradient. The flux due to diffusion is given as

$$\Gamma_\perp = N\mathbf{u}_\perp = -\frac{N\nabla p}{\sigma B^2}. \tag{8.25}$$

For a two-fluid plasma consisting of electrons and ions, we have

$$p = p_e + p_i = Nk_\mathrm{B}(T_e + T_i),$$

so that

$$\Gamma_\perp = -\underbrace{\frac{Nk_B(T_e + T_i)}{\sigma B^2}}_{D_\perp} \nabla N = -D_\perp \nabla N, \tag{8.26}$$

thus identifying the so-called *classical diffusion coefficient* for a fully ionized plasma.

Classical diffusion (magnetized, fully ionized plasma)

$$\boxed{D_\perp = \frac{Nk_B(T_e + T_i)}{\sigma B^2}}. \tag{8.27}$$

Note that this type of diffusion in a fully ionized plasma is inherently ambipolar, since we have assumed the quasi-neutrality condition ($\nabla \cdot \mathbf{J} = 0$) so that the electrons and ions stay together and diffuse together.

As can be seen, the classical diffusion coefficient has an inverse-square dependence on the magnetic field B. However, in experiments on plasma confinement $D_\perp$ has been found to have a dependence of $1/B$, leading to much faster diffusion (decay) of plasma densities than predicted by the the analysis above. This anomalous diffusion was first noted by Bohm, who came up with the following empirical formula:

Bohm (empirical) diffusion

$$\boxed{D_\perp = D_B = \frac{k_B T_e}{16 q_e B}}. \tag{8.28}$$

The *Bohm diffusion coefficient* is an experimental result. It is believed that the disparity between the Bohm and classical coefficients is caused by plasma turbulence, a subject that is beyond the scope of this text.

8.4 Summary

In this chapter we investigated diffusion, the motion of plasmas due to density and temperature gradients. We showed that such motion is proportional to the density gradient, multiplied by an appropriate diffusion coefficient derived from the fluid equation of motion and continuity equation. For partially ionized plasmas a diffusion coefficient can be calculated separately for each species, but quasi-neutrality requires that oppositely charged particles move at the same rate, yielding an effective of ambipolar diffusion rate. Fully ionized plasmas are treated with a single-fluid model that is inherently ambipolar. A key difference between diffusion in magnetized and unmagnetized plasmas is the role of collisions. In the absence of a magnetic field (or along magnetic field lines), collisions generally impede movement. In movement across a magnetic field, however, collisions facilitate transport by interrupting the indefinite gyration that would otherwise occur.

8.5 Problems

8-1. A cylindrical, fully ionized plasma column has a density distribution of

$$N = N_0 \left(1 - \frac{r^2}{a^2}\right),$$

where $a = 10$ cm and $N_0 = 10^{19}$ m^{-3}. If $k_B T_e = 100$ eV, $k_B T_i = 0$, and the axial magnetic field B_0 is 1 T: (a) What are the values of the Bohm and classical diffusion coefficients? (b) What is the ratio of the two diffusion coefficients? (c) What will be the characteristic diffusion time and loss rate from the plasma column?

8-2. At relatively high altitudes in the Earth's ionosphere, electron density varies with altitude as

$$N_e(z) \simeq 10^5 e^{-z/H} \text{ cm}^{-3},$$

where the scale height $H \simeq 300$ km and the electron temperature is approximately $T_e \simeq 3.5 \times 10^3$ K. Estimate the ambipolar electric field associated with this density profile, and estimate the associated departure from charge neutrality (i.e., find $|N_i - N_e|/N_e$, where N_i is the ion density). State all assumptions.

8-3. The electron–neutral collision cross-section for 2 eV electrons in He is $\sigma_{en} = 6\pi a_0^2$, where $a_0 \simeq 0.53 \times 10^{-8}$ cm is the radius of the first Bohr orbit of the hydrogen atom. The ion–neutral collision cross-section can be assumed to be approximately the same, i.e., $\sigma_{in} \simeq \sigma_{en}$. A very long cylindrical plasma column of radius $r_0 = 1$ cm has He pressure of 1 Torr (at room temperature), $k_B T_e = 2$ eV, and $k_B T_i = 0.1$ eV.[3] A magnetic field of magnitude $B = 0.2$ T is aligned along the column. The general solution of the diffusion equation in cylindrical coordinates is given by

[3] 1 Torr = 133.322 Pa.

$$N(r,t) = \begin{cases} N_0 e^{-t/\tau_d} J_0\left[\dfrac{r}{\sqrt{D\tau_d}}\right] & r \le r_0 \\ 0 & r > r_0 \end{cases},$$

where τ_d is the diffusion time, D is the diffusion coefficient, and $J_0(\zeta)$ is the Bessel function of the first kind and zeroth order, which is known to have a first zero at $\zeta = 2.405$ ($J_0(2.405) = 0$). (a) Considering only diffusion across the field lines, use Equation (8.15) to define the perpendicular ambipolar diffusion coefficient $D_{a\perp}$ and determine its numerical value. Since the column is very long you can assume that the normally confounding effects of simultaneous parallel diffusion can be neglected. (b) Show that $D_{a\perp}$ can be approximated by the free-electron diffusion coefficient $D_{e\perp}$.

8-4. Explain the different roles played by collisions in diffusion in fully ionized plasmas across and along the magnetic field.

8-5. Estimate the ambipolar diffusion coefficient for a partially ionized plasma with electron temperature of 1 eV, ion temperature of 0.1 eV, and collision frequency with neutrals for both ions and electrons of 3 GHz.

CHAPTER

9 Introduction to waves in plasmas

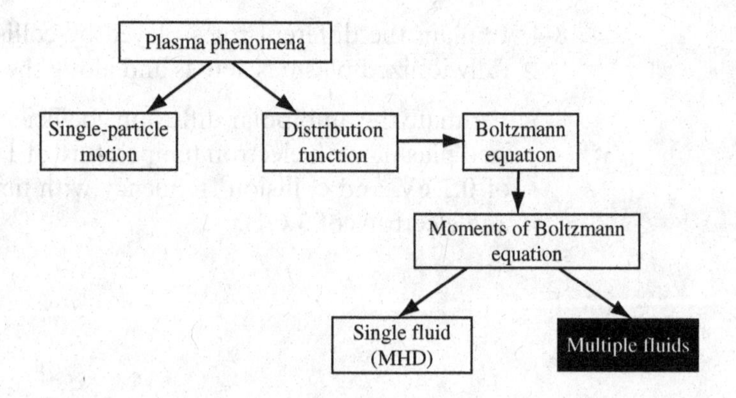

9.1 Introduction

An ionized gas is capable of a wide variety of oscillatory motions, which can in general be exceedingly complex. The subject of waves in plasmas is important not only because the natural dynamical motions of plasmas often lead to waves but also because waves are often used to excite, perturb, or probe plasmas. Waves are also important because they carry energy from the surface of a plasma (where the waves may have been excited) into the bulk of the plasma, where the waves may be absorbed and, for example, heat the plasma. The study of the propagation, linear and non-linear, of electrostatic and electromagnetic waves constitutes a large portion of the discipline of plasma physics. In our presentation, we will limit ourselves exclusively to purely sinusoidal phenomena and relatively small oscillations so that the relevant fluid equations can be "linearized" by ignoring terms which are "second-order" in the perturbed quantities. The waves we consider here are plane waves propagating in an unbounded, homogeneous, and time-independent medium. Understanding these simple oscillations will give us insight into the more complicated wave phenomena which may occur. Also, in spite of the idealized nature of such plane waves, they have been found extremely useful in interpreting experimental observations. In fact, early observations of a rich variety of waves in the Earth's ionosphere provided the stimulus for the development of many important aspects of plasma-wave theory. In this and the following two chapters, we will discuss three different types of waves in a plasma: *electrostatic waves*, *electromagnetic waves*, and *hydromagnetic waves*. The fundamental equations that we will solve for all types of waves are Maxwell's equations, from which the following wave equation can be derived:

$$\nabla^2 \mathbf{E} - \mu\epsilon \frac{\partial^2 \mathbf{E}}{\partial t^2} = \mu_0 \frac{\partial \mathbf{J}}{\partial t} + \frac{1}{\epsilon_0} \nabla \rho. \tag{9.1}$$

To determine $\mathbf{E}$, we must solve (9.1) together with the momentum transport and continuity equations derived in Chapters 5 and 6, which relate $\mathbf{E}$ and $\mathbf{B}$ to the fluid velocity $\mathbf{u}$ and the density N, which in turn are related to the current density $\mathbf{J}$ and the charge density ρ by

$$\mathbf{J}(\mathbf{r}, t) = \sum_i q_i N_i(\mathbf{r}, t) \mathbf{u}_i(\mathbf{r}, t) \tag{9.2a}$$

$$\rho(\mathbf{r}, t) = \sum_i q_i N_i(\mathbf{r}, t), \tag{9.2b}$$

where the summations are over distinct particle species. Waves in plasmas are most often treated with the multiple-fluid plasma model presented in Chapter 5. Notable exceptions are hydrodynamic waves, which are based on a single-fluid treatment (see Chapter 11), and waves in hot plasmas, which require working directly with the distribution function (see Chapter 12).

The different types of waves are obtained by applying different simplifying assumptions. If $\mathbf{E}$ and $\mathbf{J}$ are parallel to the direction of wave propagation, electrostatic restoring forces are present, and such *longitudinal* waves are called *electrostatic waves*. In electron oscillations of this type the frequency is so large that the positive ions are not affected (because of their inertia), while in positive-ion oscillations the frequency is so low that the electrons can be assumed to be in equilibrium at all times in accordance with the Maxwell–Boltzmann distribution.

When the electric field $\mathbf{E}$ is perpendicular to the wave propagation direction, we have *transverse* waves which are *electromagnetic.* The electrons in the plasma interfere with the transverse electromagnetic waves, and increase their phase velocity. In the absence of a static magnetic field, electromagnetic waves at a given frequency can only propagate through a plasma if the electron density is lower than a given value. The presence of an external magnetic field removes this requirement as new, lower-frequency modes emerge, as discussed in Chapter 10.

Hydromagnetic or MHD waves can occur only in the presence of a magnetic field, and then only for frequencies small compared to the cyclotron frequency of the ions. In a hydromagnetic wave, the positive ions provide the inertia of the oscillation, while the restoring forces are largely magnetic, resulting from the $\mathbf{J} \times \mathbf{B}$ term in the momentum equation. These oscillations may be regarded as waves in the lines of magnetic force, which behave as stretched strings and which are "loaded" with charged particles.

9.2 General properties of small-amplitude waves

Our purpose is to analyze the properties of waves of small amplitude superimposed on a background uniform and unbounded plasma. The background (or ambient) values of the quantities N, $\mathbf{u}$, p, $\mathbf{E}$, and $\mathbf{B}$ will be denoted by subscript "0" (e.g., $\mathbf{B}_0$ or N_0), while the perturbation quantities will be denoted by subscript "1" (i.e., $\mathbf{B}_1$ or N_1). However, we will exclusively consider the cases in which there is no steady fluid motion and no imposed external electric field, so that $\mathbf{u}_0 = 0$ and $\mathbf{E}_0 = 0$. It is clear from (9.2a) that we then have $\mathbf{J}_0 = 0$. Thus we do not need the subscripts for $\mathbf{u}$, $\mathbf{E}$, and $\mathbf{J}$, and we have

$$N = N_0 + N_1 \tag{9.3a}$$

$$p = p_0 + p_1 \tag{9.3b}$$

$$\mathbf{B} = \mathbf{B}_0 + \mathbf{B}_1, \tag{9.3c}$$

where N_0, p_0, and $\mathbf{B}_0$ are constants which do not vary with space or time. Neglecting collisions, and also assuming an adiabatic equation of state (4.35), we can write the linearized versions of the fluid equations (valid for each species separately, as discussed before) as

$$\frac{\partial N_1}{\partial t} + N_0 \nabla \cdot \mathbf{u} = 0 \tag{9.4a}$$

$$N_0 m \frac{\partial \mathbf{u}}{\partial t} = q N_0 (\mathbf{E} + \mathbf{u} \times \mathbf{B}_0) - \nabla p_1 \tag{9.4b}$$

$$\frac{p_1}{p_1} = \gamma \frac{N_1}{N_0} \quad \rightarrow \quad p_1 = \gamma k_B T N_1, \tag{9.4c}$$

where we have assumed each species to be a perfect isothermal gas at some temperature T (which could be different for electrons and for ions), so that $p_0 = N_0 k_B T$. Characteristics of plasma waves are determined by solving (9.4) and (9.2) together with Maxwell's equations, repeated here:

$$\nabla \times \mathbf{B}_1 = \mu_0 \mathbf{J} + \epsilon_0 \frac{\partial \mathbf{E}}{\partial t} \tag{9.5a}$$

$$\nabla \times \mathbf{E} = -\frac{\partial \mathbf{B}_1}{\partial t} \tag{9.5b}$$

$$\nabla \cdot \mathbf{E} = \frac{\rho_1}{\epsilon_0} \tag{9.5c}$$

$$\nabla \cdot \mathbf{B}_1 = 0, \tag{9.5d}$$

where the ambient value of the charge density ρ is also zero as a result of the macroscopic charge neutrality of the plasma. To study plane waves in plasmas, we search for solutions of the linearized fluid equations and Maxwell's equations in which all perturbation quantities vary proportional to

$$e^{j(\omega t - \mathbf{k} \cdot \mathbf{r})}. \tag{9.6}$$

More precisely, each physical quantity (e.g., the electric field or the electron density perturbation) is represented by a complex phasor (e.g., $\mathbf{E}(\mathbf{r})$ or $\rho(\mathbf{r})$) given by

$$\mathbf{E}(\mathbf{r}) = \mathbf{E}_c e^{-j\mathbf{k} \cdot \mathbf{r}} \quad \text{or} \quad N_1(\mathbf{r}) = N_c e^{-j\mathbf{k} \cdot \mathbf{r}}, \tag{9.7}$$

where $\mathbf{E}_c$ is a complex vector constant and N_c is a complex scalar constant. The actual physical quantities (e.g., $\overline{\mathcal{E}}(\mathbf{r}, t)$ or $\widetilde{N}_1(\mathbf{r}, t)$) can be obtained from their associated phasors by the following operation:

$$\overline{\mathcal{E}}(\mathbf{r}, t) \equiv \mathcal{R}e\left\{\mathbf{E}(\mathbf{r}) e^{j\omega t}\right\} \quad \text{or} \quad \widetilde{N}_1(\mathbf{r}) \equiv \mathcal{R}e\left\{N_1(\mathbf{r}) e^{j\omega t}\right\}. \tag{9.8}$$

Under these conditions, Maxwell's equations can be rewritten[1] as

$$\mathbf{k} \times \mathbf{B}_1 = j\mu_0 \mathbf{J} - \omega\epsilon_0\mu_0 \mathbf{E} \tag{9.9a}$$

$$\mathbf{k} \times \mathbf{E} = \omega \mathbf{B}_1 \tag{9.9b}$$

$$\mathbf{k} \cdot \mathbf{E} = \frac{j\rho_1}{\epsilon_0} \tag{9.9c}$$

$$\mathbf{k} \cdot \mathbf{B}_1 = 0, \tag{9.9d}$$

while the fluid equations become

$$j\omega N_1 - jN_0\, \mathbf{k} \cdot \mathbf{u} = 0 \tag{9.10a}$$

$$N_0 m\, j\omega\, \mathbf{u} = qN_0(\mathbf{E} + \mathbf{u} \times \mathbf{B}_0) - \nabla p_1 \tag{9.10b}$$

$$p_1 = \gamma k_B T N_1. \tag{9.10c}$$

It should be noted that the equation of continuity of electric current,

$$\nabla \cdot \overline{\mathcal{J}} + \frac{\partial \tilde{\rho}}{\partial t} = 0 \quad \text{or} \quad \mathbf{k} \cdot \mathbf{J} = \omega\rho, \tag{9.11}$$

is contained in Maxwell's equations (9.9), and is also implied by (9.4a). In studying plasma waves, we will usually not use (9.11) or (9.9c), for reasons previously mentioned, namely that charge neutrality is preserved to a very large degree. One exception to this occurs when we study longitudinal waves, i.e., waves for which $\mathbf{E}$, $\mathbf{B}_1$, $\mathbf{u}$, and $\mathbf{J}$ are all parallel to the wave propagation direction. Since we then have $\mathbf{k} \times \mathbf{E} = 0$, we must have $\mathbf{B}_1 = 0$, and (9.9d) is redundant. With $\mathbf{k} \times \mathbf{B}_1 = 0$, (9.9a) becomes equivalent to (9.9c). In such cases, it becomes more natural to use (9.9c) since this can be interpreted as Maxwell's equations having been reduced to Poisson's equation. Note that for transverse waves, the electric and magnetic fields are both perpendicular to the propagation direction $\hat{\mathbf{k}}$ ($\mathbf{B}_1$ is

[1] Note for example that

$$\mathbf{E} = \mathbf{E}_c e^{-j\mathbf{k}\cdot\mathbf{r}} \quad \to \quad \nabla \times \mathbf{E}(\mathbf{r}) = \nabla \times \mathbf{E}_c e^{-j\mathbf{k}\cdot\mathbf{r}} = -\mathbf{E}_c \times \nabla e^{-j\mathbf{k}\cdot\mathbf{r}}$$

$$= -\mathbf{E}_c \times \left[\hat{\mathbf{x}}\frac{\partial}{\partial x} + \hat{\mathbf{y}}\frac{\partial}{\partial y} + \hat{\mathbf{z}}\frac{\partial}{\partial z}\right] e^{-j(k_x x + k_y y + k_z z)}$$

$$= -\mathbf{E}_c \times \left[-j(\hat{\mathbf{x}}\, k_x + \hat{\mathbf{y}}\, k_y + \hat{\mathbf{z}}\, k_z)\right] e^{-j(k_x x + k_y y + k_z z)}$$

$$= -\mathbf{E}_c \times (-j\mathbf{k}) e^{-j\mathbf{k}\cdot\mathbf{r}} = -j\,\mathbf{k} \times \mathbf{E}_c e^{-j\mathbf{k}\cdot\mathbf{r}}$$

$$\nabla \times \mathbf{E}(\mathbf{r}) = -j\,\mathbf{k} \times \underbrace{\mathbf{E}_c e^{-j\mathbf{k}\cdot\mathbf{r}}}_{\mathbf{E}(\mathbf{r})} = -j\,\mathbf{k} \times \mathbf{E}(\mathbf{r}).$$

Similarly, it can be shown that $\nabla \cdot \mathbf{E} = -j\mathbf{k} \cdot \mathbf{E}$.

always perpendicular to **k** on account of (9.9d)). For such waves, the charge density ρ must vanish according to (9.9c). Transverse waves are similar in many ways to uniform plane electromagnetic waves in free space, except for various cutoffs and resonances, as we will see later. In general, our goal in plasma-wave analysis is to obtain a relationship between the wave number $k = |\mathbf{k}|$ and the wave frequency ω. This relationship is called the *dispersion relation* for the particular wave under study. Under the assumption of sinusoidal variation of quantities in time and space, the dispersion relation becomes the last remaining unknown. It is easy to see that solutions of the form (9.6) represent plane electromagnetic waves propagating with a *phase velocity*

$$v_p = \frac{\omega}{k}, \tag{9.12}$$

while the *group velocity*, being the velocity with which energy or information can travel, is given by

$$v_g = \frac{d\omega}{dk}. \tag{9.13}$$

9.3 Waves in non-magnetized plasmas

We first investigate waves that can propagate in non-magnetized plasmas, so that $\mathbf{B}_0 = 0$. These include plasma oscillations, which are localized and can hardly be called "waves," transverse electron plasma waves which have an electromagnetic character, and longitudinal electrostatic electron and ion waves which are similar to sound waves in a gas.

9.3.1 Plasma oscillations

We have already seen (Chapter 1) how plasma oscillations arise from the natural tendency of the plasma to maintain charge neutrality. In the context of the formalism that we now have for the study of plasma waves, we can arrive at the same conclusions as we did in Chapter 1 by adopting the cold-plasma assumptions, so that $T = 0$, and consider only the motion of the electrons, assuming the ions to be stationary. Equation (9.10b) then reduces to

$$m_e j\omega \mathbf{u}_e = q_e \mathbf{E}. \tag{9.14}$$

Taking the divergence of (9.14) and using (9.4a) (after replacing $\partial/\partial t$ with $j\omega$) we find

$$N_1 = \frac{N_0 q_e}{m_e \omega^2} \nabla \cdot \mathbf{E} \quad \rightarrow \quad \rho = N_1 q_e = \frac{N_0 q_e^2}{m_e \omega^2} \nabla \cdot \mathbf{E}. \qquad (9.15)$$

We now look for a longitudinal wave so that $\nabla \cdot \mathbf{E} \neq 0$. Combining (9.15) and (9.5c) we find that is necessary to have

$$\omega^2 = \frac{N_0 q_e^2}{m_e \epsilon_0} \equiv \omega_{pe}^2, \qquad (9.16)$$

where ω_{pe} is the electron plasma frequency. This result indicates that all quantities ($\mathbf{E}$, n_1, ρ, $\mathbf{u}$) oscillate in time at a frequency ω_{pe}. The electron gas is compressed and rarefied in a manner similar to compressional sound waves, but the restoring force is the electric field $\mathbf{E}$ set up by charge separation, rather than pressure. In arriving at (9.16), it was not necessary to assume a particular spatial form of the disturbance, i.e., the wave number $\mathbf{k}$ did not appear in the derivation, indicating that the plasma oscillates at the frequency ω_{pe} regardless of the size of the disturbed region. Since ω and k are not related, the group velocity $v_g = 0$. Thus, disturbances in the form of plasma oscillations do not propagate away from the locale in which they occur. Indeed, this phenomena is referred to as an "oscillation" rather than a "wave." We also note here that the assumption that the ions are stationary is in no way a severe restriction on our solution for plasma oscillations. In fact, as is shown in an exercise left to the reader at the end of this chapter, inclusion of ion motions simply brings in a slight correction to the oscillation frequency, namely

$$\omega^2 = \omega_p^2 = \omega_{pe}^2 + \omega_{pi}^2 = \omega_{pe}^2 \left(1 + \frac{m_e}{m_i}\right). \qquad (9.17)$$

9.3.2 Transverse electromagnetic waves

To search for transverse waves for which $\nabla \cdot \mathbf{E} = 0$, it is not possible to use (9.15) or (9.5c). However, we can find wave solutions by following the usual procedure of working with the two curl equations (9.9a) and (9.9b). The current density $\mathbf{J}$ needed in (9.9a) can be written as

$$\mathbf{J} = N_0 q_e \mathbf{u}_e = \frac{N_0 q_e^2}{m_e j\omega} \mathbf{E}, \qquad (9.18)$$

where we have used (9.14). Since transverse waves are similar in nature to uniform plane waves in ordinary dielectric media, it is convenient to define an electric flux density vector $\mathbf{D}$, so that the right-hand side of (9.9a) can be written as $-\mu_0\omega\mathbf{D}$, as in simple dielectrics. We have

$$-\omega\mu_0\mathbf{D} = j\mu_0\mathbf{J} - \omega\epsilon_0\mu_0\mathbf{E} \quad \rightarrow \quad \mathbf{D} = \epsilon_{\text{eff}}\mathbf{E}, \tag{9.19}$$

where the effective dielectric constant is

$$\epsilon_{\text{eff}} = \epsilon_0\left(1 - \frac{N_0 q_e^2}{\epsilon_0 m_e \omega^2}\right) = \epsilon_0\left(1 - \frac{\omega_{pe}^2}{\omega^2}\right). \tag{9.20}$$

With this definition the two Maxwell's equations (9.5a) and (9.5b) can be written as

$$\nabla \times \mathbf{B}_1 = j\omega\,\mu_0\,\epsilon_{\text{eff}}\,\mathbf{E} \tag{9.21a}$$

$$\nabla \times \mathbf{E} = -jw\mathbf{B}_1. \tag{9.21b}$$

Since these equations are identical to Maxwell's equations for simple dielectric media, transverse uniform plane-wave solutions for simple dielectric media must also be valid in a plasma as long as we make the substitution $\epsilon \rightarrow \epsilon_{\text{eff}}$. The refractive index of such waves is simply given as $\sqrt{\epsilon}$, so that we have[2]

$$\mu \equiv \frac{kc}{\omega} = \left(1 - \frac{\omega_{pe}^2}{\omega^2}\right)^{\frac{1}{2}}. \tag{9.22}$$

It can be noted from (9.22) that $\mu < 1$, indicating that the phase velocity of the transverse plasma waves is greater than the speed of light in free space c. However, we can easily see that the group velocity $v_g = c^2/v_p$ and is thus less than c. It can also be noted from (9.22) that for $\omega < \omega_{pe}$ the refractive index is imaginary, as is the wave number k, given by

$$k = \pm\frac{j}{c}\sqrt{\omega_{pe}^2 - \omega^2}. \tag{9.23}$$

Thus, $\omega = \omega_{pe}$ is the *cutoff frequency* at which $k = 0$, v_p is infinite, and the group velocity $v_g = 0$. Sinusoidal excitations at frequencies below ω_{pe} do not propagate in a plasma and are rapidly attenuated. Such waves are called *evanescent*. The relationship between ω and k is plotted in Figure 9.1 for real values of ω (i.e., propagating waves). Also shown is the dispersion relation for waves in free space, which follows a straight line, $w = ck$. Note that for

[2] Short of symbols, we choose to use μ for refractive index, not to be confused with magnetic moment, also denoted with μ in earlier chapters.

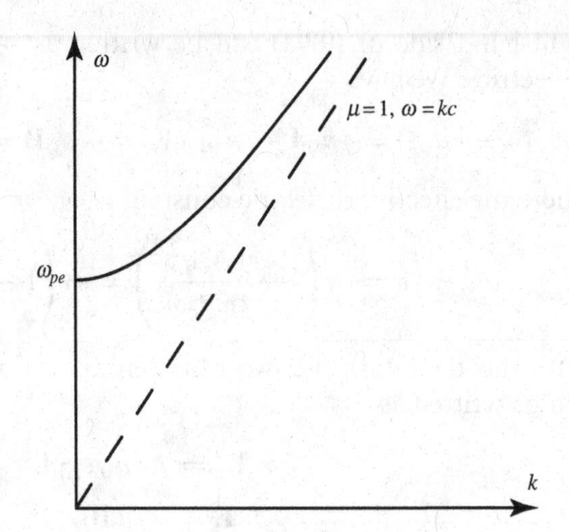

Figure 9.1 Dispersion relation for electromagnetic waves in a non-magnetized plasma (solid line) and in free space (dashed line).

high frequencies ($\omega \gg \omega_{pe}$) electromagnetic wave propagation in a plasma approaches that in free space. The physical interpretation of this convergence is that at such high frequencies the wave fields vary too fast for the plasma electrons (let alone the ions) to respond to the wave and affect its propagation. For a high-frequency wave, the plasma becomes transparent.

Total reflection from free space–plasma interface

An interesting and important consequence of the fact that uniform plane waves can only propagate in a plasma at frequencies $\omega > \omega_p$ is the total reflection of a uniform plane wave at the interface between a dielectric and a plasma medium or an ionized region. A natural example of such an interface is that between free space and the Earth's ionosphere, although the electron density in the ionosphere increases relatively gradually with height so that the interface is not a single sharp interface between two media. It is this type of reflection that makes radio waves "bounce" off the ionosphere, making long-range radio communication possible. Reflection occurs when $\omega < \omega_p$. For the ionosphere, the peak value of f_p is approximately 10 MHz; thus, AM radio broadcast frequencies are reflected from the ionospheric conducting layer. Microwave, television, and FM radio signals are typically above 40 MHz, and are thus easily transmitted through the conducting ionospheric layer with no reflection. To illustrate the basic concept of total reflection at such an interface, we consider a single sharp interface between free space and an ionized region (characterized by plasma frequency ω_p), as illustrated in Figure 9.2.

Assuming that the incident wave in Figure 9.2 has a frequency ω, and that the ionized region behaves as a dielectric with dielectric

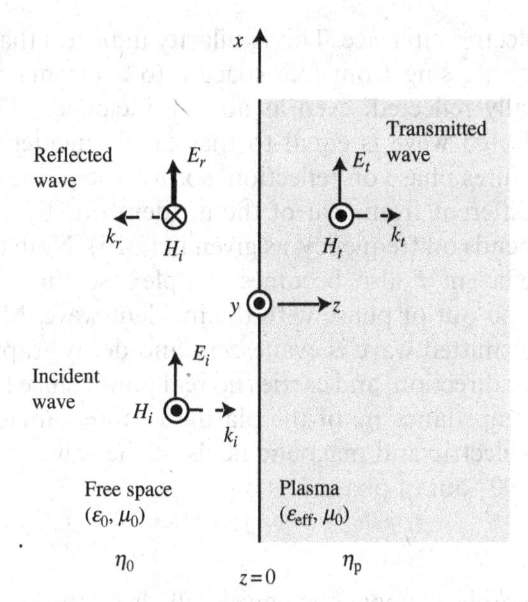

Figure 9.2 Reflection from free space–plasma interface. Normal incidence at a sharp interface between free space and an ionized medium (plasma) characterized by a plasma frequency of ω_p.

constant $\epsilon_{\text{eff}} = \epsilon\sqrt{1 - \omega_p^2/\omega^2}$, the reflection and transmission coefficients (Γ and τ, respectively) for the case of normal incidence are given by expressions similar to those for reflection from the interface between two ordinary dielectrics, namely

$$\Gamma = \frac{\sqrt{\epsilon_1} - \sqrt{\epsilon_2}}{\sqrt{\epsilon_1} + \sqrt{\epsilon_2}} = \frac{\sqrt{\epsilon_0} - \sqrt{\epsilon_0\left(1 - \omega_p^2/\omega^2\right)}}{\sqrt{\epsilon_0} + \sqrt{\epsilon_0\left(1 - \omega_p^2/\omega^2\right)}} = \frac{\omega - \sqrt{\omega^2 - \omega_p^2}}{\omega + \sqrt{\omega^2 - \omega_p^2}} \tag{9.24a}$$

$$\tau = \frac{2\sqrt{\epsilon_1}}{\sqrt{\epsilon_1} + \sqrt{\epsilon_2}} = \frac{2\omega}{\omega + \sqrt{\omega^2 - \omega_p^2}}. \tag{9.24b}$$

For $\omega > \omega_p$, we note from the above that both Γ and τ are real and that portions of the incident wave energy are accordingly reflected and transmitted. However, for $\omega < \omega_p$, Γ becomes imaginary, in which case we can write it as

$$\Gamma = \frac{\omega - j\sqrt{\left(\omega_p^2 - \omega^2\right)}}{\omega + j\sqrt{\left(\omega_p^2 - \omega^2\right)}} = 1e^{j\phi_\Gamma}. \tag{9.25}$$

We note that (9.25) is similar in form to the reflection coefficients $\Gamma_\parallel$ and $\Gamma_\perp$ for total internal reflection upon oblique incidence at a dielectric interface. This similarity indicates that an electromagnetic wave passing from free space into a plasma with $\omega < \omega_p$ is thus totally reflected, even at normal incidence. The amplitude of the reflected wave is equal to that of the incident wave but the wave acquires phase on reflection, so that the phase of the reflected wave is different from that of the incident one by an amount ϕ_Γ which depends on frequency, as given in (9.25). Note that the transmission coefficient τ also becomes complex, so that the transmitted wave is also out of phase with the incident wave. More importantly, the transmitted wave is evanescent and decays rapidly with distance in the z direction, and carries no real power since for $\omega < \omega_p$ the intrinsic impedance η_p of the plasma is purely imaginary, ensuring that the electric and magnetic fields of the wave in the ionized medium are 90° out of phase.

Effect of collisions

Some electromagnetic power will always be lost in a plasma because the electrons frequently collide with gas molecules, ions, and even other electrons. These collisions cause electromagnetic power to be transformed into heat. For $\omega > \omega_p$, the collisions cause the wave to be attenuated with distance as $e^{-\alpha_{\text{eff}} z}$. For $\omega < \omega_p$ the losses due to collisions cause total reflection (discussed above) to become partial reflection. Collisional effects may be taken into account by including the collisional friction term in the equation of motion (9.14):

$$q_e \mathbf{E} = j\omega m_e \mathbf{u}_e + m_e \nu \mathbf{u}_e = j\omega m_e \left(1 - j\frac{\nu}{\omega}\right) \mathbf{u}_e, \tag{9.26}$$

where ν is the effective collision frequency of electrons with other particles, in units of s^{-1}. With the collision term included, we can once again eliminate $\mathbf{u}_e$ to obtain

$$\nabla \times \mathbf{H} = j\omega\epsilon_0 \mathbf{E} + \frac{N q_e^2 \mathbf{E}}{j\omega m_e \left(1 - j\dfrac{\nu}{\omega}\right)}$$

or

$$\nabla \times \mathbf{H} = j\omega\epsilon_0 \left(1 - \frac{X}{1 - jZ}\right) \mathbf{E},$$

where $X = \omega_p^2/\omega^2$ and $Z = \nu/\omega$. Note that X and Z are dimensionless quantities. The effective permittivity of the plasma wlth collisions is thus

$$\epsilon_{\text{eff}} = \epsilon_0 \left(1 - \frac{X}{1 - jZ}\right) = \epsilon'_{\text{eff}} - j\epsilon''_{\text{eff}}.$$

As is the case for lossy dielectrics, the imaginary part ϵ''_{eff} represents power loss and attenuation of the wave. The expressions for uniform plane waves in a collisional plasma can be obtained by using the general form of uniform plane waves in a lossy medium represented by σ, ϵ, and μ, by replacing ϵ with ϵ'_{eff} and σ with $\omega\epsilon''_{\text{eff}}$.

9.3.3 Electrostatic electron and ion waves

The presence of a pressure gradient causes the stationary plasma oscillations to propagate away, as electrons streaming into adjacent layers of the plasma carry information about the oscillations in the disturbed region. The plasma oscillations then properly become a plasma wave. In addition, the pressure gradient causes the ions to communicate with one another (i.e., to transmit vibrations to one another), not through collisions, as is the case for acoustic waves in a gas, but through the electric field. This coordinated motion of the ions leads to waves, which are necessarily low-frequency waves, owing to the large mass of the ions. We can include the effects of pressure by means of the ∇p_1 term in (9.10b). Using (9.10a) and (9.10c) to eliminate p_1 and N_1 we have

$$\mathbf{u} = \frac{q}{mj\omega}\mathbf{E} - \frac{\gamma k_B T}{m\omega^2}\nabla(-j\mathbf{k}\cdot\mathbf{u}). \tag{9.27}$$

This new version of (9.14) is valid for both electrons and ions, which in general can have different temperatures T_e and T_i. First we note in passing that the new term in (9.27) has no effect on the transverse waves (i.e., waves with $\mathbf{k}\cdot\mathbf{E} = 0$) since for these $\mathbf{E}$, $\mathbf{J}$, and $\mathbf{u}$ are perpendicular to $\mathbf{k}$, so that, by definition, $\mathbf{k}\cdot\mathbf{u} = 0$. On the other hand, longitudinal waves are affected, since for them all of the vector quantities are parallel to $\mathbf{k}$. By the same token, we can reduce our equations to their single-dimensional forms. Noting that for such waves the magnetic field is identically zero, we can observe from (9.9a) that $E = -J/(j\epsilon_0\omega)$, and noting that $J = Nq_e(-\mathbf{u}_i + \mathbf{u}_e)$, we can write

$$E = \frac{jq_e}{\epsilon_0\omega}(-\mathbf{u}_i + \mathbf{u}_e). \tag{9.28}$$

Using (9.28) in (9.27) we find

$$\mathbf{u}_i = \frac{-q_e}{m_i j\omega}E + \frac{\gamma k_B T_i k^2}{m_i\omega^2}\mathbf{u}_i \tag{9.29a}$$

$$\mathbf{u}_i = \frac{q_e}{m_i j\omega}E + \frac{\gamma k_B T_e k^2}{m_e\omega^2}\mathbf{u}_e. \tag{9.29b}$$

Eliminating E we can write

$$\mathbf{u}_i\left(\omega^2 - \omega_{pi}^2 - c_{si}^2 k^2\right) + \omega_{pi}^2 \mathbf{u}_e = 0 \qquad (9.30a)$$

$$\mathbf{u}_e\left(\omega^2 - \omega_{pe}^2 - c_{se}^2 k^2\right) + \omega_{pe}^2 \mathbf{u}_i = 0, \qquad (9.30b)$$

where $c_s = \gamma k_B T/m$ is the speed of sound waves in an adiabatic gas at a temperature T. Eliminating $\mathbf{u}_e/\mathbf{u}_i$ from the above equations results in a dispersion relation which is quadratic in ω^2 in terms of k^2. Although this equation can be solved exactly, results simplify if we take advantage of the fact that $m_i \gg m_e$ and assume that the electron and ion temperatures are comparable, so that c_{si} and c_{se} are also of the same order. Under those conditions, one of the solutions of (9.30), characterized by $|\mathbf{u}_i| \ll |\mathbf{u}_e|$ and valid for relatively small k, is

$$\omega^2 \simeq \omega_{pe}^2 + c_{se}^2 k^2 = \omega_{pe}^2\left(1 + \gamma \lambda_D^2 k^2\right), \qquad (9.31)$$

where $\lambda_D = \sqrt{\epsilon_0 k_B T_e/(N_0 q_e^2)}$ is the Debye length. It is apparent from (9.31) that it constitutes a modification of the plasma oscillations in which electric-field and pressure-gradient restoring forces work together, resulting in a plasma oscillation combined with an acoustic wave. The second solution of (9.30), again assuming relatively small k, is

$$v_p = \frac{\omega}{k} \simeq \sqrt{\frac{\omega_{pi}^2 c_{se}^2 + \omega_{pe}^2 c_{si}^2}{\omega_{pi}^2 + \omega_{pe}^2}} \simeq \sqrt{\frac{\gamma k_B (T_e + T_i)}{m_i}}, \qquad (9.32)$$

which is characterized by $\mathbf{u}_i \simeq \mathbf{u}_e$ so that $E \simeq 0$. It appears that this is simply an acoustic wave, with the speed determined by the total pressure of the two species and the mass of the ions. It is interesting to note that this ion acoustic wave exists even when the ion temperature is zero, in which case the phase velocity is determined by the electron temperature (which determines the electric field) and ion mass. We will see later (Chapter 12) that when $\lambda_D k$ is large a new phenomenon, called Landau damping, appears and leads to the dissipation of the electron plasma waves even in the absence of collisions.

9.4 Problems

9-1. Derive the plasma frequency with ion motion included, as shown in Equation (9.17).

9-2. A pulsar emits a broad spectrum of electromagnetic radiation, which is detected with a receiver tuned to the vicinity of $f = 75$ MHz. Because of dispersion in group velocity caused by the interstellar plasma, the observed frequency during each pulse drifts at a rate given by $df/dt = -4.5$ MHz s^{-1}. (a) If the interstellar magnetic field is negligible and $\omega^2 \gg \omega_p^2$, show that $df/dt \simeq -(c/r)(f^3/f_p^2)$, where f_p is the plasma frequency, r is the distance to the pulsar, and c is the speed of light. (b) If the average electron density in interstellar space is 2×10^5 m^{-3}, how far away is the pulsar? (1 parsec = 3×10^{16} m)

9-3. Calculate the additional phase acquired by a 1420 kHz AM radio wave reflected from a free space–plasma interface if the plasma density is 10^{11} m^{-3}.

9-4. Find the reflection coefficient for reflection from a sharp plasma boundary with a density of 10^{11} m^{-3} and an effective electron collision frequency of 2 GHz.

9-5. Although during normal conditions waves with frequencies of 1 GHz can pass through the Earth's ionosphere and are thus suitable for communications with satellites, large solar flares can greatly increase the ionospheric plasma density and disrupt communications. Calculate the minumim ionospheric plasma density that would lead to a communication blackout for an Earth–satellite link operating at 1 GHz.

9-6. Calculate the phase velocity of electrostatic waves for a 10 eV equilibrium plasma assuming three degrees of freedom and relatively small k.

9-7. Find the phase and group velocities for electrostatic waves propagating in a uniform plasma with density $N_0 = 10^{16}$ m^{-3} if the wavelength and frequency of the waves are observed to be 1.2 GHz and 1.1 cm, respectively. The motion of ions can be neglected.

9-8. Consider an over-the-horizon (OTH) radar that uses reflection of waves from the ionosphere to achieve greater than line-of-site range. The radar is intended to reflect waves from the ionosheric F-region, which is at an altitude of ~300 km and has an electron density of around 10^{11} m^{-3}. (a) Assuming a spherical Earth with a radius of 6370 km, find the maximum range of the radar using a single ionospheric reflection. (b) What frequency should the radar operate at and what will be the scale of target that it will be able to detect? You can ignore the effects of ionization layers in the ionosphere at altitudes below the F-region.

CHAPTER

10 Waves in cold magnetized plasmas

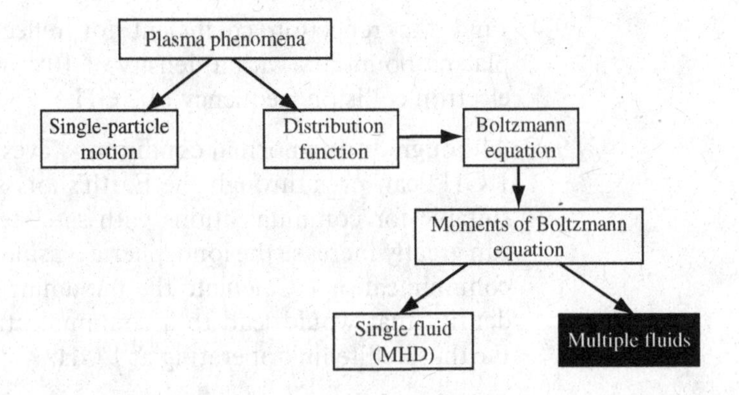

10.1 Introduction

In this chapter we continue using the multiple-fluid model to investigate wave phenomena, now examining the effect of a static magnetic field on waves in cold plasmas. The basic principles and fundamental equations on which we base our study of waves in plasmas are documented in Chapter 9. In particular, we saw that it is useful to represent the presence of the plasma medium (i.e., the plasma convection current **J** which results from the motion of the plasma fluid driven by the **E** and $\mathbf{B}_1$ fields via the momentum transport equation) in terms of an effective permittivity ϵ_{eff}, which for the case of an isotropic (non-magnetized) plasma is given by

$$\epsilon_{\text{eff}} = 1 + \frac{\sigma}{j\omega\epsilon_0}.$$

The conductivity σ is defined by the relationship between the current density and the electric field, $\mathbf{J} = \sigma\mathbf{E}$. In the more general case of the magnetized plasmas to be studied in this chapter, and based on our previous findings of the effect of a magnetic field, we expect the plasma conductivity to be a tensor, i.e., $\overleftrightarrow{\sigma}_p$, yielding an effective permittivity for the plasma which also, in general, must be a tensor, $\overleftrightarrow{\epsilon}_p$, related to the conductivity tensor by

$$\overleftrightarrow{\epsilon}_p \equiv 1 + \frac{\overleftrightarrow{\sigma}_p}{j\omega\epsilon_0}, \tag{10.1}$$

where we now drop the use of the subscript "eff" and instead use the subscript "p" to indicate that the permittivity and conductivity tensors are properties of the "plasma" medium.

10.2 The dispersion relation

Using the equivalent permittivity concept, the initial aim of our analysis of uniform plane plasma waves will be to reduce Maxwell's equations to

$$\mathbf{k} \times \mathbf{B}_1 = -\omega\mu_0\epsilon_0\overleftrightarrow{\epsilon}_p \cdot \mathbf{E} \tag{10.2a}$$

$$\mathbf{k} \times \mathbf{E} = \omega\mathbf{B}_1 \tag{10.2b}$$

$$\mathbf{k} \cdot \overleftrightarrow{\epsilon}_p \cdot \mathbf{E} = 0 \tag{10.2c}$$

$$\mathbf{k} \cdot \mathbf{B}_1 = 0, \tag{10.2d}$$

where the tensor dot-product in (10.2c) is to be carried out as shown in Chapter 4.[1] From (10.2a) and (10.2b) we can obtain the wave equation

$$\mathbf{k} \times (\mathbf{k} \times \mathbf{E}) + \frac{\omega^2}{c^2}\overleftrightarrow{\epsilon}_p \cdot \mathbf{E} = 0$$

$$\mathbf{k}\,(\mathbf{k} \cdot \mathbf{E}) - (\mathbf{k} \cdot \mathbf{k})\,\mathbf{E} + \frac{\omega^2}{c^2}\overleftrightarrow{\epsilon}_p \cdot \mathbf{E} = 0$$

$$\mathbf{k}\,(\mathbf{k} \cdot \mathbf{E}) - k^2\,\mathbf{E} + \frac{\omega^2}{c^2}\overleftrightarrow{\epsilon}_p \cdot \mathbf{E} = 0. \tag{10.3}$$

[1] In other words,

$$\mathbf{k} \cdot \overleftrightarrow{\epsilon}_p \cdot \mathbf{E} \equiv \begin{bmatrix} k_x & k_y & k_z \end{bmatrix} \begin{bmatrix} \epsilon_{11} & \epsilon_{12} & \epsilon_{31} \\ \epsilon_{21} & \epsilon_{22} & \epsilon_{23} \\ \epsilon_{31} & \epsilon_{32} & \epsilon_{33} \end{bmatrix} \begin{bmatrix} E_x \\ E_y \\ E_z \end{bmatrix}.$$

It should be noted that a similar wave equation can be written for the magnetic field perturbation $\mathbf{B}_1$; however, it is customary to work with the electric field $\mathbf{E}$ since all plasma waves have an electric field but some may have $\mathbf{B}_1 = 0$. The typical procedure is to expand (10.3) in terms of all three orthogonal electric field components. For a non-trivial wave solution, the determinant of the coefficients of this equation must vanish. This condition leads to an algebraic expression known as the *dispersion relation*:

$$D(\mathbf{k}, \omega) = 0. \tag{10.4}$$

In unbounded plasmas, (10.4) permits a continuum of normal wave modes. In bounded plasmas, only a discrete set of solutions may be possible, analogous to a discrete set of normal-mode solutions in metallic or dielectric waveguides. We will find it useful to study two principal wave modes, transverse ($\mathbf{k}\perp\mathbf{E}$) and longitudinal ($\mathbf{k} \parallel \mathbf{E}$). Transverse waves ($\mathbf{k} \cdot \mathbf{E} = 0$) are described by

$$k^2\mathbf{E} - \frac{\omega^2}{c^2} \overleftrightarrow{\epsilon}_p \cdot \mathbf{E} = 0, \tag{10.5}$$

whereas longitudinal waves ($\mathbf{k} \times \mathbf{E} = 0$) are described by

$$\overleftrightarrow{\epsilon}_p \cdot \mathbf{E} = 0 \quad \text{or} \quad \mathbf{k} \cdot \overleftrightarrow{\epsilon}_p \cdot \mathbf{k} = 0, \tag{10.6}$$

where the second term is effectively Poisson's equation (to see this note that $\mathbf{E} = j\mathbf{k}\Phi$, where Φ is the electrostatic potential). At this point it is useful to note that all of our analysis of waves in plasmas involves examination of time-harmonic uniform plane waves with all perturbation quantities varying as

$$e^{j(\omega t - \mathbf{k}\cdot\mathbf{r})}. \tag{10.7}$$

The dispersion relation (10.4) does not always give purely real time-harmonic solutions and may instead indicate damping or instability; the solution of (10.4) for a given real wave number $k = k_r + j0$ may be a complex frequency, $\omega_r + j\omega_i$, and the solution of (10.4) for a given real frequency $\omega_r + j0$ may be a complex wave number, $k_r + jk_i$. Solutions with $\omega_i < 0$ grow exponentially at all points in space, while those with $k_i < 0$ *may* grow exponentially if there is energy available for wave growth. Note that for evanescent waves in passive systems such as metallic waveguides, cases with $k_i < 0$ are automatically deemed physically unrealizable;[2] however, in a

[2] For example, for a rectangular waveguide the dispersion relation is

$$\omega_c^2 + k^2c^2 = \omega^2,$$

complicated plasma system with the possibility of free energy in the system, growth can occur up to the limits of our linearization approximations. In this and the following chapter we will concentrate on understanding the variety of possible wave modes in a plasma and will not discuss instabilities.

10.3 Waves in magnetized plasmas

We start with Maxwell's equations for time-harmonic uniform plane waves as given in (9.9), repeated here,

$$\mathbf{k} \times \mathbf{B}_1 = j\mu_0 \mathbf{J} - \omega\epsilon_0\mu_0 \mathbf{E}$$
$$\mathbf{k} \times \mathbf{E} = \omega \mathbf{B}_1$$
$$\mathbf{k} \cdot \mathbf{E} = \frac{j\rho_1}{\epsilon_0}$$
$$\mathbf{k} \cdot \mathbf{B}_1 = 0,$$

along with the momentum equation

$$m\, N_0\, j\omega\, \mathbf{u} = q\, N_0\, (\mathbf{E} + \mathbf{u} \times \mathbf{B}_0) \tag{10.8}$$

and the definition of electric current

$$\mathbf{J}(\mathbf{r}, t) = \sum_i q_i\, N_i(\mathbf{r}, t)\, \mathbf{u}_i(\mathbf{r}, t). \tag{10.9}$$

We begin by considering an infinite, cold, collisionless, and homogeneous plasma, although we will occasionally relax the first three assumptions and examine the effects of boundaries, temperature, and collisions. We will not consider inhomogeneous plasmas. In this context, note that by a homogeneous plasma we mean that the ambient density N_0 and temperature T_0 (and thus the pressure p_0) are constant as a function of space; the perturbation quantities N_1, T_1, and p_1 vary sinusoidally in the manner described by (10.7) as do all the other quantities. We also initially consider only electron motions, assuming the ions to be stationary; however, we will later (in Chapter 11) relax this assumption and account for ion motions. Our goal is to work with (9.9), (10.8), and (10.9) to write the wave equation (10.3), thus identifying $\overleftrightarrow{\epsilon}_p$. By definition we have

where ω_c is the cutoff frequency. At a frequency ω below the cutoff frequency we have

$$k_r = 0 \quad \text{and} \quad k_i \pm jc^{-2}\sqrt{\omega_c^2 - \omega^2},$$

and the solution corresponding to growth is dropped on physical grounds.

$$\overleftrightarrow{\epsilon}_p \cdot \mathbf{E} = \frac{\mathbf{J}}{j\omega\epsilon_0} + \mathbf{E}. \tag{10.10}$$

For future reference, it should be noted that if ion motions were included, the $\mathbf{J}$ above would simply be a summation $\sum \mathbf{J}$ over the different species, as indicated in (10.9). Using (10.8), (10.9), and (10.10) we have

$$\frac{\mathbf{J}}{j\omega\epsilon_0} = \frac{N_0 q_e}{j\omega\epsilon_0}\left(\frac{q_e}{j\omega m_e}\right)(\mathbf{E} + \mathbf{u}_e \times \mathbf{B}_0) = -\left(\frac{\omega_p^2}{\omega^2}\right)\mathbf{E} + \left(\frac{\omega_c}{\omega}\right)\frac{\mathbf{J} \times \hat{\mathbf{z}}}{\omega\epsilon_0}, \tag{10.11}$$

where we have assumed the static magnetic field to be in the z direction. We can expand (10.11) into its components:

$$\frac{J_x}{j\omega\epsilon_0} = -\left(\frac{\omega_p^2}{\omega^2}\right)E_x + \left(\frac{\omega_c}{\omega}\right)\frac{J_y}{\omega\epsilon_0} \tag{10.12a}$$

$$\frac{J_y}{j\omega\epsilon_0} = -\left(\frac{\omega_p^2}{\omega^2}\right)E_y - \left(\frac{\omega_c}{\omega}\right)\frac{J_x}{\omega\epsilon_0} \tag{10.12b}$$

$$\frac{J_z}{j\omega\epsilon_0} = -\left(\frac{\omega_p^2}{\omega^2}\right)E_z. \tag{10.12c}$$

Equation (10.12) can be solved to find

$$\frac{J_x}{j\omega\epsilon_0} = -\frac{\omega_p^2}{\omega^2 - \omega_c^2}E_x - j\left(\frac{\omega_c}{\omega}\right)\frac{\omega_p^2}{\omega^2 - \omega_c^2}E_y \tag{10.13a}$$

$$\frac{J_y}{j\omega\epsilon_0} = -\frac{\omega_p^2}{\omega^2 - \omega_c^2}E_y + j\left(\frac{\omega_c}{\omega}\right)\frac{\omega_p^2}{\omega^2 - \omega_c^2}E_x \tag{10.13b}$$

$$\frac{J_z}{j\omega\epsilon_0} = -\frac{\omega_p^2}{\omega^2}E_z, \tag{10.13c}$$

where the reader should recognize that the derivation is the same as that for the AC conductivity tensor in Equation (7.27), albeit with no collisions. Substitution in (10.10) gives

$$\overleftrightarrow{\epsilon}_p \cdot \mathbf{E} \equiv \begin{bmatrix} \epsilon_\perp & -j\epsilon_\times & 0 \\ j\epsilon_\times & \epsilon_\perp & 0 \\ 0 & 0 & \epsilon_\parallel \end{bmatrix}\begin{bmatrix} E_x \\ E_y \\ E_z \end{bmatrix}, \tag{10.14}$$

where

$$\epsilon_\perp = 1 - \frac{\omega_p^2}{\omega^2 - \omega_c^2}, \quad \epsilon_\times = \left(\frac{\omega_c}{\omega}\right)\frac{\omega_p^2}{\omega^2 - \omega_c^2}, \quad \epsilon_\parallel = 1 - \frac{\omega_p^2}{\omega^2}. \tag{10.15}$$

We see from (10.14) that E_x and E_y are coupled together through the static magnetic field (ω_c). Note that the E_z component is unaffected by the magnetic field, and that as $\omega_c \to 0$ we recover the equivalent permittivity for an isotropic plasma, i.e., $\epsilon_p = \epsilon_\parallel$. We now need to use (10.14) and the wave equation (10.3) to obtain the dispersion relation. As a first step, we write out the wave equation (10.3) in terms of its x, y, and z components. We have

$$k_x(k_x E_x + k_y E_y + k_z E_z) - \left(k_x^2 + k_y^2 + k_z^2\right) E_x + \frac{\omega^2}{c^2}(\epsilon_\perp E_x - j\epsilon_\times E_y) = 0 \tag{10.16a}$$

$$k_y(k_x E_x + k_y E_y + k_z E_z) - \left(k_x^2 + k_y^2 + k_z^2\right) E_y + \frac{\omega^2}{c^2}(\epsilon_\perp E_x + j\epsilon_\times E_y) = 0 \tag{10.16b}$$

$$k_z(k_x E_x + k_y E_y + k_z E_z) - \left(k_x^2 + k_y^2 + k_z^2\right) E_z + \frac{\omega^2}{c^2}\epsilon_\parallel E_z = 0. \tag{10.16c}$$

Without any loss of generality, we can orient our coordinate system as shown in Figure 10.1 so that the propagation vector lies

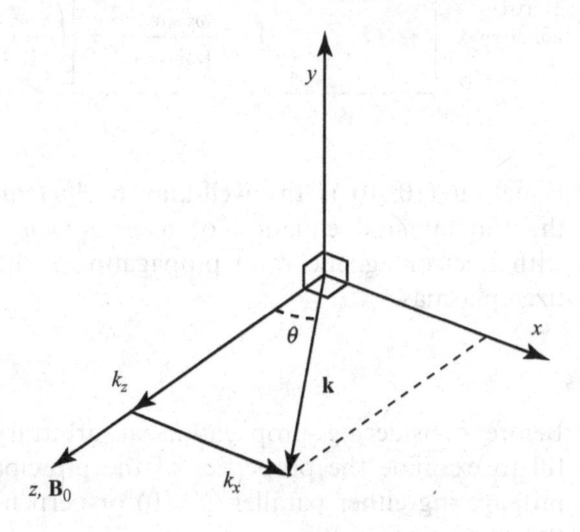

Figure 10.1 Coordinate system for propagation at an angle θ to $\mathbf{B}_0$. We choose the z axis to be aligned with $\mathbf{B}_0$ and rotate the x and y axes so that $\mathbf{k}$ is in the x–z plane, without any loss of generality.

in one of the principal planes, say the x–z plane, so that $k_y = 0$. Equation (10.16) then reduces to

$$\begin{bmatrix} \epsilon_\perp - \dfrac{k_z^2 c^2}{\omega^2} & -j\epsilon_\times & \dfrac{k_x k_z c^2}{\omega^2} \\ j\epsilon_\times & \epsilon_\perp - \left(k_x^2 + k_z^2\right)\dfrac{c^2}{\omega^2} & 0 \\ \dfrac{k_x k_z c^2}{\omega^2} & 0 & \epsilon_\parallel - \dfrac{k_x^2 c^2}{\omega^2} \end{bmatrix} \begin{bmatrix} E_x \\ E_y \\ E_z \end{bmatrix} = 0. \quad (10.17)$$

It is convenient to introduce the refractive index $\mu = kc/\omega = v/v_p$ and rewrite (10.17) as

$$\begin{bmatrix} \epsilon_\perp - \mu^2 \cos^2\theta & -j\epsilon_\times & \mu^2 \sin\theta\cos\theta \\ j\epsilon_\times & \epsilon_\perp - \mu^2 & 0 \\ \mu^2 \sin\theta\cos\theta & 0 & \epsilon_\parallel - \mu^2 \sin^2\theta \end{bmatrix} \begin{bmatrix} E_x \\ E_y \\ E_z \end{bmatrix} = 0. \quad (10.18)$$

For a non-trivial solution, i.e., $E_x, E_y, E_z \neq 0$, the determinant of the coefficient matrix must vanish. Setting determinant equal to zero gives the dispersion relation we have been looking for, now determined for propagation with refractive index μ, at an angle θ to the static magnetic field $\mathbf{B}_0$. The determinant is a biquadratic in μ and can be written in the following alternative forms:

$$\tan^2\theta = -\frac{\epsilon_\parallel\left[\mu^2 - (\epsilon_\perp + \epsilon_\times)\right]\left[\mu^2 - (\epsilon_\perp - \epsilon_\times)\right]}{\left(\mu^2 - \epsilon_\parallel\right)\left[\epsilon_\perp \mu - \left(\epsilon_\perp^2 - \epsilon_\times^2\right)\right]} \quad (10.19)$$

or

Appleton–Hartree equation

$$\boxed{\mu^2 = \frac{k^2c^2}{\omega^2} = 1 - \frac{\left(\omega_p^2/\omega^2\right)}{1 - \dfrac{\omega_c^2 \sin^2\theta}{2\left(\omega^2 - \omega_p^2\right)} \pm \left[\left(\dfrac{\omega_c^2 \sin^2\theta}{2\left(\omega^2 - \omega_p^2\right)}\right)^2 + \dfrac{\omega_c^2}{\omega^2}\cos^2\theta\right]^{1/2}}} \quad (10.20)$$

Equation (10.20) is the well-known *Appleton–Hartree* equation, the fundamental equation of *magnetoionic theory*, which deals with electromagnetic wave propagation in homogeneous magnetized plasmas.

10.3.1 Principal modes

Before considering propagation at arbitrary angles, it is useful to examine the properties of the principal modes, i.e., those propagating either parallel ($\theta = 0$) or perpendicular ($\theta = \pi/2$) to the magnetic field $\mathbf{B}_0$.

Parallel propagation ($\theta = 0$)

From (10.18) we have

$$\begin{bmatrix} \epsilon_\perp - \mu^2 & -j\epsilon_\times & 0 \\ j\epsilon_\times & \epsilon_\perp - \mu^2 & 0 \\ 0 & 0 & \epsilon_\parallel \end{bmatrix} \begin{bmatrix} E_x \\ E_y \\ E_z \end{bmatrix} = 0, \tag{10.21}$$

and (10.19) directly indicates three solutions,

$$\epsilon_\parallel = 0 \tag{10.22a}$$

$$\mu^2 = \epsilon_\perp + \epsilon_\times \tag{10.22b}$$

$$\mu^2 = \epsilon_\perp - \epsilon_\times, \tag{10.22c}$$

each of which we will examine separately. The solution indicated by (10.22a) is the plasma oscillations that we have seen before in Chapter 9. We have

$$\epsilon_\parallel E_z = \left(1 - \frac{\omega_p^2}{\omega^2}\right) E_z = 0, \tag{10.23}$$

which represents longitudinal waves at a frequency ω and arbitrary wave number k. Since the only electric field is $E_z \parallel \mathbf{k}$, it follows that $\mathbf{k} \times \mathbf{E} = 0$, implying in turn that $\mathbf{B}_1 = 0$, and that $\mathbf{E} = -\nabla\Phi$, i.e., the electric field is derivable from a scalar potential. Since $\mathbf{k} \cdot \mathbf{E} = j\rho/\epsilon_0 \neq 0$, there is space charge associated with these oscillations. Note that the group velocity is zero, so that these oscillations remain localized in space. The solutions indicated by (10.22b) and (10.22c) are polarized waves. The origin of these waves is immediately clear on examination of (10.21). The E_x and E_y components are coupled together by the magnetic field. The relation $\mu^2 = \epsilon_\perp \pm \epsilon_\times$ implies that

$$\mp\epsilon_\times E_x - j\epsilon_\times E_y = 0 \quad \rightarrow \quad \frac{E_x}{E_y} = \mp j. \tag{10.24}$$

These modes are consequently right- or left-hand circularly polarized, depending on the sign. According to the usual IEEE convention, defining polarization on the basis of the right-hand rule, with the motion of the tip of the electric field vector being counter-clockwise when viewed looking toward the $-z$ direction, the wave is said to be *right-hand* (RH) circularly polarized. An easy way to remember this convention is to use your right hand with the thumb pointing in the direction of propagation (in this case the $+z$ direction); if the electric field moves in the direction of your other

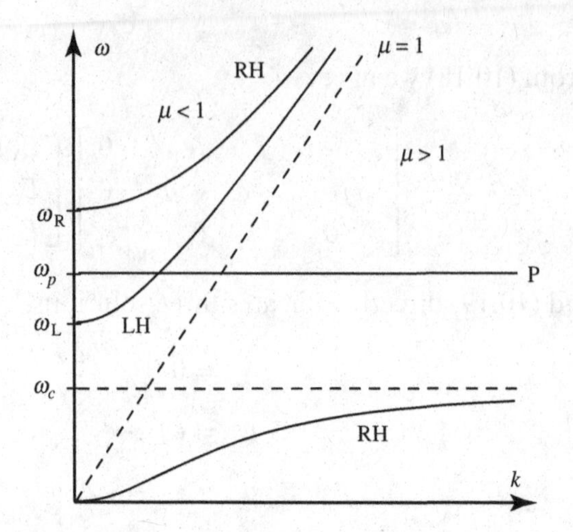

Figure 10.2 Dispersion diagram for parallel propagation of electromagnetic waves in a magnetized plasma. The branch marked P corresponds to plasma oscillations.

fingers, then the wave is polarized in the right-handed sense.[3] The dispersion relations for the polarized modes are

$$\mu^2 = \frac{k^2c^2}{\omega^2} = 1 - \frac{\omega_p^2}{\omega^2 - \omega_c^2} \pm \frac{\omega_c}{\omega}\frac{\omega_p^2}{\omega^2 - \omega_c^2} = 1 - \frac{\omega_p^2}{\omega(\omega \pm \omega_c)}. \tag{10.25}$$

The corresponding w–k diagrams are sketched in Figure 10.2. Note that the RH wave has two branches, the lower of which shows an electron-cyclotron resonance for $\omega \to \omega_c$. This branch is often termed the *whistler* mode, and is very important in the Earth's ionosphere and radiation belts as well as in certain materials-processing applications. It is also referred to as the *electron-cyclotron* mode.

Note that for both of the polarized waves we have $\mathbf{E}\perp\mathbf{k}$ ($\mathbf{k} \times \mathbf{E} \neq 0$ and $\mathbf{k} \cdot \mathbf{E} = 0$), so we can deduce that $\mathbf{B}_1 \neq 0$, but $\rho = 0$. Thus, the Poynting vector for these transverse waves is non-zero, and the wave propagates without space charge. The plasma manifests itself via $\mathbf{J}$, which flows in planes normal to $\mathbf{k}$ and $\mathbf{B}_0$.

The two cutoff frequencies (i.e., the frequencies for which $k = 0$) for the parallel propagation case indicated in Figure 10.2 can be found from (10.25) by setting the left-hand side equal to zero, and are given by

[3] The right-hand rule described here is the IEEE convention. Amazingly enough, there is considerable disagreement on how to define the sense of polarization of a wave. Most physicists, as well as scientists and engineers specializing in optics, prefer to have the thumb pointing to *where the wave is coming from*, exactly opposite to the IEEE convention. A further source of confusion is the preference of physicists for using $e^{-j\omega t}$ instead of $e^{j\omega t}$, which of course reverses the sense of rotation. In view of these ambiguities, it would be wise in any given case to carefully examine the actual sense of rotation of the electric field $\mathcal{E}(\mathbf{r}, t)$ by determining the field vector orientation in space at two specific times, such as $\omega t = 0$ and $\omega t = \pi/2$.

$$\omega_\mathrm{R} = \left(\omega_p^2 + \frac{\omega_c^2}{4}\right)^{1/2} + \frac{\omega_c}{2} \tag{10.26a}$$

$$\omega_\mathrm{L} = \left(\omega_p^2 + \frac{\omega_c^2}{4}\right)^{1/2} - \frac{\omega_c}{2}, \tag{10.26b}$$

with the subscripts signifying the fact that ω_L is the cutoff frequency for the left-hand mode, and vice versa for ω_R. It should be noted that $\omega_\mathrm{R} - \omega_\mathrm{L} = \omega_c$, and that in Figure 10.2 we have chosen the plasma parameters such that $\omega_p > \omega_c$.

Faraday rotation

The fact that the propagation constant and therefore the phase velocity is different for the RH and LH circularly polarized waves causes the plane of polarization of a linearly polarized wave to rotate as the wave propagates through the medium.[4] This result can be easily seen by decomposing the linearly polarized wave into a sum of LH and RH circularly polarized components and solving separately for the propagation of each component. At $z = 0$, consider the electric field of a linearly polarized wave propagating in the z direction to be

$$\mathbf{E} = \hat{\mathbf{x}} E_1 e^{j\omega t},$$

so that the plane of polarization includes the x axis. The complex amplitude of this wave can be written as

$$\mathbf{E} = \frac{E_1}{2}(\hat{\mathbf{x}} - j\hat{\mathbf{y}}) + \frac{E_1}{2}(\hat{\mathbf{x}} + j\hat{\mathbf{y}}),$$

where the first term is the RH circularly polarized component and the second term is the LH circularly polarized component. After the component waves propagate in the magnetoplasma for a distance d, we have, at $z = d$,

$$\mathbf{E} = \frac{E_1}{2}(\hat{\mathbf{x}} - j\hat{\mathbf{y}})e^{-jk_\mathrm{RH}d} + \frac{E_1}{2}(\hat{\mathbf{x}} + j\hat{\mathbf{y}})e^{-jk_\mathrm{LH}d},$$

[4] This phenomenon is called Faraday rotation. It was Faraday who discovered in 1845 that on passing a plane polarized ray of light through a piece of glass in the direction of the lines of force of an imposed magnetic field, the plane of polarization was rotated by an amount proportional to the thickness of the glass traversed and the strength of the magnetic field [1]. The interaction between the wave and the bound electrons within the glass material is more complex than the interaction with a free electron plasma considered here; nevertheless, the presence of a static magnetic field imposes a preferential direction for the electrons and makes the medium anisotropic.

which can be rewritten as

$$\mathbf{E} = E_1 e^{-j(k_{\rm RH}+k_{\rm LH})d/2}$$
$$\times \left\{ \hat{\mathbf{x}} \cos\left[\frac{(k_{\rm RH}-k_{\rm LH})d}{2}\right] - \hat{\mathbf{y}} \sin\left[\frac{(k_{\rm RH}-k_{\rm LH})d}{2}\right]\right\}. \quad (10.27)$$

This is a linearly polarized wave, but with the plane of polarization making an angle of θ_F with the x axis, where

$$\theta_F = \tan^{-1}\frac{E_y}{E_x} = -\tan^{-1}\left\{\tan\left[\frac{(k_{\rm RH}-k_{\rm LH})d}{2}\right]\right\}$$
$$= \frac{(k_{\rm RH}-k_{\rm LH})d}{2}. \quad (10.28)$$

Thus, it appears that the plane of polarization rotates by an amount $\theta_F = (k_{\rm RH} - k_{\rm LH})/2$ per unit distance. The wave propagates with an effective propagation constant of $(k_{\rm RH} + k_{\rm LH})/2$, as is apparent from the $e^{-j(k_{\rm RH}+k_{\rm LH})d/2}$ term in Equation (10.27), and undergoes rotation at the same time. Note that this phenomena would occur only for frequencies where both the LH and RH components would be propagating; from Figure 10.2 and the related discussion, this would happen for $\omega > \omega_R$. It is interesting to express the rotation in terms of the physical parameters of the plasma. Using the definitions of ω_p and ω_c we find

$$\theta_F = \frac{|q_e|^3 B_0}{2m_e^2 \epsilon_0 \omega^2 c} Nd,$$

where c is the speed of light in free space. If the electron density varies with distance and if B_0 is approximately constant, the net rotation is given by

$$\theta_F = \frac{|q_e|^3 B_{0z}}{2m_e^2 \epsilon_0 \omega^2 c} \int N(z)dz,$$

where the integration is along the entire propagation path. For signals in the several hundred MHz range propagating in the Earth's ionosphere, most of the Faraday rotation occurs in the 90–1000 km altitude range. Measurements of the *total electron content* (in a 1 m^2 column extending through the most highly ionized part of the ionosphere) are made regularly using satellite-to-Earth transmissions.[5]

[5] For further information, see [2].

Perpendicular propagation ($\theta = \pi/2$)

For propagation perpendicular to the magnetic field we have, from (10.18),

$$\begin{bmatrix} \epsilon_\perp & -j\epsilon_\times & 0 \\ j\epsilon_\times & \epsilon_\perp - \mu^2 & 0 \\ 0 & 0 & \epsilon_\parallel - \mu^2 \end{bmatrix} \begin{bmatrix} E_x \\ E_y \\ E_z \end{bmatrix} = 0. \tag{10.29}$$

From (10.19) we can directly see two solutions:

$$\mu^2 = \epsilon_\parallel \tag{10.30a}$$

$$\mu^2 = \frac{\epsilon_\perp^2 - \epsilon_\times^2}{\epsilon_\perp}. \tag{10.30b}$$

The mode represented by (10.30a) is called the *ordinary* mode, since its propagation is not affected by the magnetic field. E_z is completely decoupled from E_x and E_y. This means that this wave has the same properties as the transverse electromagnetic wave in a non-magnetized plasma with $\mathbf{u} \parallel \mathbf{E}$. Since the Lorentz force $\mathbf{u} \times \mathbf{B}_0 = 0$, the ordinary mode is uncoupled from the magnetic field $\mathbf{B}_0$. Note that the transverse electromagnetic wave in a non-magnetized plasma is purely transverse (i.e., $\mathbf{k} \perp \mathbf{E}$), while for the ordinary wave which occurs for $\mathbf{k} \perp \mathbf{B}_0$ we have $\mathbf{E} \parallel \mathbf{B}_0$ and thus $\mathbf{E} \perp \mathbf{k}$. Since $\mathbf{k} \perp \mathbf{E}$, we have $\mathbf{k} \times \mathbf{E} \neq 0$, and $\mathbf{k} \cdot \mathbf{E} = 0$. Thus, there is a non-zero Poynting vector, but zero space charge. The dispersion relation is shown in Figure 10.3.

The mode represented by (10.30b) has some extraordinary properties, and is thus called the *extraordinary* mode. The electric field for this mode is perpendicular to $\mathbf{B}_0$, with E_x and E_y coupled together. Since one component (E_x) is in the direction of propagation (note that $\mathbf{k}$ is perpendicular to $\mathbf{B}_0$, so $\mathbf{k} = \hat{\mathbf{x}}k_x$), we have

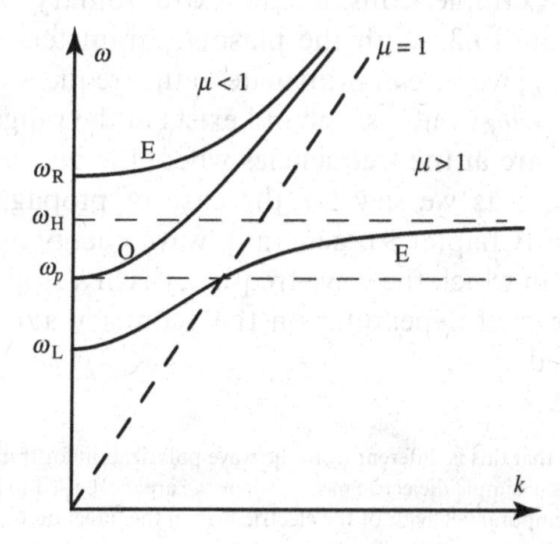

Figure 10.3 Dispersion diagram for perpendicular propagation. The ordinary (O) and extraordinary (E) branches are marked.

$\mathbf{k} \cdot \mathbf{E} \neq 0$, so that there is space charge associated with this wave. From (10.29) we can write

$$\frac{E_x}{E_y} = j\frac{\epsilon_\times}{\epsilon_\perp} = j\frac{\omega_c \omega_p^2}{\omega\left(\omega^2 - \omega_{\mathrm{H}}^2\right)}, \tag{10.31}$$

where $\omega_{\mathrm{H}} = \sqrt{\omega_p^2 + \omega_c^2}$ is the *upper hybrid frequency*. It is clear from (10.31) that the extraordinary wave is in general elliptically polarized in the plane of propagation.[6] At $\omega = \omega_{\mathrm{H}}$ we have a *resonance*, with the wave number $\mathbf{k}$ approaching infinity and hence the wavelength approaching zero. The physical meaning of a resonance is discussed below. The extraordinary wave has both transverse (i.e., E_y) and longitudinal (i.e., E_x) components, the relative magnitudes of which depend on frequency, as is clear from (10.31). When ω is very close to ω_{H}, we can see from (10.31) that $E_x \gg E_y$, so that the wave is longitudinal ($\mathbf{k} \parallel \mathbf{E}$) and electrostatic. For other frequencies, the magnitudes of the two components are comparable so that the wave is a mixed longitudinal–transverse mode, tending to be transverse electromagnetic for $\mu < 1$ and longitudinal electrostatic for $\mu > 1$. In this connection, it is useful to write (10.30b) for the extraordinary mode as

$$\mu^2 = \frac{k^2 c^2}{\omega^2} = 1 - \frac{\omega_p^2}{\omega^2}\left[\frac{\omega^2 - \omega_p^2}{\omega^2 - \left(\omega_p^2 + \omega_c^2\right)}\right]. \tag{10.32}$$

It is clear from (10.32) that $\mu = 1$ when $\omega = \omega_p$, regardless of the value of ω_c, as long as $\omega_c \neq 0$.

A few words on the nature of cutoffs and resonances are in order here. Cutoffs and resonances are important because they define the passbands and stopbands for wave propagation in a plasma. As an example, consider the extraordinary wave branches shown in Figure 10.3. With the plasma parameters chosen for this case, $\omega_{\mathrm{H}} > \omega_p$, waves can propagate in the frequency ranges $\omega_{\mathrm{L}} < \omega < \omega_{\mathrm{H}}$ and $\omega > \omega_{\mathrm{R}}$, but a stopband exists in the range $\omega_{\mathrm{H}} < \omega < \omega_{\mathrm{R}}$. The cutoffs are at the frequencies where the dispersion relation reduces to $k = 0$; as we saw for the case of propagation in an isotropic plasma (Chapter 9), generally wave energy approaching a plasma region in which the wave frequency is in a stopband is fully reflected or refracted depending on the geometry and the inhomogeneities involved.

[6] Note that this is different from the wave polarization for transverse electromagnetic waves in simple dielectric media, where we are more used to looking at polarization as the temporal behavior of the electric field in the direction transverse to $\mathbf{k}$.

Resonances are quite different than cutoffs. At a resonance, the dispersion relation reduces to $k \to \infty$, meaning that the wavelength goes to zero. When $k \to \infty$, the wave phase velocity goes to zero, and the wavefronts "pile up." Generally waves become non-propagating and purely electrostatic near resonances. We can see graphically from Figure 10.3 that as $\omega \to \omega_{\mathrm{H}}$, the group velocity ($v_g = d\omega/dk$, i.e., the slope of the dispersion curve) also goes to zero. Consider a generator sending waves into a plasma, for example by propagating down a density or magnetic field gradient so that $\omega < \omega_{\mathrm{H}} = \sqrt{\omega_p^2 + \omega_c^2}$ in the propagation region but $\omega = \omega_{\mathrm{H}}$ in the resonance region. The energy sent by the generator would continue to pile up in the region of resonance and build up wave energy density, until some process beyond the applicability of our linearized cold-plasma theory takes over. Resonances are in fact more complicated; depending on the particle distribution function, the wave can either give energy to the particles and become damped, or extract energy and momentum from the particles and grow. The fundamental point here is that, with the wavelength being nearly zero, the wave spatial scales become comparable to the spatial scales of the particles (e.g., gyroradii) so that the wave can drastically affect the orbits of individual particles.

10.3.2 Oblique propagation at an arbitrary angle θ

The case of propagation at an arbitrary angle θ is best studied using (10.20). We see that there are always two values of μ^2 for any given frequency, i.e., just two modes. In general these modes are elliptically polarized, and have strong space charge only when $\mu > 1$ and $\mathbf{k} \times \mathbf{E} \simeq 0$, i.e., when the modes are slow compared to the velocity of light and are approximately longitudinal waves.

It can be seen easily from Figure 10.4 that the principal modes for $\theta = 0$ and $\theta = \pi/2$ are the limiting cases of general propagation at an arbitrary angle θ. Using (10.20), it can be seen that the cutoff frequencies ω_{R}, ω_{L}, and ω_p are independent of θ. The cutoffs are determined by

$$0 = 1 - \frac{\left(\omega_p^2/\omega^2\right)}{1 - \dfrac{\omega_c^2 \sin^2\theta}{2\left(\omega^2 - \omega_p^2\right)} \pm \left[\left(\dfrac{\omega_c^2 \sin^2\theta}{2\left(\omega^2 - \omega_p^2\right)}\right)^2 + \dfrac{\omega_c^2}{\omega^2}\cos^2\theta\right]^{1/2}}.$$

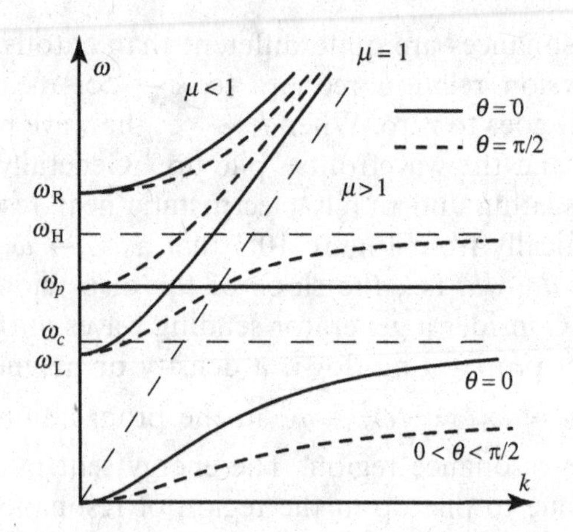

Figure 10.4 Dispersion diagram for oblique propagation. The principal mode branches corresponding to $\theta = 0$ (solid lines) and $\theta = \pi/2$ (dashed lines) are indicated. For any arbitrary angle θ, the dispersion curve lies between these two principal branches.

The resonances, on the other hand, are found from

$$0 = 1 - \frac{\omega_c^2 \sin^2\theta}{2\left(\omega^2 - \omega_p^2\right)} \pm \left[\left(\frac{\omega_c^2 \sin^2\theta}{2\left(\omega^2 - \omega_p^2\right)}\right)^2 + \frac{\omega_c^2}{\omega^2}\cos^2\theta\right]^{1/2} \tag{10.33}$$

and are clearly functions of θ. We can see from Figure 10.4 and from (10.33) that the resonance frequency of the whistler mode changes from $\omega = \omega_c$ for $\theta = 0$ to zero (actually the mode ceases to exist) for $\theta = \pi/2$. At any arbitrary propagation angle θ, the resonance frequency for the whistler mode is given by $\omega_{\text{res}} = \omega_c \cos\theta$. Alternatively, for any given wave frequency ω, the so-called *resonance-cone angle* is given by $\cos(\theta_{\text{res}}) = \omega/\omega_c$. For the dense-plasma (i.e., $\omega_p > \omega_c$) case illustrated in Figure 10.4, the resonance of the extraordinary wave moves from the upper hybrid resonance frequency for perpendicular propagation ($\theta = \pi/2$) to the plasma frequency for parallel propagation. It should be noted that, for all angles θ, we have $\mu = 1$ at $\omega = \omega_p$. In studying oblique propagation, it is sometimes convenient to simplify the Appleton–Hartree equation (10.20) by making the so-called quasi-parallel assumption, which is applicable in cases for which the angle θ is relatively small. For this case, (10.20) reduces to

$$\mu^2 \simeq 1 - \frac{\omega_p^2}{\omega(\omega \pm \omega_c \cos\theta)} \quad \text{for} \quad \frac{\omega_c^2 \sin^2\theta}{2\left(\omega^2 - \omega_p^2\right)} \ll \frac{\omega_c}{\omega}\cos\theta, \tag{10.34}$$

which can be compared directly to (10.25). Alternatively, for propagation at relatively high angles it may be useful to adopt the quasi-perpendicular approximation, in which case (10.20) reduces to

$$\mu^2 \simeq 1 - \frac{\omega_p^2}{\omega^2}\left[\frac{\omega^2 - \omega_p^2}{\omega^2 - \left(\omega_p^2 + \omega_c^2 \sin^2\theta\right)}\right] \quad \text{for}$$

$$\frac{\omega_c^2 \sin^2\theta}{2\left(\omega^2 - \omega_p^2\right)} \gg \frac{\omega_c}{\omega}\cos\theta, \tag{10.35}$$

which can be directly compared to (10.30a) and (10.32). We can note that the refractive indices for the quasi-parallel and quasi-perpendicular approximations are simply the principal modes, with B_0 replaced by its component along the wave vector **k**, respectively $B_0 \cos\theta$ or $B_0 \sin\theta$.

10.4 Summary

The presence of a static magnetic field was shown to have a profound effect on the propagation of electromagnetic waves in a plasma. The primary effect of the magnetic field is to couple the orthogonal components of the wave electric field to one another. With the exception of the ordinary mode, all electromagnetic waves in a magnetized plasma have coupled transverse electric field components, which gives rise to circular or elliptical polarization of the wave. In contrast to the single branch of the dispersion relation for non-magnetized plasmas, waves in a magnetized plasma exhibit multiple modes of propagation that depend on the direction of propagation with respect to the magnetic field and the wave polarization. The existence of multiple propagation modes for a single frequency leads to observable and scientifically useful effects such as Faraday rotation.

10.5 Problems

10-1. A transmitter satellite (Tx) and a receiver satellite (Rx) are on opposite ends of a large volume of cold plasma of unknown dimension and density. A static uniform magnetic field is as shown in the diagram overleaf. The strength of the magnetic field is such that the electron-cyclotron frequency is observed to be 6 kHz. The transmitter sends two linearly polarized EM signals at two frequencies, 68 kHz and 58 kHz. The 68 kHz signal arrives at the receiver linearly polarized but with a Faraday rotation of 5°. The 58 kHz signal arrives LH circularly polarized. Assuming that the plasma density is uniform across the volume, determine the upper and lower bounds of the plasma's lateral dimension d and density N. Justify your answer using an appropriate dispersion (ω–k) diagram.

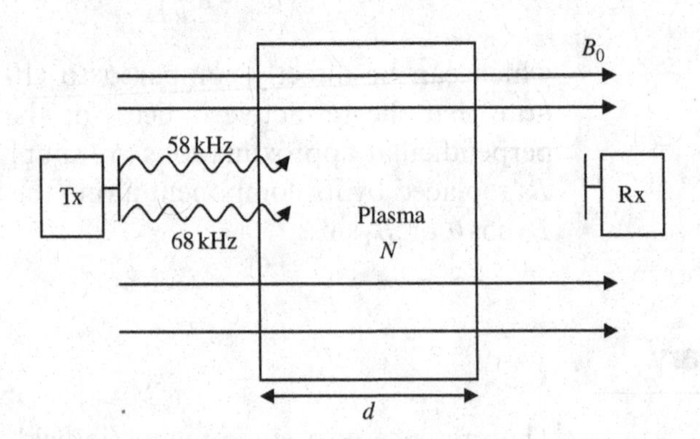

10-2. Consider wave propagation parallel to the static magnetic field in a plasma with the following parameters: $f_p = 1$ MHz, $f_{ce} = 100$ kHz. Determine the distance which a 10 MHz wave needs to travel in order for the electric field to rotate by $\pi/6$ radians.

10-3. Find the group velocity for a 3 kHz whistler-mode wave propagating parallel to the magnetic field with cyclotron frequency of 7 kHz and plasma density of 4×10^8 m^{-3}.

10-4. Find an expression for the group velocity of the whistler mode and make a plot of group velocity versus frequency. Normalize the frequency axis of your plot in units of the electron-cyclotron frequnecy w_{ce}, and the velocity axis in units of the speed of light c.

10-5. Find an analytic expression for the frequency at which the group velocity whistler mode is maximum. This is the so-called "nose frequency" of a whistler.

10-6. Microwaves of frequency 30 GHz propagating in the z direction are sent through a plasma slab that is infinite in the x–y plane and 10 cm thick in the z direction. The plasma density in the slab is $N_0 = 3.0 \times 10^{18}$ m^{-3}. The slab also has a magnetic field of 1.05 T oriented in the x direction. Calculate the number of wavelengths in the slab if the impinging waves have electric fields oriented in (a) the x direction and (b) the y direction.

10-7. LH circularly polarized waves are propagating along a uniform magnetic field into a plasma with steadily increasing density. The magnitude of the magnetic field is 0.2 T. At what density will the waves reach a cutoff if their frequency is 2.8 GHz?

References

[1] M. Faraday, On the magnetization of light and the illumination of magnetic lines of force. *Phil. Trans. R. Soc. London*, **136** (1846), 1.
[2] K. Davies, *Ionospheric Radio* (London: Peter Peregrinus, 1990).

CHAPTER

11 Effects of collisions, ions, and finite temperature on waves in magnetized plasmas

11.1 Introduction

The fundamental wave modes for a magnetized cold plasma were outlined in Chapter 10. In the application of the cold-plasma, two-fluid model, ions were assumed to be stationary and collisions and electron thermal motions were neglected. In this chapter we investigate expected wave behavior when these simplifying assumptions are relaxed: we will examine the effects of collisions, ion motion, and finite temperature, and how the previously derived modes are modified. We will show that collisions can be treated as a relatively straightforward correction factor to cold-plasma wave modes. On the other hand, for low-frequency phenomena where ion motions are included, the plasma behavior approaches that of a single fluid and new wave types are possible, appropriately called hydromagnetic or MHD waves. Likewise, it will be shown that finite temperature changes the fundamental stationary plasma oscillations into propagating waves.

11.2 Effects of collisions

We mentioned in Chapter 9 that collisions can be taken into account in a straightforward manner, simply by including the frictional damping term in the momentum equation. While the analysis in Chapter 9 was for non-magnetized plasmas, the same procedure can be used for magnetized plasmas. The time-harmonic momentum equation (9.10b) becomes

$$j\omega m N_0 \mathbf{u} = q N_0(\mathbf{E} + \mathbf{u} \times \mathbf{B}_0) - m N_0 \nu \mathbf{u}$$

$$j w m (1 - j\nu/\omega)\, \mathbf{u} = q(\mathbf{E} + \mathbf{u} \times \mathbf{B}_0), \qquad (11.1)$$

where ν is the collision frequency. The effect of the collisional drag term is thus to replace the particle mass m by $m_{\text{eff}} = m(1 - j\nu/\omega)$. If the simple substitution of $m \to m(1 - j\nu/\omega)$ is made in any of the equations derived so far, the appropriate collisional form is obtained. In general, this substitution will make all of the refractive indices complex, leading to complex values of the wave number $k = k_r - jk_i$ and to corresponding wave solutions of the type

$$e^{j(\omega t - k_r r + jk_i r)} = e^{-k_i r} e^{j(\omega t - k_r r)}, \tag{11.2}$$

that is, exponential decay with distance, as one would expect as a consequence of collisional losses. As an example, consider the RH mode for parallel propagation in the absence of ion motion. Making the substitution $m_e \to m_e(1 - j\nu/\omega)$ in (10.25) we find

$$\mu^2 = \frac{k^2c^2}{\omega^2} = 1 - \frac{\dfrac{\omega_p^2}{1 - j\nu/c}}{\omega\left(\omega - \dfrac{\omega_c}{1 - j\nu/c}\right)} = 1 - \frac{\omega_p^2}{\omega(\omega - \omega_c - j\nu)}. \tag{11.3}$$

We note that collisions are most effective near cyclotron resonance (i.e., when $\omega \simeq \omega_c$, so that the $j\nu$ term in the denominator is dominant). They are also effective near the cutoff region (i.e., $\omega \simeq \omega_R$); in fact, it can be shown that collisional damping is always strong when the group velocity is low. In general, the Appleton–Hartree refractive index expression, including the effect of collisions, is given by

$$\mu^2 = 1 - \frac{(\omega_p^2/\omega^2)}{1 - j\dfrac{\nu}{\omega} + \dfrac{\omega_c^2 \sin^2\theta}{2(\omega^2 - \omega_p^2 - j\omega\nu) \pm \left\{\left[\dfrac{\omega_c^2 \sin^2\theta}{2(\omega^2 - \omega_p^2 - j\omega\nu)}\right]^2 + \dfrac{\omega_c^2}{\omega^2}\cos^2\theta\right\}^{1/2}}}. \tag{11.4}$$

11.3 Effects of positive ions

To include the effects of positive ions, we follow the same procedure that we followed in deriving (10.14), (10.17), and (10.18), except that the current $\mathbf{J}$ is now a summation of the currents of electrons and of ions. For simplicity, we consider a two-fluid plasma, consisting of electrons and protons (i.e., $q_i = -q_e$). The definitions of the component permittivities given in (10.15) are modified additively, so that we have

$$\epsilon_{\perp} = 1 - \left[\frac{\omega_{pe}^2}{\omega^2 - \omega_{ce}^2} + \frac{\omega_{pi}^2}{\omega^2 - \omega_{ci}^2}\right] \tag{11.5a}$$

$$\epsilon_{\times} = \left(\frac{\omega_{ce}}{\omega}\right)\frac{\omega_{pe}^2}{\omega^2 - \omega_{ce}^2} - \left(\frac{\omega_{ci}}{\omega}\right)\frac{\omega_{pi}^2}{\omega^2 - \omega_{ci}^2} \tag{11.5b}$$

$$\epsilon_{\parallel} = 1 - \left[\frac{\omega_{pe}^2}{\omega^2} - \frac{\omega_{pi}^2}{\omega^2}\right]. \tag{11.5c}$$

We now investigate how these modifications will affect the principal modes.

11.3.1 Parallel propagation ($\theta = 0$)

We examine the three modes analogous to (10.22):

(a) The plasma oscillations ($\epsilon_{\parallel} = 0$) are not affected significantly, since $\omega_{pi}^2/\omega_{pe}^2 = m_e/m_i \ll 1$.

(b) Concerning the RH polarized wave, we now have

$$\mu_{\rm RH}^2 = 1 - \frac{\omega_{pe}^2}{(\omega + \omega_{ci})(\omega - \omega_{ce})}. \tag{11.6}$$

It is clear that the RH mode will only be modified for frequencies low enough to be comparable to ω_{ci}. The upper branch of the RH mode is thus unaffected, while the lower branch is modified. For $\omega \ll \omega_{ci}$, the refractive index tends to the value

$$\mu_{\rm RH}^2 \simeq 1 + \frac{\omega_{pe}^2}{\omega_{ce}\omega_{ci}} \simeq \frac{c^2}{v_{\rm A}}, \tag{11.7}$$

where $v_{\rm A}$ is the *Alfvén speed*, defined by

$$v_{\rm A} \equiv c\left(\frac{\omega_{ce}\omega_{ci}}{\omega_{pe}^2}\right)^{1/2} = \left(\frac{|q_e|B_0}{m_e}\frac{|q_e|B_0}{m_i}\frac{\epsilon_0 m_e}{N_0 q_e^2}\frac{1}{\mu_0\epsilon_0}\right)^{1/2}$$

$$= c\,\frac{\omega_{ce}}{\omega_{pe}}\left(\frac{m_e}{m_i}\right)^{1/2} = c\,\frac{\omega_{ci}}{\omega_{pi}} \quad \text{or} \quad \frac{B_0}{\sqrt{\mu_0 N_0 m_i}}, \tag{11.8}$$

where we have assumed the typical situation with $v_{\rm A} \ll c$. This wave mode is called the RH polarized *shear Alfvén wave*. Note from (11.7) that in this low-frequency range neither the numerator nor the denominator of the expression for μ^2 can go to

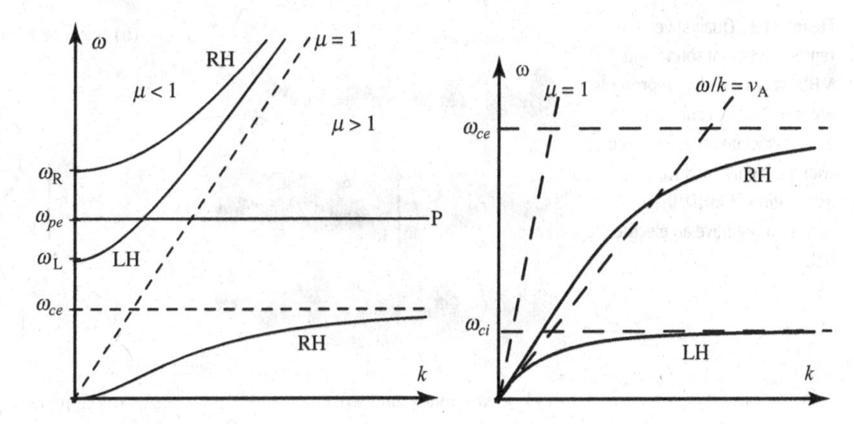

Figure 11.1 Effect of positive ions on wave propagation in a cold magnetoplasma: parallel propagation. The low-frequency end of the RH whistler mode is modified and a completely new LH ion-cyclotron wave branch appears.

zero, meaning that there are no cutoffs or resonances. This is not surprising, since the ions are left-handed, while the wave is right-handed. As we increase the frequency, the RH polarized shear Alfvén wave changes smoothly into a whistler-mode wave, with its resonance at $\omega = \omega_{ce}$. At the low-frequency end, the RH polarized shear Alfvén wave combines with the new LH polarized shear Alfvén mode described below and becomes equivalent to a simple transverse wave propagating in a medium with a large scalar dielectric constant or very low phase velocity $v_p \simeq v_A$. The new low-frequency RH branch is shown in Figure 11.1.

(c) The LH branch that we found by neglecting the effects of ions is almost unchanged since it lies above $\omega = \omega_L$, which is typically much larger than ω_{ci}. However, there is now a new branch of propagation for $\omega < \omega_{ci}$ which is called the *ion-cyclotron wave* or LH polarized shear Alfvén wave. The general dispersion relation for the LH wave is

$$\mu_{LH}^2 = 1 - \frac{\omega_{pe}^2}{(\omega - \omega_{ci})(\omega + \omega_{ce})}. \tag{11.9}$$

The low-frequency behavior is identical to that of the RH wave, as given in (11.7). The fact that the RH and LH shear Alfvén waves have the same dispersion relation means that, at these low frequencies, linearly polarized Alfvén waves can exist, and propagate, without Faraday rotation (see Figure 11.2). It should be noted, however, that the LH mode clearly has a resonance at $\omega = \omega_{ci}$, associated with LH ion-cyclotron motion. The new low-frequency LH branch is shown in Figure 11.1.

The physical nature of the shear Alfvén waves described above can be understood as follows. At the lowest frequencies ($\omega \ll \omega_{ci}$)

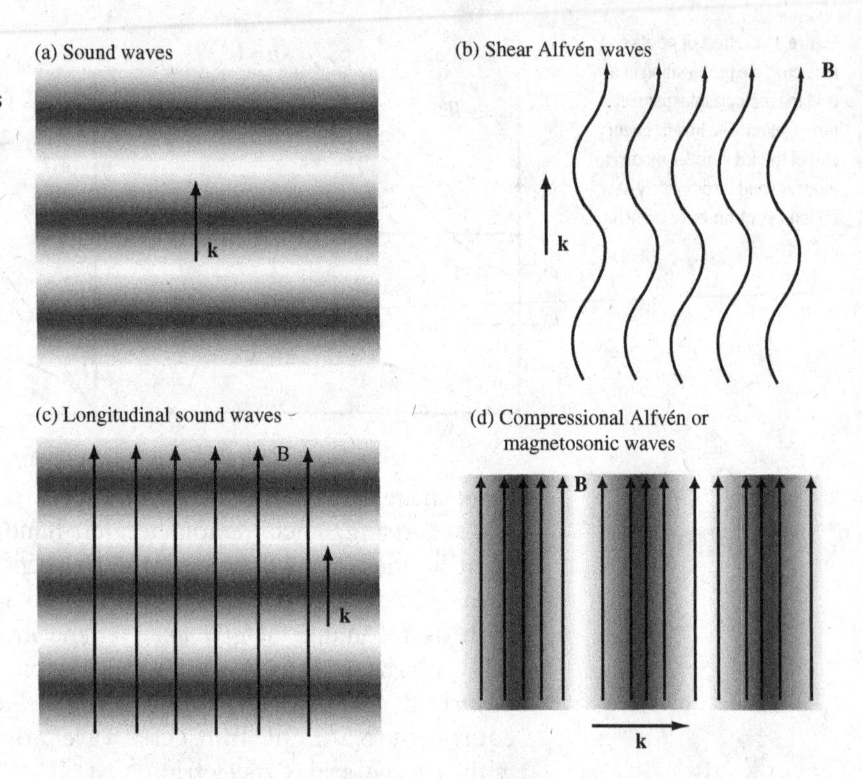

Figure 11.2 Qualitative representation of sound and MHD waves. Shading represents areas of density concentration and rarefaction. In accordance with the highly conducting fluid assumption of MHD, these waves do not have an electric field.

both the electrons and the ions execute $\mathbf{E} \times \mathbf{B}_0$ drift motion. The magnetic field lines themselves are frozen into the plasma and move with it at the same drift speed, $v_E = \mathbf{E} \times \mathbf{B}_0 / B_0^2$. The circular polarization of the wave electric field means that the field lines twist like a helix around the z axis, which is why this mode is called the shear Alfvén wave (or sometimes the torsional wave). The ions provide the inertia for this wave, causing the field lines to continue to move circularly, rather than come to rest. One characteristic of such twisting motion is that $\nabla \cdot \mathbf{u} = 0$, so that there is no compression nor any pressure perturbations (p_1); thus this wave is not affected by the finite-temperature (or pressure-gradient) term in the momentum equation.

At frequencies below ω_{ci} the traverse shear Alfvén wave is in fact an MHD wave, where the magnetic field lines behave as elastic cords under tension. In Section 11.3.4 it will be shown that this wave behavior can be obtained directly from the MHD single-fluid equations derived in Chapter 6. The wave motion is a fluctuation between plasma density changes perpendicular to the magnetic field and magnetic field line tension B_0^2/μ_0. This wave behavior is often compared to traverse vibrations of an elastic string, and is illustrated in Figure 11.2b. Under the frozen-in field concept described in Chapter 6, the magnetic field lines act as mass-loaded springs under tension, vibrating when the conducting fluid is perturbed.

11.3.2 Perpendicular propagation ($\theta = \pi/2$)

As in the absence of ions, we have ordinary and extraordinary modes, as described by (10.30), with permittivities as given by (9.8). We can make the following observations:

(a) The ordinary mode is hardly modified, since $\epsilon_{\parallel}$ is not significantly different. This mode has a cutoff frequency of $\omega = \omega_{pe}$, and is characterized by a wave electric field parallel to $\mathbf{B}_0$, i.e., $\mathbf{E} \times \mathbf{B}_0 \neq 0$. With the electric field only moving the ions along $\mathbf{B}_0$, the ion dynamics cannot produce any low-frequency effects. Thus, the ordinary branch shown in Figure 10.3 is unchanged.

(b) For the extraordinary branch, propagation near the upper hybrid frequency is not significantly affected. However, a new branch emerges, which has a resonance ($\epsilon_{\perp} = 0$) at the *lower hybrid frequency* ω_{h}, given by

$$\frac{1}{\omega_{\mathrm{h}}^2} = \frac{1}{\omega_{pi}^2 + \omega_{ci}^2} + \frac{1}{\omega_{ce}\omega_{ci}}$$

$$\rightarrow \quad \omega_{\mathrm{h}}^2 = \frac{\omega_{ce}\omega_{ci}\left(\omega_{pi}^2 + \omega_{ci}^2\right)}{\omega_{pi}^2 + \omega_{ci}^2 + \omega_{ci}\omega_{ce}} \simeq \frac{\omega_{ce}\omega_{ci}\left(\omega_{pi}^2 + \omega_{ci}^2\right)}{\omega_{pi}^2 + \omega_{ci}\omega_{ce}}. \tag{11.10}$$

For $\omega \ll \omega_{ci}$, this mode has a phase velocity given by the Alfvén speed. It has associated space charge, and finite $\mathbf{k} \cdot \mathbf{u}$, so that it compresses the plasma, which is why it is often called a *compressional Alfvén wave*. Once again, since the plasma is frozen into the field lines at the low hydromagnetic frequencies considered here, the magnetic field is also compressed. The wave propagates across the magnetic field, alternately compressing and expanding it like the pressure in a sound wave, which is why this mode is also often called the *magnetosonic* mode. For the same reason, we expect this wave to be affected by finite temperature effects, which we will study in the next section. The new extraordinary wave branch corresponding to this mode is shown in Figure 11.3. The basic propagation characteristics of this MHD wave are shown in Figure 11.2d.

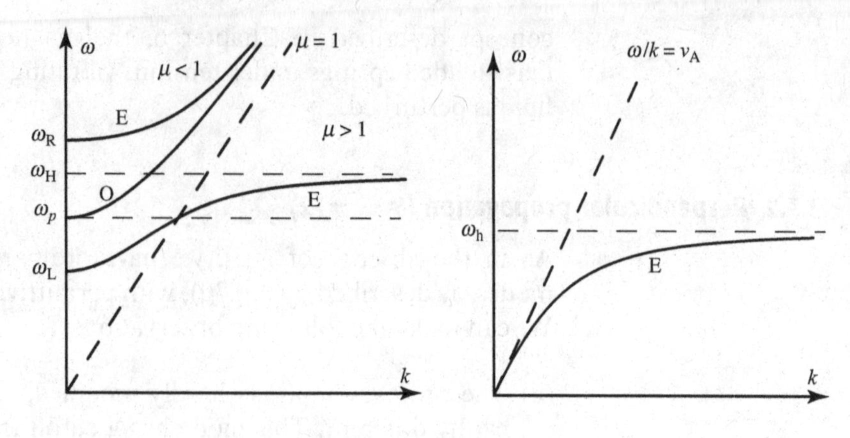

Figure 11.3 Effect of positive ions on wave propagation in a cold magnetoplasma: perpendicular propagation. The ordinary mode is not affected but a completely new extraordinary magnetosonic wave (or compressional Alfvén wave) branch appears.

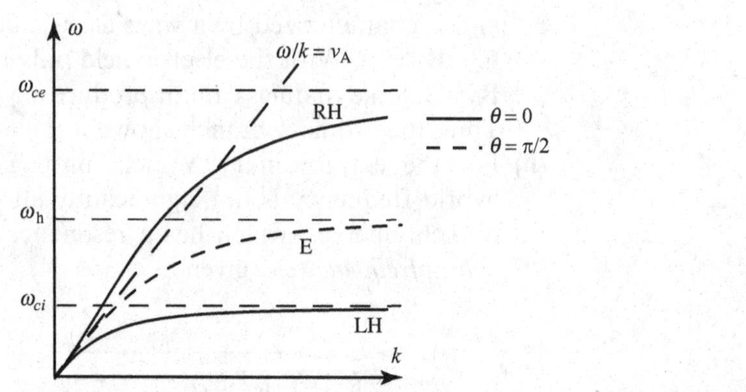

Figure 11.4 Effect of positive ions on wave propagation in a cold magnetoplasma: oblique propagation.

11.3.3 Oblique propagation (arbitrary θ)

The precise nature of the dispersion diagrams for oblique propagation depends on the magnitude of ω_{pe}/ω_{ce}; however, the general behavior is as shown in Figure 11.4. In the low-frequency limit, i.e., $\omega \ll \omega_{ci}$, one branch has phase velocity v_{A} at all angles, while the other, the ion-cyclotron branch, has $v_p = v_{\mathrm{A}} \cos\theta$.

11.3.4 Hydromagnetic (MHD) waves

The Alfvén waves that we encountered with the inclusion of ion motions are in fact natural solutions of the magnetohydrodynamic equations (6.39) and (6.40). Such waves are physically caused by the magnetic forces discussed in Chapter 6, including the component of the magnetic force that tends to straighten out curved magnetic field lines. To the degree that the MHD fluid is perfectly conducting, the electric field within it is zero, so that the wave action (i.e., the generation of one quantity by the other and vice versa) in such waves is between the fluid velocity $\mathbf{u}_{\mathrm{m}}$ and the magnetic field $\mathbf{B}_1$ rather than between electric field $\mathbf{E}$ and $\mathbf{B}_1$. Consider an unbounded conductive fluid embedded in a homogeneous magnetic field of magnitude $\mathbf{B}_0$. If we disturb this fluid by moving a column in a given direction with velocity $\mathbf{u}$, currents arise in order to set up the charge distribution necessary to maintain a zero electric field. These currents interact with the magnetic field and produce forces which impede the motion of the column but also move adjacent layers, thus leading to wave motion. We now show that these waves can be directly obtained from the MHD equations (6.39) and (6.40).

Using (6.42) in (6.40) we have

$$\rho_\mathrm{m}\frac{\partial \mathbf{u}_\mathrm{m}}{\partial t} = -\nabla\left[p + \frac{B^2}{2\mu_0}\right] + \frac{1}{\mu_0}(\mathbf{B}\cdot\nabla)\mathbf{B}$$

$$= \underbrace{-\nabla\left[p + \frac{(B_0+B_1)^2}{2\mu_0}\right] + \frac{1}{\mu_0}(\mathbf{B}_0\cdot\nabla)\mathbf{B}_1}_{\text{linearized form}}, \qquad (11.11)$$

while from (6.39), for the case when flow dominates (nearly infinite conductivity), we have

$$\frac{\partial \mathbf{B}}{\partial t} = \nabla\times(\mathbf{u}_\mathrm{m}\times\mathbf{B}) \quad\rightarrow\quad \underbrace{\frac{\partial \mathbf{B}_1}{\partial t} = \nabla\times(\mathbf{u}_\mathrm{m}\times\mathbf{B}_0)}_{\text{linearized form}}. \qquad (11.12)$$

For simplicity, we will consider incompressible fluids,[1] with

$$\nabla\cdot\mathbf{u}_\mathrm{m} = 0 \quad\rightarrow\quad \frac{d\rho_\mathrm{m}}{dt} = 0 \quad\rightarrow\quad \rho_\mathrm{m} = \rho_\mathrm{m0} = \text{constant}. \qquad (11.13)$$

We can now expand (11.12) using a vector identity:

$$\nabla\times(\mathbf{u}_\mathrm{m}\times\mathbf{B}) \equiv (\mathbf{B}_0\cdot\nabla)\mathbf{u}_\mathrm{m} - \underbrace{(\mathbf{u}_\mathrm{m}\cdot\nabla)\mathbf{B}_0}_{=0\ (\mathbf{B}_0=\text{constant})} + \mathbf{u}_\mathrm{m}\underbrace{(\nabla\cdot\mathbf{B}_0)}_{=0} - \mathbf{B}_0\underbrace{(\nabla\cdot\mathbf{u}_\mathrm{m})}_{=0\text{ from (11.13)}} \qquad (11.14)$$

We thus have

$$\frac{\partial \mathbf{B}_1}{\partial t} = (\mathbf{B}_0\cdot\nabla)\mathbf{u}_\mathrm{m}. \qquad (11.15)$$

Without further discussion, the potential for hydromagnetic wave action can be seen by inspection of (11.11) and (11.15), where we see

[1] Inclusion of compressibility complicates matters only slightly, since it can be shown that hydromagnetic plane-wave equations for compressible fluids conveniently split into parts that are parallel and perpendicular to $\mathbf{B}$, the latter representing transverse Alfvén waves while the former represents ordinary sound waves.

that the time derivatives of the two quantities $\mathbf{B}_1$ and $\mathbf{u}_{\rm m}$ are related to spatial derivatives of one another, very similar to (for example) Maxwell's equations, in which the time derivatives of electric and magnetic fields are related to the spatial derivatives (in that case curls) of one another. To proceed further, we select our coordinate system so that (as before) $\mathbf{B}_0 = \hat{\mathbf{z}}\, B_0$, and seek uniform plane waves propagating along $\mathbf{B}_0$, i.e.,

$$e^{j(\omega t - \mathbf{k}\cdot\mathbf{r})} = e^{j(\omega t - kz)}, \tag{11.16}$$

where $k = |\mathbf{k}|$. Note that by definition (i.e., for a *uniform* plane wave) we have $\partial/\partial x = 0$ and $\partial/\partial y = 0$, so that (11.11) and (11.15) reduce to

$$\rho_{\rm m0}\frac{\partial \mathbf{u}_{\rm m}}{\partial t} = -\hat{\mathbf{z}}\frac{\partial}{\partial z}\left[p + \frac{(B_0 + B_1)^2}{2\mu_0}\right] + \frac{1}{\mu_0}B_0\frac{\partial \mathbf{B}_1}{\partial z} \tag{11.17}$$

$$\frac{\partial \mathbf{B}_1}{\partial t} = B_0\frac{\partial \mathbf{u}_{\rm m}}{\partial z}. \tag{11.18}$$

Noting the divergence-free nature of both the magnetic field and the fluid velocity (11.13), we must have

$$\frac{\partial \mathbf{u}_{{\rm m}z}}{\partial z} = 0 \quad \text{and} \quad \frac{\partial \mathbf{B}_{1z}}{\partial z} = 0, \tag{11.19}$$

indicating that both of these quantities are uniform in space. Since we are looking for wave-like solutions, we can take both of these to be zero, i.e., $\mathbf{u}_{{\rm m}z} = 0$ and $\mathbf{B}_{1z} = 0$. Re-examination of (11.17) indicates that the left-hand term and the second term on the right-hand side do not have any components in the z direction, so that we must have

$$-\frac{\partial}{\partial z}\left[p + \frac{(B_0 + B_1)^2}{2\mu_0}\right] = 0.$$

Our linearized equations (11.17) and (11.18) can now be written as

$$\frac{\partial \mathbf{u}_{\rm m}}{\partial t} = \frac{B_0}{\rho_{\rm m0}\mu_0}\frac{\partial \mathbf{B}_1}{\partial z} \tag{11.20}$$

$$\frac{\partial \mathbf{B}_1}{\partial t} = B_0\frac{\partial \mathbf{u}_{\rm m}}{\partial z}. \tag{11.21}$$

Combining after differentiation, we obtain wave equations which can be contrasted with similar equations for uniform plane electromagnetic waves propagating in the z direction:

$$\frac{\partial^2 \overline{\mathcal{E}}}{\partial t^2} = \frac{1}{\mu\epsilon}\frac{\partial^2 \overline{\mathcal{E}}}{\partial z^2} \qquad \frac{\partial^2 \mathbf{u}_\text{m}}{\partial t^2} = \frac{B_0^2}{\rho_{\text{m}0}\mu_0}\frac{\partial^2 \mathbf{u}_\text{m}}{\partial z^2} \tag{11.22}$$

$$\frac{\partial^2 \overline{\mathcal{H}}}{\partial t^2} = \frac{1}{\mu\epsilon}\frac{\partial^2 \overline{\mathcal{H}}}{\partial z^2} \qquad \frac{\partial^2 \mathbf{B}_1}{\partial t^2} = \frac{B_0^2}{\rho_{\text{m}0}\mu_0}\frac{\partial^2 \mathbf{B}_1}{\partial z^2} \tag{11.23}$$

$$v_p = \frac{1}{\sqrt{\mu\epsilon}}, \quad k = \omega\sqrt{\mu\epsilon} \qquad v_\text{A} = \frac{B_0}{\sqrt{\mu_0\rho_{\text{m}0}}}, \quad k_\text{A} = \frac{\omega\sqrt{\mu_0\rho_{\text{m}0}}}{B_0}, \tag{11.24}$$

where $\overline{\mathcal{E}}$ and $\overline{\mathcal{H}}$ are respectively the total electric and total magnetic fields. We see that hydromagnetic waves propagating along $\mathbf{B}_0$ are very similar to uniform plane electromagnetic waves (also known as transverse electromagnetic or TEM waves), with a phase velocity independent of frequency and given by

$$\text{Alfvén speed} \quad \boxed{v_\text{A} = \frac{B_0}{\sqrt{\mu_0\rho_{\text{m}0}}}}, \tag{11.25}$$

which, of course, is identical to the Alfvén speed for shear Alfvén waves as given in (11.8). These hydromagnetic waves are indeed one and the same with the shear Alfvén waves that we found in Section 11.3 by taking the limit of the more general refractive index for $\omega \ll \omega_i$. Note that there is no electric field for these waves, since the electric field within the highly conducting fluid is nearly zero. The wave action arises from a give-and-take between $\mathbf{B}_1$ and $\mathbf{u}_\text{m}$, rather than that between $\overline{\mathcal{E}}$ and $\overline{\mathcal{H}}$ for an electromagnetic wave (see Figure 11.2). Note that these waves are transverse, just like TEM waves, since $\mathbf{B}_1$ and $\mathbf{u}_\text{m}$ do not have z components (see (11.19)). However, it is clear from (11.20) (or (11.21)) that $\mathbf{B}_1$ and $\mathbf{u}_\text{m}$ are polarized in the same direction, as opposed to $\overline{\mathcal{E}}$ and $\overline{\mathcal{H}}$ for TEM waves, which are perpendicular to one another. Note, however, that the current $\mathbf{J}$ is perpendicular to both $\mathbf{B}_1$ and $\mathbf{k}$, since we have

$$\nabla \times \mathbf{B}_1 = \frac{1}{\mu_0}\mathbf{J} \quad \rightarrow \quad \mathbf{k} \times \mathbf{B}_1 = \frac{j}{\mu_0}\mathbf{J}. \tag{11.26}$$

Noting that both $\mathbf{B}_1$ and $\mathbf{k}$ have the form given in (11.16), we can rewrite (11.20) as

$$\omega\mathbf{B}_1 = B_0\, k\, \mathbf{u}_\text{m} \quad \rightarrow \quad \mathbf{B}_1 = \sqrt{\mu_0\rho_{\text{m}0}}\, \mathbf{u}_\text{m}, \tag{11.27}$$

which is indeed similar to $|\overline{\mathcal{H}}| = \left[\sqrt{\mu/\epsilon}\right]|\overline{\mathcal{E}}|$, so that the quantity $\sqrt{\mu_0\rho_{\text{m}0}}$ may be loosely viewed as the intrinsic "impedance" of the hydromagnetic fluid for shear (longitudinal) Alfvén waves.

The above analysis assumes that the conductivity of the plasma fluid is infinite, so that the $(\mu_0\sigma)^{-1}\nabla^2\mathbf{B}$ term in (6.39) can be left out of (11.12). The inclusion of finite but very large conductivity leads to damping of the Alfvén waves, as would be expected. One interesting difference between TEM waves and hydromagnetic waves is that higher conductivity means less damping for the latter, while TEM waves in a good conductor attenuate exponentially with distance in accordance with a skin depth of $\delta = (\mu_0\sigma\omega)^{-1/2}$.

11.4 Effects of temperature

Following up on our general discussions in Chapter 9, finite temperature effects can be included by using the pressure-gradient term in the momentum equation. The relevant time-harmonic equations are (9.10b) and (9.10c), repeated here:

$$N_0 m\, j\omega\, \mathbf{u} = q N_0(\mathbf{E} + \mathbf{u} \times \mathbf{B}_0) - \nabla p_1 \tag{11.28a}$$

$$p_1 = \gamma k_B T N_1; \tag{11.28b}$$

substituting (11.28b) in (11.28a) and rearranging gives

$$j\omega\mathbf{u} + \frac{\gamma k_B T_0}{N_0 m}\nabla N_1 = \frac{q}{m}(\mathbf{E} + \mathbf{u} \times \mathbf{B}_0), \tag{11.29}$$

valid, as before, for each plasma species. We do not expect cold-plasma modes with zero space charge ($N_1 = 0$) to be greatly affected by the inclusion of temperature and shall therefore concentrate on the modes that are effectively longitudinal (i.e., $\mathbf{k} \parallel \mathbf{E}$). We consider only the principal modes, to determine how they are modified and whether there are any new branches of propagation.

11.4.1 Parallel propagation ($\theta = 0$)

The only longitudinal mode is plasma oscillations with $\mathbf{E} \parallel \mathbf{B}_0$, given by $\epsilon_\parallel = 0$. We see from (11.29) that the equation of motion does not couple motions perpendicular to $\mathbf{B}_0$. The z component for electrons is given by

$$j\omega u_z + \frac{c_{se}^2}{N_0}(-jkN_1) = \frac{q_e}{m_e}E_z, \quad c_s = \sqrt{\frac{\gamma_e k_B T_e}{m_e}}, \tag{11.30}$$

where c_s is the speed of sound waves in the electron gas at temperature T_e. This is identical to the equation of motion (9.14)

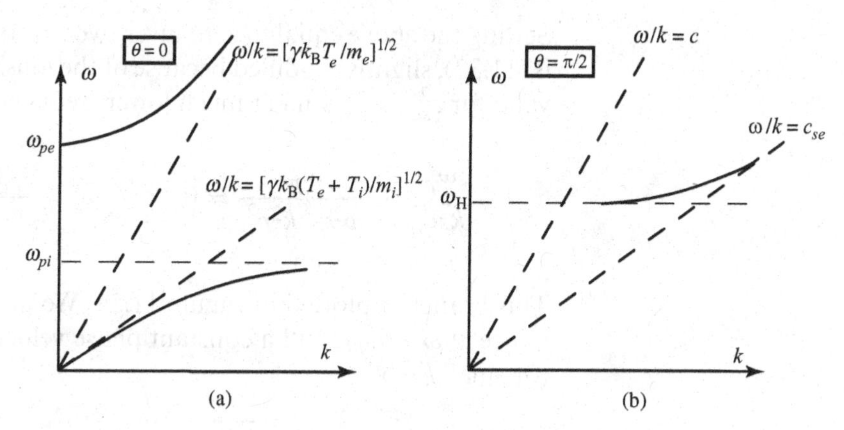

Figure 11.5 Effects of temperature. (a) Parallel propagation. (b) Perpendicular propagation. Both diagrams are for longitudinal waves (**k** ∥ **E**) and should not be confused with the dispersion diagrams for transverse waves (**k**⊥**E**).

for plasma oscillations in a cold plasma if we make the substitution $m_e \to m_e\left(1 - k^2 c_{se}^2/\omega^2\right)$. Thus, the modified expression for $\epsilon_\parallel = 0$ is

$$\epsilon_\parallel = 1 - \frac{\omega_{pe}^2}{\omega^2 - k^2 c_{se}^2} = 0, \tag{11.31}$$

which can be rearranged to give the dispersion relation

$$\omega^2 = \omega_{pe}^2 + k^2 c_{se}^2 = \omega_{pe}^2\left(1 + \gamma_e k^2 \lambda_{\mathrm{D}}^2\right). \tag{11.32}$$

This is identical to (9.31) from Chapter 9, with λ_{D} being the Debye length. We thus see that the solutions for a non-magnetized plasma obtained in Chapter 9 are simply special cases of the more general magnetized-plasma case. This dispersion relation is plotted in Figure 11.5a, showing how the plasma oscillations branch is modified for finite temperature. Note that the asymptotic phase velocity at $\omega \to \infty$ is $v_p = \omega/k = c_{se}$. Also note that the inclusion of finite temperature effects makes the propagation k-dependent, and that the group velocity is no longer zero. For a cold plasma, it was mentioned in Section 11.3 that inclusion of positive-ion motions had no significant effect on the plasma oscillations. However, we know from our discussions that finite temperature allows ions to communicate with one another, leading to low-frequency ion acoustic waves. This wave mode is a new branch that is brought about by finite temperature. As before, the dispersion relation for the case which includes ion motions can be obtained by simple additive modification of that for the electron-only case. We thus have

$$\epsilon_\parallel = 1 - \left[\frac{\omega_{pe}^2}{\omega^2 - k^2 c_{se}^2} + \frac{\omega_{pi}^2}{\omega^2 - k^2 c_{si}^2}\right]. \tag{11.33}$$

Setting the above equal to zero gives two solutions, the first of which is (11.32), slightly modified because of the ions. The second solution, valid for $c_{se}^2 \gg c_{si}^2$ and at much lower frequencies, is

$$1 + \frac{\omega_{pe}^2}{k^2 c_{se}^2} - \frac{\omega_{pi}^2}{\omega^2 - k^2 c_{si}^2} = 0 \quad \text{or} \quad \omega^2 = \frac{\omega_{pi}^2 k^2 c_{se}^2 + \omega_{pe}^2 k^2 c_{si}^2}{\omega_{pe}^2}. \tag{11.34}$$

This branch is plotted in Figure 11.5a. We note that there is a resonance at $\omega = \omega_{pi}$, and a constant phase velocity, at low frequencies (or small k), of

$$v_s = \frac{\omega}{k} \simeq \sqrt{\frac{\omega_{pi}^2 c_{se}^2 + \omega_{pe}^2 c_{si}^2}{\omega_{pe}^2}} \simeq \sqrt{\frac{\gamma k_B (T_e + T_i)}{m_i}}, \tag{11.35}$$

which is identical to (9.32). Since this behavior is qualitatively similar to sound waves in an isothermal gas, these waves are referred to as ion sound waves or ion acoustic waves (see Figure 11.2a).

11.4.2 Perpendicular propagation ($\theta = \pi/2$)

The only branch that exhibits space-charge effects is the extraordinary branch near the upper hybrid frequency. The version of $\epsilon_\perp$ modified as a result of finite temperature is

$$\epsilon_\perp = 1 - \frac{\omega_{pe}^2}{\omega^2 - k^2 c_{se}^2 - \left(\omega_{ce}^2 + \omega_{pe}^2\right)} = 0, \tag{11.36}$$

so that the dispersion relation for this branch becomes

$$\omega^2 = \left(\omega_{pe}^2 + \omega_{ce}^2\right) + k^2 c_{se}^2; \tag{11.37}$$

this is sketched in Figure 11.5b. Inclusion of ion motions for perpendicular propagation becomes complicated and needs to be studied separately for different plasma regimes. One important result is that the compressional Alfvén wave (i.e., the magnetosonic wave), which has space charge associated with it, has its phase velocity modified from v_A to $\sqrt{v_A^2 + v_s^2}$, where v_s is given by (11.35):

Finite-temperature magnetosonic speed $\boxed{v_m = \sqrt{v_A^2 + v_s^2}}$. (11.38)

11.5 Summary

In this chapter we examined the effects of collisions, positive ions, and finite temperature on waves in magnetized plasmas. The effect of collisions can be handled in a straightforward way by defining an effective particle mass $m_{\text{eff}} = m(1 - j\nu/\omega)$ that can be directly substituted into previously derived equations for collisionless cold-plasma wave modes. The inclusion of ions leads to new wave modes at low frequencies near and below the ion-cyclotron frequency. In the low-frequency limit, $\omega \ll \omega_{ci}$, the two-fluid plasma treatment with ion motion yields the same results as that derived from the single-fluid MHD equations. In such MHD waves the wave action is a give-and-take between the background magnetic field and the conducting fluid motion, instead of between electric and magnetic fields as in electromagnetic waves.

11.6 Problems

11-1. It has been proposed that a solar power station be built in space with huge panels of solar cells collecting sunlight continuously and transmitting the power to Earth in a microwave beam at 30 cm wavelength. However, losses in the microwave beam as it passes through the ionosphere are likely to be problematic. Treating the ionosphere as a weakly ionized gas with constant electron–neutral collision frequency, what fraction of the beam power would be lost in traversing 100 km of plasma with $N_e = 10^{11}$ m^{-3} and $N_n = 10^{16}$ m^{-3}? Assume the Earth's magnetic field intensity at the altitudes of interest to be ~50 μT. Also assume an electron temperature of $T_e \simeq 3.5 \times 10^3$ K and an electron–neutral collision cross-section of $\sigma_{\text{en}} \simeq 10^{-19}$ m^2. State all other assumptions.

11-2. Derive an expression for Faraday rotation for electromagnetic waves propagating parallel to the static magnetic field at frequencies below the ion-cyclotron frequency. Use the derived expression to determine the distance which a 1 Hz wave needs to travel in order for the electric field to rotate by $\pi/6$ radians if the plasma parameters are $f_p = 1$ MHz and $f_{ce} = 100$ kHz.

11-3. Calculate the attenuation in dB km^{-1} of a RH circularly polarized "whistler-mode" wave at high latitudes in the Earth's ionosphere. Assume that the wave propagates parallel to the Earth's magnetic field, which is here close to vertical.

The strength of the geomagnetic field at this altitude is 50 μT, the electron density is 10^9 m^{-3}, and the electron–neutral collision frequency is 10 MHz.

11-4. Calculate the speed of a shear Alfvén wave for interstellar space, where the electron density is $N_e = 10^7$ m^{-3}, and for a magnetic field of $B = 10^{-7}$ T.

11-5. Acoustic sound waves like those shown in Figure 11.2a obey the adiabatic equation of state between pressure and density, such that

$$\nabla p = \gamma k_B T \nabla N = v_s^2 m \nabla N,$$

where p, m, and T are assumed to be bulk "total" quantities. For magnetosonic waves the total pressure is given by $p + B_0^2/(2\mu_0)$. Use the adiabatic relationship between total pressure and density to show that the propagation speed for magnetosonic waves at finite temperature is $v_m = \sqrt{v_A^2 + v_s^2}$, as given in Equation (11.38).

11-6. An ambitious but naive student claims to be be able to construct a simple communication system that can transfer information faster than the speed of light. The student proposes extending a tight rope (or solid bar) over many hundreds of miles. When a person at one end tugs the bar or rope, the receiver on the other end will instantaneously feel the tug and register the communication signal with no delay. Explain why the student's logic is flawed and find the actual speed of communication for such a system as a fraction of the speed of light. Make reasonable assumptions about the material used for the rope or bar.

CHAPTER

12 Waves in hot plasmas

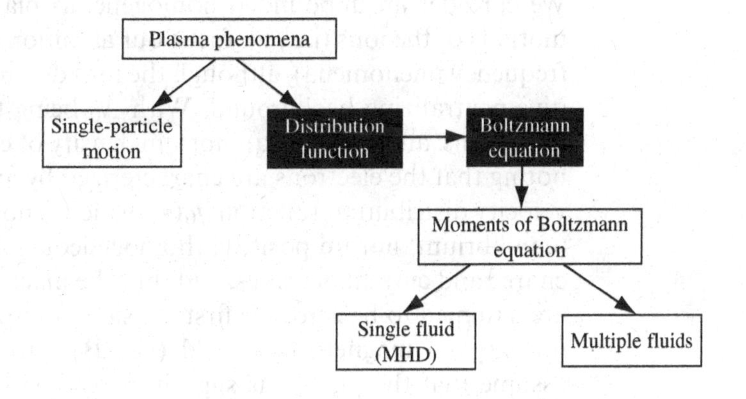

12.1 Introduction

Up to now, we have primarily been working with fluid equations, derived by taking the first two moments of the Boltzmann equation. All of our analysis of plasma wave phenomena in Chapters 9–11 was based on either the cold- or the warm-plasma approximations, sometimes with collisions also included. However, in the most general cases when the plasmas cannot be adequately described by moments and equations of state, it is necessary to solve Maxwell's equations simultaneously with the Boltzmann equation. Plasmas that are far from an equilibrium state are called "hot" plasmas, the simplest examples of which are beams. While there is some disagreement among plasma researchers about which specific conditions deserve to be denoted as "hot," the preceding definition is the simplest and most fundamental. In this context, we reiterate that in this text we consider cold plasmas to be those for which the temperature is strictly zero, and warm plasmas to have non-zero temperatures but still be treatable with a finite number of moment equations.

In this chapter we provide a preliminary discussion of waves in collisionless hot plasmas, which means that we can replace the equation of motion (which was used in previous chapters to relate the **E** field to the fluid velocity **u**) by the Vlasov equation. This type of analysis is sometimes called the Vlasov theory of plasma waves and involves working directly with the velocity distribution function and the Boltzmann equation.

12.2 Waves in a hot isotropic plasma

We consider an unbounded homogeneous plasma and ignore the motions of the ions (i.e., we focus our attention on reasonably high-frequency phenomena), although the ions do provide necessary positive neutralizing background. With N_0 being the ambient density of the ions, and with an equilibrium density of electrons, we start by noting that the electrons are characterized by an initial equilibrium velocity distribution function $f_0(\mathbf{v})$, which is not dependent on time (equilibrium) nor on position (homogeneous). Initially, the electric charge and current densities, and thus the macroscopic electric field, are assumed to be zero. We first consider an isotropic plasma, with no external magnetostatic field (i.e., $\mathbf{B}_0 = 0$). At time $t = 0$, we assume that the plasma is slightly perturbed from its equilibrium state so that at later times $t > 0$ the electrons are characterized by a modified distribution function $f(\mathbf{r}, \mathbf{v}, t)$ given by

$$f(\mathbf{r}, \mathbf{v}, t) = f_0(\mathbf{v}) + f_1(\mathbf{r}, \mathbf{v}, t), \tag{12.1}$$

where f_1 is always small compared to f_0 (i.e., $|f_1| \ll f_0$) and $f_1(\mathbf{r}, \mathbf{v}, t) = 0$ for $t < 0$. Because of this perturbation, non-zero electric charge and current densities arise which in turn give rise to macroscopic electric and magnetic fields. The perturbation $f_1(\mathbf{r}, \mathbf{v}, t)$ and the associated electromagnetic fields may decay in time (with characteristic decay times and natural frequencies), in which case the original state of the equilibrium is said to be *stable*. Alternatively, it is possible that the perturbation and the associated fields may increase in time, with characteristic growth rates, until a new state of equilibrium is reached, or even indefinitely in time, in which case the plasma is said to be *unstable*. As before, we will seek time-harmonic plane-wave solutions for which the perturbation quantities can vary as

$$e^{j(\omega t - \mathbf{k}\cdot\mathbf{r})} \tag{12.2}$$

but we will allow the frequency ω to be complex, to represent the decay of perturbations (ω has a small negative imaginary part) or growth of the wave in time (ω has small positive imaginary part). To start with, we note that the velocity distribution function $f(\mathbf{r}, \mathbf{v}, t)$ must satisfy the Vlasov equation:

$$\frac{\partial f}{\partial t} + (\mathbf{v} \cdot \nabla) f + \frac{q_e}{m_e}[\mathbf{E} + \mathbf{v} \times \mathbf{B}] \cdot \nabla_{\mathbf{v}} f = 0. \tag{12.3}$$

Note that $\mathbf{E}$, $\mathbf{B}$, and f_1 are perturbation quantities, and neglecting second-order terms we have the linearized Vlasov equation:

$$\frac{\partial f_1}{\partial t} + (\mathbf{v} \cdot \nabla) f_1 + \frac{q_e}{m_e}(\mathbf{E} + \mathbf{v} \times \mathbf{B}) \cdot \nabla_{\mathbf{v}} f_0 = 0. \tag{12.4}$$

Assuming that the equilibrium distribution function f_0 is isotropic, its velocity-space gradient (i.e., $\nabla_{\mathbf{v}} f_0$) is in the direction of velocity $\mathbf{v}$,[1] i.e.,

$$\nabla_{\mathbf{v}} f_0(\mathbf{v}) = \frac{\mathbf{v}}{v} \frac{d f_0(\mathbf{v})}{dv}, \tag{12.5}$$

so that we have

$$[\mathbf{v} \times \mathbf{B}] \cdot \nabla_{\mathbf{v}} f_0 = 0, \tag{12.6}$$

indicating that, in the linear regime, the wave magnetic field has no effect on the particle distribution. The final form of the linearized Vlasov equation is thus

$$\frac{\partial f_1}{\partial t} + (\mathbf{v} \cdot \nabla) f_1 + \frac{q_e}{m_e}\mathbf{E} \cdot \nabla_{\mathbf{v}} f_0 = 0. \tag{12.7}$$

Note that the electric field $\mathbf{E}$ is a first-order perturbation quantity and can be calculated from f_1 with the help of Poisson's equation:

$$\nabla \cdot \mathbf{E} = \frac{\rho}{\epsilon_0} = \frac{1}{\epsilon_0} q_e \int_v f_1 d^3 v, \tag{12.8}$$

[1] To see this, consider each component, e.g.,

$$\frac{\partial f_0(\mathbf{v})}{\partial v_x} = \underbrace{\frac{\partial f_0(\mathbf{v})}{dv} \frac{\partial v}{\partial v_x}}_{\text{chain rule}} = \frac{v_x}{v} \frac{d f_0(\mathbf{v})}{dv},$$

noting that $v = \left(v_x^2 + v_y^2 + v_z^2\right)^{\frac{1}{2}}$, so that

$$\frac{\partial v}{\partial v_x} = \frac{1}{2}\left(v_x^2 + v_y^2 + v_z^2\right)^{-\frac{1}{2}} (2v_x) = \frac{v_x}{v},$$

the same being true for the other velocity components. Note that (12.5) is only valid for an isotropic distribution.

where we have noted that the perturbation charge density is given by

$$\rho(\mathbf{r}, t) = q_e N_0 + q_e \int_v f(\mathbf{r}, \mathbf{v}, t) d^3v = q_e \int_v f_1(\mathbf{r}, \mathbf{v}, t) d^3v. \quad (12.9)$$

Similarly the perturbation current density is given by

$$\mathbf{J}(\mathbf{r}, t) = q_e \int_v \mathbf{v} f_1(\mathbf{r}, \mathbf{v}, t)\, d^3v. \quad (12.10)$$

Note that, since the equilibrium distribution function f_0 is known (in most cases it is a Maxwellian distribution for a given temperature), Equations (12.7) and (12.8) are fully sufficient to determine the two unknowns $\mathbf{E}$ and f_1. We now look for perturbation solutions of the form (12.2) so that all quantities ($\mathbf{E}$, f_1, etc.) vary in this manner. The linearized Vlasov equation then becomes

$$j\omega f_1(\mathbf{v}) - j\mathbf{k} \cdot \mathbf{v} f_1(\mathbf{v}) + \frac{q_e}{m_e} \mathbf{E} \cdot \nabla_{\mathbf{v}} f_0(\mathbf{v}) = 0, \quad (12.11)$$

from which we can write the perturbed distribution function as

$$f_1 = \frac{j}{\omega - \mathbf{k} \cdot \mathbf{v}} \frac{q_e}{m_e} \mathbf{E} \cdot \nabla_{\mathbf{v}} f_0, \quad (12.12)$$

which describes the dependence of f_1 on the electric field $\mathbf{E}$. The notable aspects of this dependence are the critical role played by the velocity-space gradients of the unperturbed distribution function f_0 and the denominator, which shows the potential for resonance effects, to be discussed later. We also note that Poisson's equation (12.8) can be rewritten using (12.2) as

$$j\mathbf{k} \cdot \mathbf{E} = \frac{q_e}{\epsilon_0} \int_v f_1\, d^3v. \quad (12.13)$$

12.2.1 Longitudinal waves ($\mathbf{k} \parallel \mathbf{E}$)

We now exclusively consider longitudinal plasma waves with $\mathbf{k} \parallel \mathbf{E}$, so that $\mathbf{k} \times \mathbf{E} = 0$, indicating that $\mathbf{B} = 0$, so that the waves considered are electrostatic. Integrating (12.12) over velocity space and noting that $\mathbf{E} = (\mathbf{k}/k)E$, we have

$$\int_v f_1 d^3v = j \frac{q_e}{m_e} \frac{E}{k} \mathbf{k} \cdot \int_v \frac{\nabla_{\mathbf{v}} f_0}{\omega - \mathbf{k} \cdot \mathbf{v}}\, d^3v. \quad (12.14)$$

Using (12.13) in (12.14) we have

Longitudinal wave dispersion relation

$$\boxed{1+\frac{q_e^2}{m_e\epsilon_0}\frac{1}{k^2}\mathbf{k}\cdot\int_v\frac{\nabla_{\mathbf{v}}f_0}{\omega-\mathbf{k}\cdot\mathbf{v}}\,d^3v=0}. \quad (12.15)$$

For a given $f_0(\mathbf{v})$, Equation (12.15) is a function of $\mathbf{k}$ and ω, i.e., a *dispersion relation* relating the wave number to the wave frequency. Unfortunately, the denominator of the integrand appears to vanish for certain values of $\mathbf{v}$, so that the integral is not well defined. Choosing, for example, a real value of $\mathbf{k}$, this equation may yield different values of ω, unless a prescription is given as to how to integrate around the pole. Note that, physically, this difficulty is caused by electrons moving with exactly the phase velocity of the wave, since for those electrons we have $\mathbf{k}\cdot\mathbf{v}=\omega$. In general, it is necessary to reformulate our approach from a search for normal-mode solutions of the type given in (12.2) to an initial-value problem which might be solved using Laplace transform methods and contour integration in the complex plane. In this text, we will be largely content with discussing the results of such an analysis, which indicates that plasma oscillations in a Maxwellian plasma are damped.

Cold-plasma model

We can first examine the dispersion relation (12.15) to see whether the results we obtained in previous chapters are borne out by this more general solution. For this purpose, it is useful to rewrite (12.15) by choosing our coordinate system so that the z axis is aligned with the electric field $\mathbf{E}$ (which is aligned with $\mathbf{k}$ and $\mathbf{J}$). Thus, for a nontrivial solution we must have

$$1+\frac{q_e^2}{\epsilon_0 m_e k}\int_v\frac{\partial f_0/\partial v_z}{\omega-k_z v_z}\,dv_z dv_y dv_x=0. \quad (12.16)$$

Consider, for example, a cold plasma, for which we have

$$f_0(\mathbf{v})=N_0\delta(\mathbf{v})=N_0\delta(v_x)\delta(v_y)\delta(v_z). \quad (12.17)$$

At this point, it is useful to note that

$$\frac{\partial}{\partial v_z}\left(\frac{f_0}{\omega-\mathbf{k}\cdot\mathbf{v}}\right)=\underbrace{\frac{\partial f_0/\partial v_z}{\omega-\mathbf{k}\cdot\mathbf{v}}+\frac{k_z f_0}{(\omega-\mathbf{k}\cdot\mathbf{v})^2}}_{\text{derivative of } f_0 \text{ times } (w-\mathbf{k}\cdot\mathbf{v})^{-1}}, \quad (12.18)$$

where we have simply applied the rule of differentiation for the product of two functions, noting that $\mathbf{k}\cdot\mathbf{v}=k_xv_x+k_yv_y+k_zv_z$ so that $\partial(\mathbf{k}\cdot\mathbf{v})/\partial v_z = k_z$. If we now integrate (12.18) with respect to v_z from $-\infty$ to $+\infty$, we have

$$\int_{-\infty}^{\infty} \frac{\partial}{\partial v_z}\left(\frac{f_0}{\omega - \mathbf{k}\cdot\mathbf{v}}\right) dv_z = \int_{\infty}^{\infty} \left[\frac{\partial f_0/\partial v_z}{\omega - \mathbf{k}\cdot\mathbf{v}} + \frac{k_z f_0}{(\omega - \mathbf{k}\cdot\mathbf{v})^2}\right] dv_z$$

$$\left[\frac{\partial}{\partial v_z}\left(\frac{f_0}{\omega - \mathbf{k}\cdot\mathbf{v}}\right)\right]_{v_z=-\infty}^{\infty} = \int_{-\infty}^{\infty} \left[\frac{\partial f_0/\partial v_z}{\omega - \mathbf{k}\cdot\mathbf{v}} + \frac{k_z f_0}{(\omega - \mathbf{k}\cdot\mathbf{v})^2}\right] dv_z$$

$$0 = \int_{-\infty}^{\infty} \frac{\partial f_0/\partial v_z}{\omega - \mathbf{k}\cdot\mathbf{v}} dv_z + \int_{-\infty}^{\infty} \frac{k_z f_0}{(\omega - \mathbf{k}\cdot\mathbf{v})^2} dv_z$$

$$= \int_{-\infty}^{\infty} \frac{\partial f_0/\partial v_z}{\omega - \mathbf{k}\cdot\mathbf{v}} dv_z + \int_{-\infty}^{\infty} \frac{k_z f_0}{(\omega - \mathbf{k}\cdot\mathbf{v})^2} dv_z,$$

where the left-hand side evaluates out to zero, both because f_0 tends to zero (e.g., a Maxwellian distribution) and because kv_z tends to $\pm\infty$ at $v_z = \pm\infty$. We thus have

$$\int_{-\infty}^{\infty} \frac{\partial f_0/\partial v_z}{\omega - \mathbf{k}\cdot\mathbf{v}} dv_z = -\int_{-\infty}^{\infty} \frac{k_z f_0}{(\omega - \mathbf{k}\cdot\mathbf{v})^2} dv_z$$

and can rewrite (12.16) as

$$1 - \frac{q_e^2}{\epsilon_0 m_e} \int_v \frac{f_0(\mathbf{v})}{(\omega - \mathbf{k}\cdot\mathbf{v})^2} dv_z dv_y dv_x. \quad (12.19)$$

Substituting (12.17) for f_0, and noting that $\int_{-\infty}^{\infty} \delta(\zeta) d\zeta = 1$, we then have

$$1 - \frac{N_0 q_e^2}{\epsilon_0 m_e \omega^2} = 0 \quad \rightarrow \quad \omega = \omega_{pe}, \quad (12.20)$$

showing that the Vlasov theory fully reproduces the result we previously obtained using cold-plasma theory.

Warm-plasma model

As the next step, we introduce an electron temperature low enough so that no electrons can be found to travel above a certain speed. In other words, a cutoff is introduced in the function $f_0(\mathbf{v})$ at a maximum speed $v_{\max}$. In this case, for sufficiently small wave numbers such that $\omega/k > v_{\max}$, the integrand in the dispersion relation is well behaved. Once again we find it convenient to use (12.19) and expand it in $(\mathbf{k}\cdot\mathbf{v})/w$ (appropriate since we want an approximation valid for small values of k) to find

$$1 = \frac{q_e^2}{m_e\epsilon_0\omega^2}\int_v \left[1 + 2\left(\frac{\mathbf{k}\cdot\mathbf{v}}{\omega}\right) + 3\left(\frac{\mathbf{k}\cdot\mathbf{v}}{\omega}\right)^2 + \cdots\right] f_0(\mathbf{v})\, d^3v. \tag{12.21}$$

Each of the terms on the right can be evaluated for a given distribution function. The first term yields N_0, so that the result in (12.20) is recovered. For an isotropic distribution, the integral with the second term vanishes. Assuming an isotropic distribution and retaining only the first three terms of (12.21) we find

$$\begin{aligned} 1 &= \frac{\omega_p^2}{\omega^2}\left[1 + 3\frac{(\langle\mathbf{k}\cdot\mathbf{v}\rangle)^2}{\omega^2}\right] \\ &= \frac{\omega_p^2}{\omega^2}\left[1 + \frac{k^2\langle v^2\rangle}{\omega^2}\right] \quad \to \quad [\omega^2]^2 - \omega_p^2\,\omega^2 - \omega_p^2 k^2\langle v^2\rangle = 0. \end{aligned} \tag{12.22}$$

This is a quadratic equation in ω^2, the solution of which is

$$\begin{aligned} \omega^2 &= \frac{\omega_p^2 \pm \left[\omega_p^4 + 4\omega_p^2 k^2\langle v^2\rangle\right]^{1/2}}{2} \\ &= \frac{\omega_p^2}{2} \pm \frac{\omega_p^2}{2}\underbrace{\left[1 + \frac{4k^2\langle v^2\rangle}{\omega_p^2}\right]^{1/2}}_{1+2\frac{k^2\langle v^2\rangle}{\omega_p^2} - \underbrace{\frac{1}{2}\left[\frac{k^2\langle v^2\rangle}{\omega_p^2}\right]^2 + \cdots}_{\simeq 0}} \simeq \omega_p^2 + \langle v^2\rangle\, k^2, \end{aligned} \tag{12.23}$$

where we have assumed that $k^2\langle v^2\rangle \ll \omega_p^2$ and used the "+" in the "±" to find the solution corresponding to electron plasma waves. Note that the choice of the "−" solution will give the dispersion relation valid for frequencies in the vicinity of ω_{pi}. For a Maxwellian plasma with an electron temperature T_e we have

$$\omega^2 = \omega_p^2 + \frac{\gamma k_B T_e}{m_e} k^2 = \omega_p^2 + c_{se}^2 k^2, \tag{12.24}$$

which is identical to (11.32) and (9.31), obtained earlier for warm-plasma electron acoustic waves. Thus, we see once again that the Vlasov formulation fully entails what we previously found with our cold- and warm-plasma approximations. The phase velocity of these waves is very large, since we obtained the result by restricting to small values of k. This is necessary so that there are no particles in the perturbed distribution which travel with speeds close to the wave phase velocity, thus ensuring that the integrand in (12.15) is well behaved. Note, however, that the group velocity for these waves, given by

$$v_g = \frac{d\omega}{dk} = \frac{\gamma k_B T_e}{m_e}\frac{k}{\omega} = \frac{v_{th}^2}{v_p}, \tag{12.25}$$

is much smaller than the thermal speed. Thus, the velocity with which perturbations propagate (i.e., v_g) is quite small.

Relative permittivity

As before, it is useful to express the effects of the plasma as an effective permittivity for the plasma medium. For this purpose, we note that the electric displacement $\mathbf{D}$, the volume polarization density $\mathbf{P}$, and the relative dielectric permittivity are related by

$$\mathbf{D} = \epsilon_0\mathbf{E} + \mathbf{P} = \epsilon_{eff}\epsilon_0\mathbf{E}. \tag{12.26}$$

Taking the divergence of (12.26) we find

$$\epsilon_{eff} = 1 + \frac{\nabla\cdot\mathbf{P}}{\epsilon_0\,\nabla\cdot\mathbf{E}}. \tag{12.27}$$

The polarization density $\mathbf{P}$ is produced as the response of the plasma to the applied field. It is related to volume polarization charge density by

$$\nabla\cdot\mathbf{P} = -\rho_p.$$

However, since the plasma has no net free charge under equilibrium conditions, the polarization charge density is simply the local charge density associated with the waves, i.e., ρ in (12.8). Using (12.8) and (12.12) we can write

$$\rho = \rho_p = q_e\int_v f_1 d^3v = \frac{jq_e^2}{m_e}E\int_v \frac{\partial f_0/\partial v_z}{\omega - kv_z}\,dv_z dv_y dv_x. \tag{12.28}$$

Substituting this expression for ρ_p in (12.27) and noting that $\nabla\cdot\mathbf{E} = -j\mathbf{k}\cdot\mathbf{E}$ we find

$$\epsilon_{eff}(\omega, k) = 1 + \frac{q_e^2}{\epsilon_0 m_e k}\int_v \frac{\partial f_0/\partial v_z}{\omega - kv_z}\,dv_z dv_y dv_x \tag{12.29}$$

as the basic kinetic expression for the relative dielectric permittivity of a plasma, for an electrostatic wave with frequency ω and wave number k. Comparing (12.29) and (12.16), we see that electrostatic normal modes satisfy the equation

$$\epsilon_{\text{eff}}(\omega, k) = 0. \tag{12.30}$$

The physical interpretation of (12.30) and (12.26) is that although the normal modes involve non-zero electric field $\mathbf{E}$ they produce no overall electric displacement $\mathbf{D}$ in the plasma. This very general result is entirely equivalent to our earliest result concerning electrostatic plasma oscillations, namely that they were defined by the relation $\epsilon_{\|} = 0$. Thus, the kinetic approach is a powerful extension of our previous analyses, providing results fully consistent with our earlier ones. The most striking aspect of kinetic theory, however, is the fact that it predicts and entails plasma behavior that is not brought out by the fluid formulations. One of the most important results is the so-called *Landau damping*, which is investigated in the next section.

Landau damping

Consider the effective permittivity derived above (Equation (12.29)):

$$\epsilon_{\text{eff}}(\omega, k) = 1 + \frac{q_e^2}{\epsilon_0 m_e k} \int_v \frac{\partial f_0/\partial v_z}{\omega - k v_z} dv_z dv_y dv_x. \tag{12.29}$$

We know that electrostatic normal modes satisfy the relationship

$$\epsilon_{\text{eff}}(\omega, k) = 0,$$

which is what we evaluated for the cold- and warm-plasma cases treated in previous sections. In the general case, when the plasma may have a non-trivial zeroth-order equilibrium distribution which may contain particles with velocities close to the wave phase velocity, the integral in (12.29) must be evaluated using complex analysis, i.e., contour integration in the complex plane. The additional contribution to ϵ_{eff} due to the pole at $v_z = \omega/k$ is an imaginary term,[2]

$$j\epsilon_{\text{eff}}^i(\omega, k) = -\frac{j\pi q_e^2}{\epsilon_0 m_e k^2} \int \left(\frac{\partial f_0}{\partial v_z}\right)_{v_z=\omega/k} dv_x\, dv_y, \tag{12.31}$$

which is in addition to the real part $\epsilon^r(\omega, k)$, which corresponds to expressions such as (12.21). In general, the normal modes are described by

$$\epsilon_{\text{eff}} = \epsilon^r(\omega, k) + j\epsilon_{\text{eff}}^i(\omega, k) = 0. \tag{12.32}$$

[2] The complex integration analysis required to obtain this expression involves application of the residue theorem and an appropriate choice of integration contour, which is beyond the scope of this text. Interested readers are pointed to discussions by other authors [1, 2].

Typically, we can assume that $|\epsilon^i_{\text{eff}}| \ll |\epsilon^r|$, so that the wave characteristics are predominantly determined by $\epsilon^r(\omega, k)$. Our goal is to solve (12.32) for ω, given the wave number k. we expect that ω will have a real part and a small imaginary part. We can write

$$\omega = \omega_r + j\gamma, \quad |\gamma| \ll \omega_r. \tag{12.33}$$

The time dependence of our field solutions then becomes

$$e^{j\omega t} = e^{j\omega_r t} e^{-\gamma t}, \tag{12.34}$$

so that a positive value of γ would correspond to damping (attenuation in time) of the wave that is at frequency ω_r. We now expand (12.32) in a Taylor series around $\omega \simeq \omega_r$, noting that $|\epsilon^i_{\text{eff}}| \ll |\epsilon^r|$:

$$\epsilon_{\text{eff}}(\omega, k) = \epsilon^r_{\text{eff}}(\omega_r, k) + j\epsilon^i_{\text{eff}} - j\gamma \left[\frac{\partial \epsilon^r_{\text{eff}}(\omega, k)}{\partial \omega} \right]_{\omega=\omega_r} + \cdots = 0. \tag{12.35}$$

Separating the real and imaginary parts of (12.35) we find

$$\epsilon^r_{\text{eff}}(\omega_r, k) = 0, \tag{12.36}$$

which gives solutions of the type obtained previously for propagation in a warm plasma, and

$$\gamma = \frac{\epsilon^i_{\text{eff}}(\omega_r, k)}{\left[\dfrac{\partial \epsilon^r_{\text{eff}}(\omega, k)}{\partial \omega} \right]_{\omega=\omega_r}}. \tag{12.37}$$

Substituting the particular expression for ϵ^i_{eff} for longitudinal waves in an isotropic plasma, i.e., (12.31), we find

$$\gamma = \frac{-\pi q_e^2}{\epsilon_0 m_e k^2 \left[\dfrac{\partial \epsilon^r_{\text{eff}}(\omega, k)}{\partial \omega} \right]_{\omega=\omega_r}} \int \left(\frac{\partial f_0}{\partial v_z} \right)_{v_z=\omega/k} dv_x \, dv_y. \tag{12.38}$$

We note that the sign and magnitude of γ are determined by the velocity-space gradient of f_0 at $v_z = \omega_r/k$. For a Maxwellian distribution, the slope of the distribution is always negative, so that γ is positive and longitudinal oscillations are necessarily damped. This phenomenon is known as *Landau damping*, since it was first shown by Landau in 1946. Note that this damping is not due to collisions, but originates physically from the wave–particle interaction energy exchange between a wave propagating at a velocity ω_r/k and particles traveling at speeds $v_z \simeq \omega_r/k$. Since the particles and wave move together, there is an opportunity for significant energy exchange. Particles moving slightly slower than the wave gain energy and accelerate, while particles moving slightly faster than the wave lose energy and give it up to the wave. Whether the wave is damped or amplified will depend on the relative number of slower and faster particles near the phase velocity, which is described by the slope of the distribution function at this velocity $v_z = \omega_r/k$, as mentioned above. For a Maxwellian plasma, with $f_0(\mathbf{v})$ as given in (4.4), the damping rate can be shown (by substitution in (12.38)) to be

$$\gamma_{\text{Maxwellian}} = \left(\frac{\pi}{8}\right)^{\frac{1}{2}} e^{-3/2} \frac{\omega_p}{(k\lambda_D)^3} \exp\left[-\frac{1}{2(k\lambda_D)^2}\right], \tag{12.39}$$

where λ_D is the Debye length.

A common analogy with the Landau effect is the interaction of a surfer riding an ocean wave. To harness wave energy a surfer must first achieve a velocity close to that of the wave. This "resonance" or "catching the wave" requirement is the reason surfers begin by paddling with their hands while laying flat on the board. Once the surfer and the wave are moving together, the surfer is carried and accelerated by the wave. The acceleration (and also gravity) make the surfer actually move ahead of the wave, and surfers can be seen to make sharp turns to travel across the wave in order to remain in resonance. In making such a maneuver the surfer actually gives some kinetic energy back to the water, thus exhibiting both the damping and amplification aspects of the resonant interaction. Further discussion of Landau damping, and its physical interpretation, are to be found in texts by Schmidt [3] and Stix [1].

12.2.2 Transverse waves

Up to now we have considered kinetic theory (or the Vlasov theory) of plasma waves applicable to small perturbations of the equilibrium distribution of a "hot" plasma. Our starting point was to seek time-harmonic plane-wave solutions of the form (Equation (12.2))

$$e^{j(\omega t - \mathbf{k}\cdot\mathbf{r})},$$

allowing the possibility that the frequency ω may be complex, to represent the decay of perturbations (ω has a small negative imaginary part) or growth of the wave in time (ω has a small positive imaginary part). We then wrote down the linearized version of the Vlasov equation (Equation (12.3)):

$$\frac{\partial f_1}{\partial t} + (\mathbf{v}\cdot\nabla)f_1 + \frac{q_e}{m_e}(\mathbf{E} + \mathbf{v}\times\mathbf{B})\cdot\nabla_{\mathbf{v}} f_0 = 0,$$

which was solved simultaneously with Poisson's equation in order to determine the relationship between ω and k, i.e., the dispersion relation. This was appropriate since we confined our attention exclusively to longitudinal waves with $\mathbf{k} \parallel \mathbf{E}$, so that $\mathbf{k} \times \mathbf{E} = 0$, but $\mathbf{k} \cdot \mathbf{E} \neq 0$, making Poisson's equation the appropriate relationship with which to relate the electric field and the perturbed distribution f_1. In this section, we consider transverse waves with $\mathbf{k} \perp \mathbf{E}$ so that $\mathbf{k} \cdot \mathbf{E} = 0$, but $\mathbf{k} \times \mathbf{E} \neq 0$. As we know from Chapter 9 (Section 9.3.2), in a cold plasma, transverse electromagnetic waves propagate as long as the wave frequency is above ω_p, and the dispersion relation is (Equation (9.22))

$$\mu^2 \equiv \frac{k^2 c^2}{\omega^2} = \left(1 - \frac{\omega_{pe}^2}{\omega^2}\right).$$

Our starting point in considering transverse waves is the linearized Vlasov equation (12.3). We retain the magnetic field term in order to allow for equilibrium distributions f_0 which may be anisotropic. In this connection, note once again that (12.5) is only valid for isotropic distributions. For transverse waves, it is not possible to use Poisson's equation as a second equation relating the electric and magnetic fields to f_1, since $\mathbf{k} \cdot \mathbf{E} = 0$. Instead, we work with Maxwell's equations,

$$\nabla \times \mathbf{B} = \mu_0 \mathbf{J} + \mu_0 \epsilon_0 \frac{\partial \mathbf{E}}{\partial t} \tag{12.40}$$

$$\nabla \times \mathbf{E} = -\frac{\partial \mathbf{B}}{\partial t}. \tag{12.41}$$

Taking the time derivative of (12.40) we have

$$\nabla \times \frac{\partial \mathbf{B}}{\partial t} = \mu_0 \frac{\partial \mathbf{J}}{\partial t} + \mu_0 \epsilon_0 \mu_0 \frac{\partial^2 \mathbf{E}}{\partial t^2}. \tag{12.42}$$

Substituting from (12.41) we find

$$\nabla \times (\nabla \times \mathbf{E}) = \nabla(\nabla \cdot \mathbf{E}) - \nabla^2 \mathbf{E} = -\mu_0 \frac{\partial \mathbf{J}}{\partial t} - \epsilon_0 \mu_0 \frac{\partial^2 \mathbf{E}}{\partial t^2}. \tag{12.43}$$

Using the assumed time and space dependence of the wave quantities as given in (12.2), we can split (12.43) into two equations,

$$-\mathbf{k} k\, E_\parallel + \mathbf{k} k\, E_\parallel = 0 = -\mu_0 \frac{\partial \mathbf{J}_\parallel}{\partial t} - \epsilon_0 \mu_0 \frac{\partial^2 \mathbf{E}_\parallel}{\partial t^2} \tag{12.44}$$

$$k^2 \mathbf{E}_\perp = -\mu_0 \frac{\partial \mathbf{J}_\perp}{\partial t} - \epsilon_0 \mu_0 \frac{\partial^2 \mathbf{E}_\perp}{\partial t^2}, \tag{12.45}$$

where we note that we are considering an isotropic plasma with no static magnetic field, so that the $\|$ and $\perp$ designations refer simply to the direction of the wave vector $\mathbf{k}$. Equation (12.44) clearly describes the longitudinal electrostatic oscillations considered in the previous section, whereas (12.45) describes the transverse waves. We can select k to be in the z direction (i.e., $\mathbf{k} = \hat{\mathbf{z}}k$) and take the perpendicular electric field to be in the x direction, i.e., $\mathbf{E}_\perp = \hat{\mathbf{x}}E$. Noting that the current $\mathbf{J}_\perp$ is then given by

$$\mathbf{J}_\perp = q_e \int v_x f_1 d^3v, \tag{12.46}$$

and replacing time derivatives with $j\omega$, we find

$$\left(k^2c^2 - \omega^2\right) E = -j\frac{q_e\omega}{\epsilon_0} \int v_x f_1 d^3v, \tag{12.47}$$

where we have dropped the subscript $\perp$, so that the electric field is understood to be the transverse field. Equation (12.47) relates the electric field to the perturbed distribution function f_1. We can now write the linearized Vlasov equation (12.3) by imposing the particular solution type we seek, as given by (12.2). We find

$$j(\omega - \mathbf{k}\cdot\mathbf{v})f_1 + \frac{q_e}{m_e}(\mathbf{E} + \mathbf{v}\times\mathbf{B})\cdot\nabla_{\mathbf{v}} f_0 = 0. \tag{12.48}$$

We can also write (12.41) as

$$\mathbf{k}\times\mathbf{E} = \omega\mathbf{B}. \tag{12.49}$$

Substituting (12.49) into (12.48) and expressing the vectors in their component form (noting that $\mathbf{E} = \hat{\mathbf{x}}E$ and that $\mathbf{k} = \hat{\mathbf{z}}k$) we find

$$\begin{aligned} f_1 &= j\frac{q_e}{m_e}\frac{\left[\mathbf{E} + \omega^{-1}\mathbf{v}\times(\mathbf{k}\times\mathbf{E})\right]\cdot\nabla_{\mathbf{v}} f_0}{\omega - \mathbf{k}\cdot\mathbf{v}} \\ &= j\frac{q_e}{m_e}\frac{E\partial f_0/\partial v_x + (kE/\omega)\left[v_x\partial f_0/\partial v_z - v_z\partial f_0/\partial v_x\right]}{\omega - kv_z}. \end{aligned} \tag{12.50}$$

Equation (12.50) describes the dependence of the perturbed particle distribution on the wave electric field, and is equivalent to (12.12) for longitudinal waves. Inserting (12.50) into (12.47) eliminates f_1 and E and reveals the dispersion relation:

Transverse wave dispersion relation

$$\boxed{k^2c^2 = \omega^2 - \frac{q_e^2}{m_e\epsilon_0}\int\left[\frac{(\omega - kv_z)v_x\partial f_0/\partial v_x}{\omega - kv_z} + \frac{kv_x^2\partial f_0/\partial v_z}{\omega - kv_z}\right]d^3v}. \tag{12.51}$$

If the equilibrium distribution does not have particles traveling at speeds close to the phase velocity of the wave, (12.51) provides real solutions (real values of ω) for any given real k, as long as $\omega > \omega_p$. In fact, the cold-plasma dispersion relation of (9.22) can be obtained from (12.51) by simply adopting a cold-plasma distribution as given by (12.17). For this purpose, we can follow the same procedure that we did for longitudinal waves, i.e., first integrate (12.51) by parts and then substitute $f_0(\mathbf{v}) = N_0\delta(\mathbf{v}) = N_0\delta(v_x)\delta(v_y)\delta(v_z)$. If $f_0(\mathbf{v})$ contains particles traveling with velocity $v_z = \omega/k$, the integrand diverges and we have the same type of problem as we had for longitudinal waves. In principle, a solution along the same lines can be sought, and we might expect Landau damping of the waves. However, this problem is much less serious for the transverse wave than for the longitudinal wave. From our cold-plasma analysis of the transverse mode, and directly from (9.22), we know that the phase velocity for the transverse mode is $v_p = \omega/k > c$. Most typical plasmas will not have an appreciable number of particles at velocities near c, and in fact the theory of relativity bounds v_z by c. However, resonant particles propagating near the phase velocity of the wave are not the only possible source of wave growth or damping, i.e., of complex frequency ω with a small imaginary part. Instabilities of the transverse mode can be caused by a large class of non-thermal distribution functions. To study such instabilities, we need to solve for ω from (12.51) for given real k. For this purpose, and as was done for the longitudinal wave (see (12.18)), it is useful to integrate (12.51) by parts to find

$$k^2c^2 = \omega^2 - \omega_p^2 - \frac{q_e^2 k^2}{m_e \epsilon_0} \int \frac{v_x^2 f_0 d^3v}{(\omega - kv_z)^2}. \tag{12.52}$$

Consider, for example, an anisotropic distribution function

$$f_0(\mathbf{v}) = \delta(v_z)\varphi(x, y), \tag{12.53}$$

where $\varphi(x, y)$ is an arbitrary function. Substituting (12.53) in (12.52) gives the dispersion relation:

$$\omega^2 - \omega_p^2 \left[1 + \frac{k^2 \left\langle v_x^2 \right\rangle}{\omega^2} \right] = k^2c^2$$

$$\omega^4 - \left(k^2c^2 + \omega_p^2\right)\omega^2 - k^2\omega^2\left\langle v_x^2 \right\rangle = 0$$

$$\rightarrow \quad \omega^2 = \frac{1}{2}\left\{ k^2c^2 + \omega_p^2 \pm \left[\left(k^2c^2 + \omega_p^2\right)^2 + 4k^2\omega_p^2\left\langle v_x^2 \right\rangle \right]^{1/2} \right\}. \tag{12.54}$$

We see that one of the roots is always negative for all values of k, so that the corresponding values of ω represent exponentially growing solutions. In general, these unstable transverse waves occur when the velocity distribution f_0 is anisotropic. If the distribution is smeared out in v_z, only a limited set of k values leads to growing waves. When the square average of the velocity v_z approaches the square average of the velocity v_x, the instability disappears, as the plasma becomes isotropic.

12.2.3 The two-stream instability

A particularly good example of wave growth (instability) which occurs for non-Maxwellian equilibrium velocity distributions is the case of two interpenetrating particle streams, either two oppositely directed electron beams or an electron beam moving through a stationary background of ions. To illustrate this instability, consider the dispersion relation for longitudinal modes in a non-magnetized plasma, i.e., Equation (12.16), repeated here:

$$1 + \frac{q_e^2}{\epsilon_0 m_e k} \int_v \frac{\partial f_0/\partial v_z}{\omega - k v_z} dv_z dv_y dv_x = 0, \tag{12.55}$$

or its alternative form,

$$1 - \frac{q_e^2}{\epsilon_0 m_e} \int_{-\infty}^{\infty} \frac{f_0(\mathbf{v})}{(\omega - k v_z)^2} dv_z dv_y dv_x = 0. \tag{12.56}$$

For simplicity, let us assume two identical but oppositely directed cold electron streams with velocities $\pm v_0$. We can express such a distribution as

$$f_0(\mathbf{v}) = f_0(v_x, v_y, v_z) = \tfrac{1}{2} N_0[\delta(v_z - v_0) + \delta(v_z + v_0)]\delta(v_x)\delta(v_y). \tag{12.57}$$

Substituting (12.57) in (12.56) we find

$$D(\omega, k) = 1 - \frac{1}{2}\left[\frac{\omega_p^2}{(\omega - k v_0)^2} + \frac{\omega_p^2}{(\omega + k v_0)^2}\right] = 0. \tag{12.58}$$

The function $D(\omega, k)$ is plotted versus ω in Figure 12.1. In general, the function is a quartic and always has four roots in ω (for given k). For short waves, $|k| > \sqrt{2}(\omega_p/v_0)$, illustrated in Figure 12.1 as Case A, there are four real roots $(\omega_1, \omega_2, \omega_3, \omega_4)$ so that the solutions are purely oscillatory. For long waves, $|k| < \sqrt{2}(\omega_p/v_0)$, illustrated as Case B in Figure 12.1, there are only two real roots (ω_1, ω_4) and two complex conjugate roots (ω_2, ω_3) one of which must have a negative imaginary part and must thus lead to wave growth or instability. The condition for instability

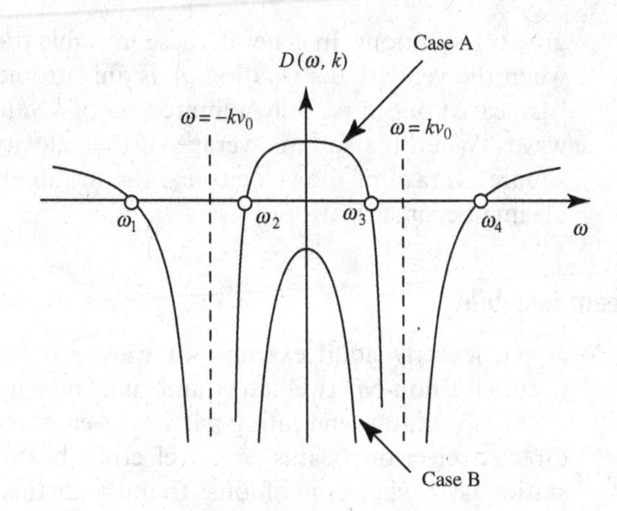

Figure 12.1 The roots of the dispersion relation for the two-stream instability.

is $|k| < \sqrt{2}(\omega_p/v_0)$. It can be shown that the maximum growth rate is $\omega_p/2$ (see Problem 12-1 at the end of this chapter). The phenomenon of exponentially growing oscillations in two streams in relative motion is known as the *two-stream instability*. This instability prevents two oppositely directed uniform beams of electrons from passing through each other, even if the electrons are neutralized by a background of ions. The instability produces strong spatial inhomogeneities, in which electrons are "bunched" together, ultimately causing the energy of the electron beams to be significantly dissipated into plasma waves.

12.3 Waves in a hot magnetized plasma

In the previous sections of this chapter, we derived the dispersion relation for non-magnetized hot plasmas by simultaneous solution of the linearized Vlasov equation and Maxwell's equations (either the $\mathbf{k} \cdot \mathbf{E}$ equation for longitudinal waves or the $\mathbf{k} \times \mathbf{E}$ equation for transverse waves). For magnetized plasmas, we can basically follow the same procedure, except that the relationship between the current density $\mathbf{J}$ and the electric field $\mathbf{E}$ is now anisotropic. The equivalent dielectric tensor that was obtained in magnetoionic theory (i.e., the cold-plasma model) must now be determined from the Vlasov equation instead of the simple Lorentz force equation of motion. The solution of the general problem is a difficult task indeed; accordingly, we restrict our attention to waves propagating parallel to the static magnetic field, i.e., the case of $\theta = 0$. Our starting point is the Vlasov equation. First, we realize that in the absence of any perturbations we have an equilibrium distribution $f_0(\mathbf{v})$ which must satisfy the zeroth-order Vlasov equation,

$$\frac{\partial f_0}{\partial t} + (\mathbf{v} \cdot \nabla_\mathbf{r}) f_0 + \frac{q_e}{m_e}[0 + \mathbf{v} \times \mathbf{B}_0] \cdot \nabla_\mathbf{v} f_0 = 0$$

$$0 + 0 + (\mathbf{v} \times \mathbf{B}_0) \cdot \nabla_\mathbf{v} f_0 = 0. \qquad (12.59)$$

Equation (12.59) indicates that the velocity-space gradient of f_0 contains no components in the $\mathbf{v} \times \mathbf{B}_0$ direction. Since this is true for any velocity vector $\mathbf{v}$, the equilibrium distribution must be constant on circles defined by $v_\perp$ = constant, so that its dependence on the two transverse coordinates v_x and v_y (assuming $\mathbf{B} = \hat{\mathbf{z}} B_0$) can be accounted for by its dependence on the single scalar quantity $v_\perp = \sqrt{v_x^2 + v_y^2}$. Thus, we have

$$f_0(\mathbf{v}) = f_0(v_x, v_y, v_z) = f_0(v_\perp, v_\parallel). \qquad (12.60)$$

The fact that the equilibrium distribution is only a function of $v_\parallel$ and $v_\perp$ can also be deduced from the facts that these quantities are constants of particle motion in a uniform magnetic field, and that, from Section 3.7, *any* function of the constants of motion is a solution of the Vlasov equation. With (12.59) in mind, the linearized Vlasov equation for a magnetized plasma can be written as

$$\frac{\partial f_1}{\partial t} + (\mathbf{v} \cdot \nabla_\mathbf{r}) f_1 + \frac{q_e}{m_e}[\mathbf{E} + \mathbf{v} \times \mathbf{B}] \cdot \nabla_\mathbf{v} f_0 + \frac{q_e}{m_e}(\mathbf{v} \times \mathbf{B}) \cdot \nabla_\mathbf{v} f_1 = 0. \qquad (12.61)$$

Imposing the time-harmonic uniform plane-wave dependence of the quantities (12.61) reduces this to

$$-j(\mathbf{k} \cdot \mathbf{v} - \omega) f_1 + \frac{q_e}{m_e}[\mathbf{E} + \mathbf{v} \times \mathbf{B}] \cdot \nabla_\mathbf{v} f_0 + \frac{q_e}{m_e}(\mathbf{v} \times \mathbf{B}) \cdot \nabla_\mathbf{v} f_1 = 0. \qquad (12.62)$$

Solving (12.62) together with Maxwell's equations

$$\mathbf{k} \times \mathbf{E} = \omega \mathbf{B}$$

$$-j\mathbf{k} \times \mathbf{B} = \mu_0 \mathbf{J} + j\omega\mu_0\epsilon_0 \mathbf{E},$$

while also noting that

$$\mathbf{J} = q_e \int \mathbf{v} f_1 d^3 v,$$

yields the following dispersion relation for parallel propagating transverse waves ($\mathbf{k} \parallel \mathbf{B}_0$ and $\mathbf{k} \perp \mathbf{E}$):

$$\boxed{k^2c^2 = \omega^2 + \frac{q_e^2\pi}{m_e\epsilon_0}\int_{-\infty}^{\infty}\int_0^{\infty}\frac{(\omega - kv_\parallel)\partial f_0/\partial v_\perp + kv_\perp \partial f_0/\partial v_\parallel}{(\omega - kv_\parallel) \pm \omega_c} v_\perp^2 dv_\perp dv_\parallel}, \tag{12.63}$$

where the upper and lower signs correspond respectively to the right- and left-hand modes. Integrating the right-hand side by parts yields the following alternative form:

$$k^2c^2 = \omega^2 - \frac{q_e^2}{m_e\epsilon_0}\int\left[\frac{(\omega - kv_\parallel)}{\omega - kv_\parallel \pm \omega_c} + \frac{\frac{1}{2}k^2v_\perp^2}{(\omega - kv_\parallel \pm \omega_c)^2}\right] f_0 d^3v. \tag{12.64}$$

Note that, for a magnetized plasma, the denominator of the dispersion relation diverges if there are resonant particles which feel the Doppler-shifted wave frequency as the cyclotron frequency in their coordinate system. In such a case, the integration should be carried out in the complex plane to bring out the contribution to the integral from the vicinity of the pole.

As in previous cases, any given equilibrium distribution function can be substituted into (12.63) or (12.64), to obtain the dispersion relation between wave number k and frequency ω. For isotropic distributions $f_0(\mathbf{v})$, we know that $(\mathbf{v} \times \mathbf{B}) \cdot \nabla_{\mathbf{v}} f_0 = 0$, so that the magnetic field of the wave has no influence on the particle motions in the linear approximation. In such a case, it can be shown that all terms in the numerators of (12.63) and (12.64) that contain k disappear, so that

$$k^2c^2 = \omega^2 - \frac{q_e^2}{m_e\epsilon_0}\int\frac{\omega}{\omega - kv_\parallel \pm \omega_c} f_0 d^3v. \tag{12.65}$$

In the absence of resonant particles, (12.65) yields a real value of ω for a given value of k. The presence of resonant particles, on the other hand, leads to damping in a manner quite similar to Landau damping. Instabilities (i.e., wave growth) in a hot magnetoplasma can occur as a result of anisotropic particle distributions. To see this, consider an equilibrium distribution given by

$$f_0(\mathbf{v}) = \delta(v_\parallel)\varphi(v_\perp), \tag{12.66}$$

where φ is any arbitrary function. Substituting (12.66) into (12.64) we find

$$k^2c^2 = \omega^2 - \omega_p^2 \frac{\omega}{\omega \pm \omega_c} + \omega_p^2 \frac{\frac{1}{2}k^2 \langle v_\perp^2 \rangle}{(\omega \pm \omega_c)^2}. \qquad (12.67)$$

Solving for k^2 from (12.67) we find

$$k^2 = \frac{\omega^2(\omega \pm \omega_c)^2 - \omega_p^2 \omega(\omega \pm \omega_c)}{c^2(\omega \pm \omega_c)^2 + \frac{1}{2}\omega_p^2 \langle v_\perp^2 \rangle}. \qquad (12.68)$$

For $k^2 \to \infty$ (i.e., in the vicinity of the resonance at $\omega \simeq \omega_c$) we have

$$\omega^2 \pm 2\omega\omega_c + \omega_c^2 + \frac{\omega_p^2 \langle v_\perp^2 \rangle}{2c^2} = 0$$

or

$$\text{RH:} \quad \omega = -\omega_c \pm \left[\omega_c^2 - \left(\omega_c^2 + \frac{\omega_p^2 \langle v_\perp^2 \rangle}{2c^2}\right)\right]^{1/2} = -\omega_c \pm j \frac{\omega_p \sqrt{\langle v_\perp^2 \rangle}}{c\sqrt{2}}$$

$$\text{LH:} \quad \omega = +\omega_c \pm \left[\omega_c^2 - \left(\omega_c^2 + \frac{\omega_p^2 \langle v_\perp^2 \rangle}{2c^2}\right)\right]^{1/2} = +\omega_c \pm j \frac{\omega_p \sqrt{\langle v_\perp^2 \rangle}}{c\sqrt{2}}.$$

We find that the wave frequencies for both of the modes have imaginary parts of both signs, the negative of which leads to wave growth and instability. We note that growth and instability occur when the real part of the wave frequency is near ω_c, i.e., resonance, as expected based on earlier discussions (e.g., see the discussion at the end of Section 10.3.1). Note that the presence of such effects (e.g., damping or growth) was not visible to us within the context of the cold- and warm-plasma models. We thus see once again the generality of the Vlasov theory of plasma waves.

Two-stream instability for magnetized plasma

The parallel propagating mode in a magnetized plasma is also unstable in the presence of currents through the plasma along the magnetic field, for example in the form of particle beams. Consider two counterstreaming beams containing particles with charges of opposite polarity, moving with velocity $v_z = \pm v_0$. Neglecting the thermal speeds and thus taking the particle distributions to be delta functions as given in (12.57), and also making the simplifying assumption that the masses of the positively and negatively charged particles are the same (i.e., $\omega_p^+ = \omega_p^-$ and $\omega_c^+ = -\omega_c^-$), we can show from (12.64) that

$$D(\omega, k) = k^2c^2 - \omega^2 + \omega_p^2 \left[\frac{\omega + kv_0}{\omega + kv_0 - \omega_c} + \frac{\omega - kv_0}{\omega - kv_0 - \omega_c} \right] = 0. \tag{12.69}$$

As in the case of the two-stream instability for non-magnetized plasma, $D(\omega, k)$ in general has four roots, two of which are always real, while the other two can be complex, if

$$k^2 < \frac{\omega_p^2}{c^2} \frac{2kv_0}{\omega_c - kv_0}; \tag{12.70}$$

thus (12.70) is the condition for stability. It can be shown that the system is always stable if $k > \omega_c/v_0$.

12.4 More on collisions in plasmas

Up to now, we have not discussed the effects of collisions on plasma dynamics in any great detail, although it was understood (Chapter 3) that collisions are the means by which thermal equilibrium distributions (e.g., Maxwellian) are established and maintained. We also noted in Chapter 6 that the magnetohydrodynamic treatment of plasmas implicitly assumes collision-dominated conditions, and in Chapters 7 and 8 that collisions are the underlying facilitating physical process for transport effects in plasmas, whether that be via resistivity or diffusion.

In studying waves in cold and warm plasmas using fluid theory (Chapters 9–11), we accounted for collisions in a phenomenological manner, using a friction term in the momentum equations, with an *effective* collision frequency. To include collisional effects in hot plasmas, we would need to use the Boltzmann equation as our starting point:

$$\frac{\partial f}{\partial t} + (\mathbf{v} \cdot \nabla_{\mathbf{r}}) f + \frac{q_e}{m_e}[\mathbf{E} + \mathbf{v} \times \mathbf{B}] \cdot \nabla_{\mathbf{v}} f = \left(\frac{\partial f}{\partial t} \right)_{\text{coll}}, \tag{12.71}$$

where the term $(\partial f/\partial t)_{\text{coll}}$ represents the resultant gain or loss of particles due to interactions between particles. Since collisions occur between particles that are close to one another, their dynamics is not likely to be accurately represented by the macroscopic averaged self-consistent fields **E** and **B**. In fact, the actual constitution of the $(\partial f/\partial t)_{\text{coll}}$ term can be quite involved, and is substantially different for various types of collisions, such as electron–electron,

electron–ion, electron–neutral, etc. Most expressions for $(\partial f/\partial t)_{\text{coll}}$ involve integral functionals of f itself, so that (12.71) is actually an integro-differential equation. Furthermore, the collision term in the Boltzmann equation for one-plasma species, e.g., electrons, includes the distribution functions of other species, thus coupling the Boltzmann equations for different plasma species. Note that this coupling is in addition to the coupling of the dynamics of the different charged species via Maxwell's equations and the macroscopic fields **E** and **B**, the result of collective effects due to all species.

If the form of $(\partial f/\partial t)_{\text{coll}}$ is known, the procedure one would follow in analyzing the plasma dynamics (e.g., waves) is similar to what we have used up to now. We assume a small perturbation to be superimposed on an equilibrium distribution, i.e., $f = f_0 + f_1$, where f_0 satisfies the zeroth-order equation in the absence of any fields. (Note that this in fact dictates that the f_0 be a uniform Maxwellian distribution.) We then consider time-harmonic perturbations for f_1, **E**, and **B**, and solve for a dispersion relation in the same manner. Unfortunately, such a solution is prohibitively complicated for any properly derived form of $(\partial f/\partial t)_{\text{coll}}$.

Fortunately, though, solution of (12.71) with the collision term is only warranted under rather unusual circumstances. In studying high-frequency electromagnetic wave propagation in both natural and laboratory plasmas, the thermal speed is usually unimportant since the wave speed is high enough so that the magnetoionic theory, with the collisions included via the momentum transport equation, is sufficient. Note that the use of magnetoionic theory amounts to neglecting the $(\mathbf{v} \cdot \nabla_{\mathbf{r}})f$ term in (12.71). For low-frequency, large-spatial-scale phenomena, collisions may well be dominant so that a hydrodynamic treatment applies, in which case the left-hand side of (12.71) is a small perturbation. For rarefied plasmas, collisions occur seldom enough that the right-hand side is negligible, reducing (12.71) to the Vlasov equation, so that all of our analyses in previous sections of this chapter are valid, with the primary features being Landau damping and cyclotron resonance effects. Circumstances in which all of the terms in (12.71) are of comparable order arise very seldom and, when they do, they often do not require results of great accuracy. In most cases, all that is needed is to interpolate roughly between hydrodynamic and collisionless regimes to see how features of one give way to those of another as the collision frequency is changed. Such analyses are usually facilitated by using simpler formulas for $(\partial f/\partial t)_{\text{coll}}$, arranged to have a similar mathematical form (and order of magnitude) as the real Boltzmann collision integral. We now discuss one such "model" for the collision term.

12.4.1 The Krook collision model

The simplest model for the collision term, referred to as the *Krook collision term* or the BGK model [4, 5], is

$$\left(\frac{\partial f}{\partial t}\right)_{\text{coll}} = -\nu_{\text{eff}}(f - f_{\text{m}}), \tag{12.72}$$

where f_{m} is a suitable Maxwellian distribution and ν_{eff} is an empirical effective collision frequency which can in general be velocity-dependent but is usually taken to be a constant. The idea behind (12.72) is that $-\nu_{\text{eff}} f$ represents the rate of loss of particles, due to collisions, from a differential element of phase space, while $+\nu_{\text{eff}} f_{\text{m}}$ represents the corresponding rate of gain of particles as the end product of collisions. The main approximation involved in writing (12.72), especially when ν_{eff} is held constant, is that the detailed statistics and dynamics of collisions are disregarded. Also disregarded is the fact that the particle velocity after a collision is correlated with that before.

The Maxwellian distribution f_{m} is specified by fixing the density, mean velocity, and temperature, which depend in some way on f itself as evaluated at the same point of physical space and at the same time. Thus, (12.72) represents a purely local effect in which particles are transferred abruptly across velocity space only, at a rate ν_{eff}. The Krook collision model thus simulates the effect of close binary collisions in which there is a substantial change in velocity, particularly applicable for electron–neutral collisions in weakly ionized plasmas. Thinking of the collisions as a Poisson process, occurring with probability $\nu_{\text{eff}} dt$ in the time interval between t and $t + dt$, (12.72) tends to establish a Maxwellian distribution in a time on the order of a few multiples of ν_{eff}^{-1}. When several species are involved, we can extend (12.72) as follows:

$$\left(\frac{\partial f^i}{\partial t}\right)_{\text{coll}} = -\sum_i \nu_{\text{eff}}^{ij}\left(f^i - f_{\text{m}}^{ij}\right), \tag{12.73}$$

where f^i is the distribution function for the ith species, ν_{eff}^{ij} is the effective collision rate between particles of the ith and jth type, and f_{m}^{ij} is a Maxwellian distribution of the ith species, the parameters of which are determined by the local values of f^i and f^j. Note that f_{m}^{ij} is known once its density, drift velocity, and temperature, i.e., N^{ij}, $\mathbf{u}^{ij}$, and T^{ij}, are specified.

To analyze plasma dynamics in the presence of collisions, we need to use the Krook collision model in the Boltzmann equation which is to be solved simultaneously with Maxwell's equations. As before, we will look for time-harmonic perturbations. In principle, we must carry out such an analysis for every species, charged or neutral, with the equations being coupled through the collision terms. As a simple illustrative example, we consider the solution of (12.71) only for electrons, with the positive ions forming a fixed background, and with the only collisions being those between electrons and neutral species. For simplicity, we take $(f - f_m)$ in (12.72) to have the same density as f, zero mean velocity, and temperature equal to the ambient temperature. Taking $f = f_0 + f_1$, where f_0 is the unperturbed Maxwellian distribution, we have

$$f_m(\mathbf{r}, \mathbf{v}, t) = \frac{N(\mathbf{r}, t) f_0(\mathbf{v})}{N_0}, \tag{12.74}$$

where $N = N_0 + N_1$ as before. The net effect of (12.74) is that electrons lose momentum at a rate $N m_e \nu_{\text{eff}} \mathbf{u}$ per unit volume, exactly as assumed in cold magnetoionic theory. After linearization we have

$$\left(\frac{\partial f}{\partial t}\right)_{\text{coll}} = -\nu \left[f_1(\mathbf{r}.\mathbf{v}, t) - \frac{N_1(\mathbf{r}, t) f_0(\mathbf{v})}{N_0} \right], \tag{12.75}$$

where, for convenience, we have dropped the subscript "eff" from ν.

The first term on the right-hand side of (12.75) is very easy to incorporate into our earlier analysis of waves in hot plasmas since it only involves the replacement of angular frequency ω by $\omega - j\nu$. However, it is necessary to solve the full equation since, without the last term, (12.75) does not conserve particles locally. Solving the full equation (12.71), with (12.75) as the collision term and for time-harmonic plane waves, we have

$$j(\omega - j\nu - \mathbf{k} \cdot \mathbf{v}) f_1 + \frac{q_e}{m_e} (\mathbf{v} \times \mathbf{B}_0) \cdot \nabla_{\mathbf{v}} f_1 + \frac{q_e}{m_e} \mathbf{E} \cdot \nabla_{\mathbf{v}} f_0$$

$$+ \underbrace{\frac{q_e}{m_e} (\mathbf{v} \times \mathbf{B}) \cdot \nabla_{\mathbf{v}} f_0}_{=0} = \nu \frac{N_1}{N_0} f_0$$

$$j(\omega - j\nu - \mathbf{k} \cdot \mathbf{v}) f_1 + \frac{q_e}{m_e} (\mathbf{v} \times \mathbf{B}_0) \cdot \nabla_{\mathbf{v}} f_1 + \frac{q_e}{m_e} \mathbf{E} \cdot \nabla_{\mathbf{v}} f_0 +$$

$$= \nu \frac{N_1}{N_0} f_0, \tag{12.76}$$

where the wave magnetic field term reduces to zero since the distribution f_0 is necessarily isotropic, and the velocity-space gradient of an isotropic distribution is in the direction of $\mathbf{v}$, as shown previously.

Considering for simplicity a non-magnetized plasma (i.e., $\mathbf{B}_0 = 0$), we can solve for f_1 as

$$f_1 = \frac{-\dfrac{q_e}{m_e}\mathbf{E} \cdot \nabla_\mathbf{v} f_0 + \nu \dfrac{N_1}{N_0} f_0}{j(\omega - j\nu - \mathbf{k} \cdot \mathbf{v})},$$

which can then be solved simultaneously with Maxwell's equations (i.e., either the $j\mathbf{k} \cdot \mathbf{E} = \rho/\epsilon_0$ or the $\mathbf{k} \times \mathbf{E} = \omega\mathbf{B}$ equations) in order to obtain the dispersion relation between k and ω. The net effect of collisions with a stationary background is to introduce collisional damping, in addition to Landau damping. For longitudinal waves, plasma oscillations are heavily damped if ν is comparable to ω_p; for transverse waves, the cyclotron resonance effect is lost if ν is comparable to ω_c.

12.5 Summary

In this chapter we presented the basic methodology for describing wave phenomena in hot or non-equilibrium plasmas. It is important to note that this is a different approach than that in previous chapters (9–11), which was built upon the fluid equations (single- or multiple-fluid versions) as derived from the moments of the Boltzmann equation and including a truncating equation of state. Unfortunately, plasmas far from equilibrium are not accurately described by such bulk averages over the distribution function, and it becomes necessary to work directly with the unintegrated Boltzmann equation and Maxwell's equations. This approach, called the Vlasov theory of plasmas, is general and powerful. Wave behavior is still described by a dispersion relation as before, but this expression includes an appropriate integral over the distribution function which can be specified. All the results from previous chapters can be reproduced by plugging in the appropriate distribution function, as was shown for the cases of a simple cold or warm plasma.

The power of the Vlasov approach is that for more complicated distributions it exposes new physical effects that result chiefly from particles being in resonance with the wave, for example moving at the wave phase velocity. Wave growth and damping can be accurately described, as integration over the poles of the distribution leads to complex values of frequency. One of the many remarkable results is Landau damping, a non-collisional wave attenuation. It is worth mentioning that, in contrast to many other discoveries in plasmas (and other fields), Landau damping was first described theoretically and only later verified by experiment. This triumph of applied mathematics is a testament to the versatility of the Vlasov theory of plasmas.

12.6 Problems

12-1. Solve the dispersion relation for electrostatic oscillations for two interpenetrating cold electron streams and show that the roots of ω are either real or imaginary, both coming in $\pm$ pairs. At which value of the wave number does the electrostatic two-stream instability grow fastest? Prove that the growth rate of the fastest instability is $\omega_p/2$.

12-2. Consider an unmagnetized plasma with a fixed neutralizing ion background. The one-dimensional electron distribution is given by

$$f_0(v) = N_p \frac{a}{\pi} \frac{1}{v^2 + a^2} + N_b \delta(v - v_0),$$

where $N_0 = N_p + N_b$ and $N_b \ll N_p$. (a) Derive the dispersion relation for high-frequency longitudinal perturbations. (b) In the limit $a \gg \omega/k$, show that a solution exists in which the imaginary part of ω is positive, i.e., growing oscillations.

12-3. A "bump-on-tail" instability occurs when a Maxwellian plasma distribution is modified as shown below. Discuss the implications for wave growth and damping for longitudinal waves with phase velocities near v_{bump}.

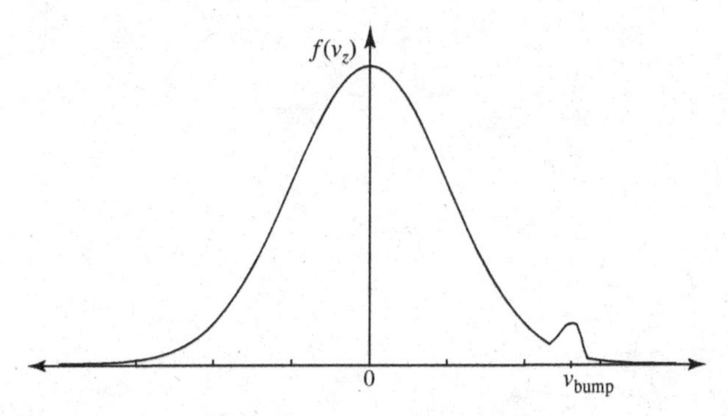

12-4. Explain why Landau resonance for longitudinal waves propagating in a Maxwellian plasma always leads to damping and not to wave growth.

12-5. A longitudinal wave is excited in a 10 eV plasma with density $N_0 = 2 \times 10^{15}$ m^{-3}. After the excitation is removed the wave motion is observed to attenuate rapidly by Landau damping.

Calculate how long it will take the amplitude to fall by a factor of e.

12-6. Estimate the cyclotron damping that will occur for a RH polarized 2 kHz wave with frequency in the vicinity of the cyclotron frequency if the plasma has a temperature of 1 keV and a density of $10^{-6}\ \mathrm{m}^{-3}$.

References

[1] T. H. Stix, *Waves in Plasmas* (New York: American Institute of Physics, 1992), 171–5.

[2] R. J. Goldston, *Introduction to Plasma Physics* (Bristol: Institute of Physics Publishing, 1995), 409–22.

[3] G. Schmidt, *Physics of High Temperature Plasmas* (New York & London: Academic Press, 1966), Chapter VII.

[4] P. L. Bhatnagar, E. P. Gross, and M. Krook, A model for collision processes in gases, I. Small-amplitude processes in charged and neutral one-component systems. *Phys. Rev.* **94** (1954), 511–25.

[5] E. P. Gross and M. Krook, A model for collision processes in gases, II. Small-amplitude oscillations of charged two-component systems. *Phys. Rev.* **102** (1956), 593–604.

CHAPTER

13 The plasma sheath and the Langmuir probe

13.1 Introduction

Practical plasmas are typically contained in a vacuum chamber of finite size. When ions and electrons hit the walls, they recombine and are lost. Since electrons have much higher thermal velocities than the ions, one may at first expect that they would be lost faster, leaving the otherwise neutral plasma with a net positive charge. However, because of this tendency, the plasma develops a positive potential with respect to the wall, which prevents most of the electrons from reaching the wall. In equilibrium, a potential gradient arises near the wall so that most of the electrons are reflected back into the plasma, with the number reaching the wall being equal to the corresponding number of positive ions reaching the wall. This potential gradient cannot extend over large sections of the plasma, since Debye shielding (Chapter 1) confines potential variations in a plasma to a layer of the order of several Debye lengths in thickness. This layer of potential gradient, which must exist on all walls with which the plasma is in contact, is called the *sheath*. Within the sheath, charge neutrality is not preserved, not even approximately, since $q\Phi$ changes through the sheath by an amount comparable to $k_B T$. In this chapter, we provide a simple treatment of the formation and structure of the plasma sheath and discuss a simple but very useful device, called the *Langmuir probe*, which is commonly used to measure electron density and temperature.

13.2 Particle flux

In applications involving a material body immersed in a plasma, or in studying the interaction between the walls of a container and a plasma, it is desirable to know the flux of particles arising from their random motions. The particle flux is defined as the number of particles impinging on unit area of the body per unit time. If the velocity distribution function is isotropic, the particle flux is independent of the orientation of the area. We can thus find an expression for the directed flux in any direction and know that it is the same in any other direction. Consider, for example, the directed flux in the $+z$ direction, given by $\Gamma = N\langle v_z \rangle$, where the average is taken over $v_z > 0$ only. Noting that in spherical coordinates $v_z = v\cos\theta$, where v is the particle speed, and recalling the expression (4.4) for the distribution function for a Maxwellian plasma, we have

$$\Gamma_z = N\left(\frac{m}{2\pi k_B T}\right)^{\frac{3}{2}} \int_0^{2\pi}\int_0^{\frac{\pi}{2}}\int_0^{\infty} v$$

$$\times \cos\theta\, e^{-mv^2/(2k_B T)} \underbrace{v^2 \sin\theta\, dv d\theta d\phi}_{dv_x dv_y dv_z}$$

$$= \frac{1}{4}N\langle v\rangle = \frac{1}{4}N\left(\frac{8k_B T}{\pi m}\right)^{\frac{1}{2}}, \qquad (13.1)$$

where we have noted from Chapter 3 that the average speed for a Maxwellian distribution is $\langle v\rangle = \sqrt{8k_B T/\pi m}$.

13.3 Sheath characteristics

We begin with a qualitative discussion of the physical mechanism of the formation of the plasma sheath. If the average energies of the electrons and ions were equal, i.e., if

$$\frac{1}{2}m_e\left\langle v_e^2\right\rangle = \frac{1}{2}m_i\left\langle v_i^2\right\rangle, \qquad (13.2)$$

then the average thermal speed of the electrons would be much greater, by at least a factor of ~1800. However, in general, the average thermal energy of the electrons is much larger than that for the positive ions. The reason for this is that whenever there is an electric field in the plasma, the electrons are rapidly accelerated and lose a very small fraction (typically about $2m_e/m_i$) of their energy in collisions with ions or neutrals. When the electric field is removed, the electrons and ions should eventually relax to an equilibrium state in which the electron and ion temperatures (and hence the average energies) are equal. However, the relaxation time is typically very long so that the ions are almost never in thermal equilibrium.

In practical laboratory plasmas, the energy to sustain the plasma is generally the heating of the electrons by the plasma source, while the ions are in thermal equilibrium with the background gas. The electron temperature is then typically a few electron volts, while the ions can be considered cold. In summary, the thermal speed of electrons is typically larger than that of the ions, by a factor much greater than $\sim\sqrt{1800}$. The particle fluxes impinging on the container walls are given, from (13.1), by

$$\Gamma_e = \frac{1}{4} N_e \langle v_e \rangle, \quad \Gamma_i = \frac{1}{4} N_i \langle v_i \rangle. \tag{13.3}$$

It is apparent that, initially, the particle flux toward the wall contributed by the electrons is much greater than that due to the ions. As this large number of electrons impinges on the wall and is lost from the plasma, the wall acquires a negative potential with respect to the plasma. The negative potential works to stop the large electron current: it repels the electrons while attracting the ions. Eventually the wall potential acquires a value such that the electron and ion fluxes are equalized, so that the net current is zero and the walls, now charged to a floating negative potential, reach a dynamical equilibrium with the plasma.

We now examine the sheath problem quantitatively by determining the wall potential Φ_{wall}. For simplicity, we assume a planar or one-dimensional sheath, assuming the wall to be on the z–y (i.e., $x = 0$) plane, bounding the plasma in the region $x > 0$. Let the potential distribution near the wall be $\Phi(x)$, such that

$$\Phi(0) = \Phi_{\text{wall}} \quad \text{and} \quad \Phi(\infty) = 0. \tag{13.4}$$

Assuming that the electrons and ions are in thermal equilibrium under the action of a conservative electric field $\mathbf{E} = -\nabla\Phi$, and assuming (for simplicity) that $T_e \simeq T_i$, the electron and ion number densities are given by (1.6), i.e.,

$$N_e(x) = N_0 e^{-q_e \Phi(x)/(k_B T)} \tag{13.5a}$$

$$N_i(x) = N_0 e^{-q_i \Phi(x)/(k_B T)}. \tag{13.5b}$$

Note that these equations were also discussed in Chapter 3 and represent unperturbed quantities (N_0) multiplied by a Boltzmann factor resulting from the potential. Also note that, far away from the wall ($x \to \infty$), (13.5) gives $N_e = N_i = N_0$, consistent with a charge-neutral equilibrium plasma. One word of caution is in order with respect to (13.5a) and (13.5b). The average velocity associated with the Maxwell–Boltzmann distribution on which these equations are based is zero, yet in view of the negative wall potential, there must be a steady stream of particles toward the wall. Thus Equations (13.5) do not take into account the drift velocity of the particles. To properly solve the sheath problem, the particles can be taken to have shifted Maxwellian distributions with $u_y, u_z = 0$, as defined in Chapter 3. However, the resulting expressions become so complicated that only numerical solutions are possible. On the other hand, results which hold remarkably well can be obtained via an approximate analysis. To begin with, we note that the thermal energy of the electrons is much larger than the energy associated with the electron drift motion in the x direction, so that (13.5a) is well justified. Equation (13.5b) does not hold, since the ions can be assumed to be cold ($T_i = 0$) and simply drifting (as a monoenergetic beam) with a velocity u_0, so that their motion is more suitably treated using fluid equations. In view of the qualifications mentioned above, we will use two different approaches to analyze the sheath potential: (i) taking (13.5b) to be true, and (ii) using fluid equations to model the ion number density.

Taking (13.5a) and (13.5b) to be true, we can proceed to evaluate Φ_{wall}. Under equilibrium conditions, we have $\Gamma_e = \Gamma_i$ at the wall (i.e., at $x = 0$), so that using (13.3), (13.4), (13.5a), (13.5b), and (13.1), and noting that $q_i = -q_e$, we find

$$\frac{e^{-q_e\Phi_{\text{wall}}/(k_B T)}}{m_e^{1/2}} = \frac{e^{q_e\Phi_{\text{wall}}/(k_B T)}}{m_i^{1/2}} \quad \rightarrow \quad e^{2q_e\Phi_{\text{wall}}/(k_B T)} = \left(\frac{m_i}{m_e}\right)^{\frac{1}{2}} \tag{13.6}$$

$$\rightarrow \quad \Phi_{\text{wall}} = \frac{k_B T}{4q_e} \ln\left(\frac{m_i}{m_e}\right). \tag{13.7}$$

Near the wall the ratio of the magnitude of the potential energy $|q_e\Phi_{\text{wall}}|$ to the thermal energy $k_B T$ of a particle is given by

$$\frac{|q_e\Phi_{\text{wall}}|}{k_B T} = \frac{1}{4} \ln\left(\frac{m_i}{m_e}\right). \tag{13.8}$$

For a hydrogen ion (proton), this quantity is approximately 2, so that the wall potential is of the same order of magnitude as the average thermal energy of the plasma particles in electron volts.

As noted above, an improved estimate of the ion density can be obtained by replacing (13.5b) with appropriate fluid equations, noting that the ions move in the form of a monoenergetic beam with a velocity u_x. Under steady-state conditions, the continuity equation reduces to

$$\frac{d}{dx}(N_i u_x) = N_i \frac{du_x}{dx} + u_x \frac{dN_i}{dx}. \tag{13.9}$$

In writing down the momentum transport equation, we neglect viscosity effects (i.e., $\nabla \cdot \Psi = \nabla p$), magnetic field effects, and collisions, and assume an isothermal plasma. Collisions can be neglected since the plasma sheath thickness is much less than the mean free path, and magnetic field can be neglected since the sheath thickness is typically much smaller than the gyroradii. We then have

$$m_i u_x \frac{du_x}{dx} = -\frac{k_B T_i}{N_i}\frac{dN_i}{dx} + q_e \frac{d\Phi(x)}{dx}, \tag{13.10}$$

where T_i is the ion temperature. We can further assume that the ions are cold, so that $T_i \simeq 0$; thus the first term on the right-hand side of (13.10) can be neglected. More generally, using (13.9), we can compare the magnitudes of the two terms on the right to conclude that this ratio is $(k_B T/(m_i u_x^2) \ll 1$, indicating that the first term is negligible. We thus have

$$m_i u_x \frac{du_x}{dx} = q_e \frac{d\Phi(x)}{dx}. \tag{13.11}$$

Integrating (13.9) and (13.11) we find

$$\frac{1}{2} m_i u_x^2 - q_e \Phi(x) = C_1 \quad \text{and} \quad N_i u_x = C_2, \tag{13.12}$$

where C_1 and C_2 are constants. We also note the boundary conditions, namely that at $x = \infty$ we have $\Phi(\infty) = 0$, $N_i = N_0$, and $u_x = u_0$. With these we can find C_1 and C_2 and eliminate u_x in (13.12) to relate N_i to $\Phi(x)$:

$$N_i(x) = N_0 \left(1 + \frac{2q_e \Phi(x)}{m_i u_0^2}\right)^{-\frac{1}{2}}. \tag{13.13}$$

Using (13.5a) and (13.13), we can write Poisson's equation within the sheath

$$\nabla^2 \Phi = -\frac{\rho}{\epsilon_0} = -\frac{q_e(N_e - N_i)}{\epsilon_0}$$

$$\frac{d^2\Phi(x)}{dx^2} = -\frac{N_0 q_e}{\epsilon_0}\Bigg[\underbrace{e^{-q_e\Phi/(k_B T)}}_{N_e(x)\ \text{term}} - \underbrace{\left(1 + \frac{2q_e\Phi(x)}{m_i u_0^2}\right)^{-\frac{1}{2}}\Bigg]}_{N_i(x)\ \text{term}}, \tag{13.14}$$

where the ion drift velocity far away from the sheath (i.e., u_0) is yet to be determined. Equation (13.14) is non-linear and in general has to be solved numerically. However, based on the approximate treatment in (13.8) we know that the potential energy $|q_e\Phi(x)|$ ranges from zero in the plasma to a value of order $k_B T$ on the wall. Thus, near the outer edge of the sheath (i.e., the edge close to the plasma), we must have a region in which $|q_e\Phi(x)| \ll k_B T$, as well as $|q_e\Phi(x)| \ll m_i u_x^2$. In this case, using a first-order Taylor expansion, the sheath equation (13.14) reduces to

$$\frac{d^2\Phi(x)}{dx^2} = \frac{\Phi(x)}{h}, \quad h = \lambda_D\left(1 - \frac{k_B T}{m_i u_0^2}\right)^{-\frac{1}{2}}, \tag{13.15}$$

where λ_D is the Debye length. The solution of (13.15) consistent with the boundary conditions (13.4) is

$$\Phi(x) = \Phi_{\text{wall}} e^{-x/h}, \tag{13.16}$$

where it is implied that the solution which is strictly valid near the edge of the plasma sheath applies throughout the sheath. However, as we will see below, the nature of the sheath potential near the wall is quite different. Nevertheless, it is apparent from (13.15) that we must have

$$k_B T \langle m_i u_0^2, \quad \text{or} \quad u_0 \rangle \sqrt{\frac{k_B T_e}{m_i}} \tag{13.17}$$

in order for h to be real so that the potential increases monotonically toward the wall. This condition is known as the *Bohm sheath criterion*. In words, it states that the ions must enter the sheath region with a minimum directed velocity, sometimes referred to as the critical drift velocity. To give this directed velocity to the ions, there must be a non-zero electric field in the plasma. Since the electron density falls off exponentially with $q_e\Phi/(k_B T)$ (i.e., (13.5a)), the electron density can be neglected in the region of large potential near the wall, in which case (13.14) can be written without the electron density term:

$$\frac{d^2\Phi(x)}{dx^2} \simeq -\frac{N_0 q_e}{\epsilon_0}\left(1 + \frac{2q_e\Phi(x)}{m_i u_0^2}\right)^{-\frac{1}{2}}. \tag{13.18}$$

Upon manipulation it can be shown that

$$u_0 \simeq \frac{k_B T_e}{m_i}\frac{4\sqrt{2}}{9}\frac{(q_e\Phi_{\text{wall}})^{\frac{3}{2}}}{(k_B T_e)^{\frac{3}{2}} d^2}\lambda_D^2, \tag{13.19}$$

where d is the approximate extent of the electron-free region. Upon substitution we can write the ion current $J_i = -q_e N_0 u_0$ into the wall as

$$J_i \simeq \frac{4}{9}\left(\frac{2|q_e|}{m_i}\right)^{\frac{1}{2}} \frac{\epsilon_0 |\Phi_{\text{wall}}|^{\frac{3}{2}}}{d^2}, \tag{13.20}$$

which is the well-known Child–Langmuir law for a space-charge-limited current in a planar diode. We thus see that the potential variation in a plasma–wall system can be divided into three parts. Nearest the wall, there is an electron-free region whose thickness d is determined by (13.20). In (13.20) the ion current J is typically determined by the ion production rate and Φ_{wall} is determined by the equality of electron and ion fluxes. Next is a region with scale length of about a Debye length, in which the electron density is appreciable. Finally there is a region, typically with much larger scale, which is known as the *pre-sheath* region, in which the ions are accelerated to the required velocity u_0 by a potential drop of $\frac{1}{2}|q_e|^{-1}k_B T_e$. Depending on the application, the scale length of the pre-sheath region may be set by the size of the plasma, the mean free path, or ionization (plasma production) mechanisms.

13.4 The Langmuir probe

The Langmuir or plasma probe is a practical instrument that has been widely used to investigate plasmas in the laboratory and in the space environment. This electrostatic device, first developed by Langmuir and Mott-Smith, gives a measure of the temperature and density of a plasma. A conducting probe, or electrode, is inserted into the plasma and the current that flows through it is measured for various potentials applied to the probe. When the surface of the probe is a plane, the current as a function of potential yields a curve like that shown in Figure 13.1.

When the probe is placed in the plasma, the plasma reacts by shielding the probe potential, thus forming a sheath with thickness on the order of a Debye length. When no current flows through the probe, it stays at the negative floating potential Φ_w, which is the wall potential (Φ_{wall}) discussed above. Under these equilibrium conditions, the number of electrons reaching the probe per unit time is equal to the number of ions reaching the probe per unit time, yielding no net current. If the probe potential is made more negative than Φ_w, electrons will be repelled and ions will be attracted to the probe. Under the convention that current flowing

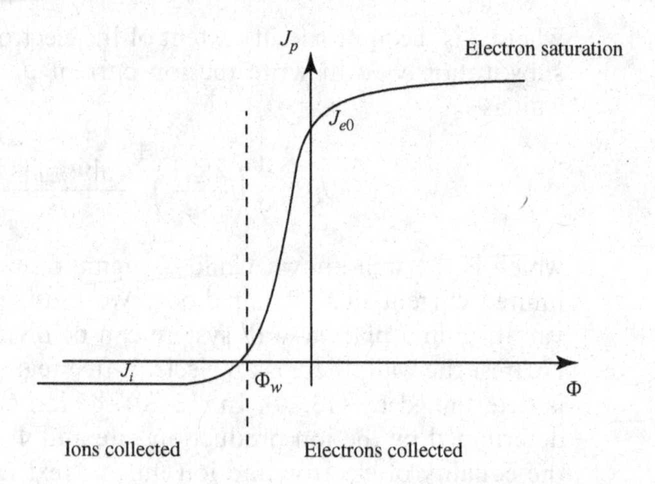

Figure 13.1 Characteristic current–potential curve for a Langmuir probe.

away from the probe is positive, an applied potential less than Φ_w leads to a negative current J_i that is carried only by ions. The ions that make up J_i are those that find themselves at the sheath boundary and their number remains constant even if the potential is made even more negative, since the Debye length remains unchanged.

If the applied potential is increased from the negative value of Φ_w, the repulsion force on the electrons is decreased and the number of electrons reaching the probe becomes greater than the number of ions. When the probe potential is zero, meaning that there is no electric field between the probe and the plasma, there is a net positive current J_{e0} made up primarily of electrons, which have much higher thermal velocities than the ions. As the potential is increased further, the electron current eventually saturates in a reverse of the previous situation, in that all of the electrons at the sheath boundary are collected by the probe. This region is known as the saturation region.

Away from the saturation region the current due to electrons can be expressed as

$$J_e = J_{e0} e^{-q_e \Phi/(k_B T)}, \tag{13.21}$$

using Equation (13.5a); J_{e0} is the electron current for $\Phi = 0$. Using (13.1) we can express J_{e0} as

$$J_{e0} = -q_e N \left(\frac{k_B T}{2\pi m} \right)^{\frac{1}{2}}. \tag{13.22}$$

When $\Phi < 0$ the ions continue to fall into the negative potential of the probe, making the ion current density J_i constant in the negative potential region. Therefore, the probe current density in this region can be expressed as

$$J_p = J_{e0} e^{-q_e \Phi/(k_B T)} - J_i \quad \text{for } \Phi < 0, \tag{13.23}$$

which yields

$$T = \frac{-q_e}{k_B} \left[\frac{d}{d\Phi} \left[\ln(J_p + J_i) \right] \right]^{-1}. \tag{13.24}$$

Equation (13.24) permits the following procedure for determining the electron temperature. First, a sufficiently large negative potential is applied to the probe that J_i can be determined directly.[1] Second, the current–potential characteristic of the probe is obtained and a plot is made of $\ln(J_p + J_i)$ as a function of Φ. This curve has a straight-line portion that can be used to obtain the value of $\frac{d}{d\Phi}\left[\ln(J_p + J_i)\right]$, and the temperature can be calculated from Equation (13.24). The value of J_{e0} is then determined by observing the current at $\Phi = 0$, or at slightly higher potentials since the current in the saturation region is fairly constant. With the temperature and J_{e0} determined, Equation (13.22) can be rearranged to give the density:

$$N = \frac{-J_{e0}}{q_e} \left(\frac{2\pi m}{k_B T} \right)^{\frac{1}{2}}. \tag{13.25}$$

13.5 Problems

13-1. Show how Equation (13.15) is obtained from (13.14).

13-2. Estimate the ion current for a 0.1 mm sheath in a 1.2 eV electron and ion plasma.

13-3. Estimate the sheath distance for a 40 W fluorescent lamp operated at 120 V with a plasma density of 5×10^{17} m^{-3} and a temperature of 1 eV. Assume that 99% of the current in the lamp is carried by electrons.

13-4. The current–voltage characteristics of a plasma probe inserted into a plasma contained by a magnetic mirror machine are given below. The collecting area of the probe is 0.644 cm^2. All voltages are measured with respect to a fixed reference potential. Determine the electron temperature

[1] The observed total current in the probe is $I_p = J_p A_p$, where A_p is the probe area.

in the plasma, the electron number density, and the floating potential of the probe.

I–V characteristics	
Voltage	Current
−30.0	−0.9
−18.5	−0.8
−8.0	−0.6
−3.0	−0.4
−0.5	−0.2
0.5	0.0
1.0	0.1
1.8	0.5
3.0	0.8
5.5	1.2
32.0	2.3
72.0	2.5

13-5. A plasma probe is connected to a circuit that can measure the first and second derivatives of total current $I_p = J_p A$ with respect to potential ϕ, where A is the probe area. Show how these two measurements can be used to estimate the electron temperature directly.

13-6. Explain why a plasma probe with no voltage applied will acquire a negative potential.

13-7. A satellite with very large solar collectors of area 1 km^2 is in geostationary orbit in a 1 eV hydrogen plasma of density 10^6 m^{-3}. During a solar event the satellite is bombarded by energetic electrons which charge it to a potential of −2 kV. Find the resulting damaging flux of ions on the solar collector.

Appendix A Derivation of the second moment of the Boltzmann equation

The second moment is obtained by multiplying the Boltzmann equation (3.9) by $\frac{1}{2}mv^2$ and integrating over velocity space:

$$\frac{m}{2}\int v^2 \frac{\partial f}{\partial t}\, d\mathbf{v} + \frac{m}{2}\int v^2 (\mathbf{v}\cdot\nabla_{\mathbf{r}}) f\, d\mathbf{v} + \frac{q}{2}\int v^2 [(\mathbf{E}+\mathbf{v}\times\mathbf{B})\cdot\nabla_{\mathbf{v}}] f\, d\mathbf{v}$$

$$= \int \frac{m}{2} v^2 \left(\frac{\partial f}{\partial t}\right)_{\text{coll}} d\mathbf{v}. \tag{A.1}$$

The first term is

$$\frac{m}{2}\int v^2 \frac{\partial f}{\partial t}\, d\mathbf{v}$$

$$= \frac{m}{2}\left[\frac{\partial}{\partial t}\int v^2 f\, d\mathbf{v} - \int f \frac{\partial (v^2)}{\partial t} d\mathbf{v}\right] \tag{A.2a}$$

$$= \frac{\partial}{\partial t}\left[N\frac{1}{2}m\langle v^2\rangle\right] \tag{A.2b}$$

$$= \frac{\partial}{\partial t}\left[N\frac{1}{2}mu^2\right], \tag{A.2c}$$

where the last term in (A.2a) is zero since v^2 is independent of t. The second term is

$$\frac{m}{2}\left[\int v^2 \mathbf{v}\cdot\nabla f d\mathbf{v}\right]$$

$$= \frac{m}{2}\left[\nabla\cdot\left(\int \mathbf{v}v^2 f d\mathbf{v}\right) - \int f\mathbf{v}\cdot\nabla v^2 d\mathbf{v} - \int f v^2 \nabla\cdot\mathbf{v} d\mathbf{v}\right] \tag{A.3a}$$

$$= \nabla \cdot \left(N \frac{1}{2} m \left\langle v^2 \mathbf{v} \right\rangle \right) \tag{A.3b}$$

$$= \nabla \cdot \left(N \frac{1}{2} m \left\langle u^2 \mathbf{u} \right\rangle \right), \tag{A.3c}$$

where we have used the fact that the $\nabla \cdot \mathbf{v}$ and ∇v^2 terms are zero since velocity and spatial variables are independent. The third term is

$$\frac{q}{2} \int v^2 \left[(\mathbf{E} + \mathbf{v} \times \mathbf{B}) \cdot \nabla_{\mathbf{v}} \right] f \, d\mathbf{v}$$

$$= \frac{q}{2} \left[\int \nabla_{\mathbf{v}} \cdot \left[(\mathbf{E} + \mathbf{v} \times \mathbf{B}) v^2 f \right] d\mathbf{v} - \int f (\mathbf{E} + \mathbf{v} \times \mathbf{B}) \cdot \nabla_{\mathbf{v}} v^2 d\mathbf{v} \right.$$

$$\left. - \int f v^2 \nabla_{\mathbf{v}} \cdot (\mathbf{E} + \mathbf{v} \times \mathbf{B}) d\mathbf{v} \right] \tag{A.4a}$$

$$= -\frac{q}{2} \int f (\mathbf{E} + \mathbf{v} \times \mathbf{B}) \cdot \nabla_{\mathbf{v}} v^2 d\mathbf{v} \tag{A.4b}$$

$$= -\frac{q}{2} N \left\langle (\mathbf{E} + \mathbf{v} \times \mathbf{B}) \cdot \nabla_{\mathbf{v}} v^2 \right\rangle \tag{A.4c}$$

$$= -\frac{q}{2} N \left\langle (\mathbf{E} + \mathbf{v} \times \mathbf{B}) \cdot \nabla_{\mathbf{v}} (\mathbf{v} \cdot \mathbf{v}) \right\rangle \tag{A.4d}$$

$$= -qN \left\langle (\mathbf{E} + \mathbf{v} \times \mathbf{B}) \cdot (\mathbf{v} \cdot \nabla_{\mathbf{v}}) \mathbf{v} \right\rangle \tag{A.4e}$$

$$= -qN \left\langle (\mathbf{E} + \mathbf{v} \times \mathbf{B}) \cdot \mathbf{v} \right\rangle \tag{A.4f}$$

$$= -qN \left\langle \mathbf{E} \cdot \mathbf{v} \right\rangle \tag{A.4g}$$

$$= -qN \left\langle \mathbf{E} \cdot \mathbf{u} \right\rangle, \tag{A.4h}$$

where the last integral in (A.4a) is zero since any Lorentz force component F_i is independent of the corresponding velocity component v_i. The first integral in (A.4a) is also zero since the distribution function is zero at infinite velocity (see discussion of Equation (4.7)).

Combining all terms leads to (4.26):

$$\frac{\partial}{\partial t} \left[N \frac{1}{2} m u^2 \right] + \nabla \cdot \left[N \frac{1}{2} m \langle u^2 \mathbf{u} \rangle \right] - Nq \langle \mathbf{E} \cdot \mathbf{u} \rangle$$

$$= \frac{m}{2} \int u^2 \left(\frac{\partial f}{\partial t} \right)_{\text{coll}} d\mathbf{u}.$$

Appendix B Useful vector relations

B.1 Definitions and identities

For vectors **A**, **B**, **C**, and **D**:

$$\mathbf{A} \cdot \mathbf{B} = \mathbf{B} \cdot \mathbf{A} = A_x B_x + A_y B_y + A_z B_z$$

$$\mathbf{A} \times \mathbf{B} = -\mathbf{B} \times \mathbf{A} = \begin{vmatrix} \hat{\mathbf{x}} & \hat{\mathbf{y}} & \hat{\mathbf{z}} \\ A_x & A_y & A_z \\ B_x & B_y & B_z \end{vmatrix}$$

$$= (A_y B_z - A_z B_y)\hat{\mathbf{x}} + (A_x B_z - A_z B_x)\hat{\mathbf{y}} + (A_y B_x - A_x B_y)\hat{\mathbf{z}}$$

$$\mathbf{A} \cdot (\mathbf{B} \times \mathbf{C}) = (\mathbf{A} \times \mathbf{B}) \cdot \mathbf{C} = (\mathbf{C} \times \mathbf{A}) \cdot \mathbf{B}$$

$$\mathbf{A} \times (\mathbf{B} \times \mathbf{C}) = (\mathbf{A} \cdot \mathbf{C})\mathbf{B} - (\mathbf{A} \cdot \mathbf{B})\mathbf{C}$$

$$(\mathbf{A} \times \mathbf{B}) \times \mathbf{C}) = (\mathbf{A} \cdot \mathbf{C})\mathbf{B} - (\mathbf{B} \cdot \mathbf{C})\mathbf{A}$$

$$\nabla \cdot (\mathbf{A} \times \mathbf{B}) = \mathbf{B} \cdot (\nabla \times \mathbf{A}) - \mathbf{A} \cdot (\nabla \times \mathbf{B})$$

$$\nabla(\mathbf{A} \cdot \mathbf{B}) = (\mathbf{A} \cdot \nabla)\mathbf{B} + (\mathbf{B} \cdot \nabla)\mathbf{A} + \mathbf{A} \times (\nabla \times \mathbf{B}) + \mathbf{B} \times (\nabla \times \mathbf{A})$$

$$\nabla \times (\mathbf{A} \times \mathbf{B}) = \mathbf{A}(\nabla \cdot \mathbf{B}) + (\mathbf{B} \cdot \nabla)\mathbf{A} - \mathbf{B}(\nabla \cdot \mathbf{A}) - (\mathbf{A} \cdot \nabla)\mathbf{B}$$

$$\nabla \times (\nabla \times \mathbf{A}) = \nabla(\nabla \cdot \mathbf{A}) - (\nabla \cdot \nabla)\mathbf{A}$$

$$\nabla \cdot (\nabla \cdot \mathbf{A}) = 0$$

B.2 Relations in Cartesian coordinates

Orthogonal unit vectors:

$$\hat{\mathbf{x}}, \quad \hat{\mathbf{y}}, \quad \hat{\mathbf{z}}$$

Orthogonal line elements:

$$dx, \quad dy, \quad dz$$

Components of gradient of a scalar function ψ:

$$(\nabla\psi)_x = \frac{\partial\psi}{\partial x}$$

$$(\nabla\psi)_y = \frac{\partial\psi}{\partial y}$$

$$(\nabla\psi)_z = \frac{\partial\psi}{\partial z}$$

Divergence of a vector function $\mathbf{A}$:

$$\nabla\cdot\mathbf{A} = \frac{\partial A_x}{\partial x} + \frac{\partial A_y}{\partial y} + \frac{\partial A_z}{\partial z}$$

Components of curl of a vector function $\mathbf{A}$:

$$(\nabla\times\mathbf{A})_x = \frac{\partial A_z}{\partial y} - \frac{\partial A_y}{\partial z}$$

$$(\nabla\times\mathbf{A})_y = \frac{\partial A_x}{\partial z} - \frac{\partial A_z}{\partial x}$$

$$(\nabla\times\mathbf{A})_z = \frac{\partial A_y}{\partial x} - \frac{\partial A_x}{\partial y}$$

Laplacian of a scalar function ψ:

$$\nabla^2\psi = \frac{\partial^2\psi}{\partial x^2} + \frac{\partial^2\psi}{\partial y^2} + \frac{\partial^2\psi}{\partial z^2}$$

B.3 Relations in cylindrical coordinates

Orthogonal unit vectors:

$$\hat{\mathbf{r}}, \quad \hat{\phi}, \quad \hat{\mathbf{z}}$$

Orthogonal line elements:

$$dr, \quad r\,d\phi, \quad dz$$

Components of gradient of a scalar function ψ:

$$(\nabla\psi)_r = \frac{\partial\psi}{\partial r}$$

$$(\nabla\psi)_\phi = \frac{1}{r}\frac{\partial\psi}{\partial\phi}$$

$$(\nabla\psi)_z = \frac{\partial\psi}{\partial z}$$

Divergence of a vector function $\mathbf{A}$:

$$\nabla\cdot\mathbf{A} = \frac{1}{r}\frac{\partial(rA_r)}{\partial r} + \frac{1}{r}\frac{\partial A_\phi}{\partial\phi} + \frac{\partial A_z}{\partial z}$$

Components of curl of a vector function $\mathbf{A}$:

$$(\nabla\times\mathbf{A})_r = \frac{1}{r}\frac{\partial A_z}{\partial\phi} - \frac{\partial A_\phi}{\partial z}$$

$$(\nabla\times\mathbf{A})_\phi = \frac{\partial A_r}{\partial z} - \frac{\partial A_z}{\partial r}$$

$$(\nabla\times\mathbf{A})_z = \frac{1}{r}\frac{\partial(rA_\phi)}{\partial r} - \frac{1}{r}\frac{\partial A_r}{\partial\phi}$$

Laplacian of a scalar function ψ:

$$\nabla^2\psi = \frac{1}{r}\frac{\partial}{\partial r}\left(r\frac{\partial\psi}{\partial r}\right) + \frac{1}{r^2}\frac{\partial^2\psi}{\partial\phi^2} + \frac{\partial^2\psi}{\partial z^2}$$

B.4 Relations in spherical coordinates

Orthogonal unit vectors:

$$\hat{\mathbf{r}},\quad \hat{\theta},\quad \hat{\phi}$$

Orthogonal line elements:

$$dr,\quad r\,d\theta,\quad r\sin\theta\,d\phi$$

Components of gradient of a scalar function ψ:

$$(\nabla\psi)_r = \frac{\partial\psi}{\partial r}$$

$$(\nabla\psi)_\theta = \frac{1}{r}\frac{\partial\psi}{\partial\theta}$$

$$(\nabla\psi)_\phi = \frac{1}{r\sin\theta}\frac{\partial\psi}{\partial\phi}$$

Divergence of a vector function $\mathbf{A}$:

$$\nabla\cdot\mathbf{A} = \frac{1}{r^2}\frac{\partial r^2 A_r}{\partial r} + \frac{1}{r\sin\theta}\frac{\partial(\sin\theta\, A_\theta)}{\partial\theta} + \frac{1}{r\sin\theta}\frac{\partial A_\phi}{\partial\phi}$$

Components of curl of a vector function $\mathbf{A}$:

$$(\nabla\times\mathbf{A})_r = \frac{1}{r\sin\theta}\frac{\partial(\sin\theta\, A_\phi)}{\partial\theta} - \frac{1}{r\sin\theta}\frac{\partial A_\theta}{\partial\phi}$$

$$(\nabla\times\mathbf{A})_\theta = \frac{1}{r\sin\theta}\frac{\partial A_r}{\partial\phi} - \frac{1}{r}\frac{\partial(r\,A_\phi)}{\partial r}$$

$$(\nabla\times\mathbf{A})_\phi = \frac{1}{r}\frac{\partial r\,A_\theta}{\partial r} - \frac{1}{r}\frac{\partial A_r}{\partial\theta}$$

Laplacian of a scalar function ψ:

$$\nabla^2\psi = \frac{1}{r^2}\frac{\partial}{\partial r}\left(r^2\frac{\partial\psi}{\partial r}\right) + \frac{1}{r^2\sin\theta}\frac{\partial}{\partial\theta}\left(\sin\theta\frac{\partial\psi}{\partial\theta}\right) + \frac{1}{r^2\sin^2\theta}\frac{\partial^2\psi}{\partial\phi^2}$$

Index

Printed in the United States
By Bookmasters